Recent Trends in Advanced High-Strength Steels

Recent Trends in Advanced High-Strength Steels

Editors

Ricardo Branco
Filippo Berto
Andrei Kotousov

MDPI • Basel • Beijing • Wuhan • Barcelona • Belgrade • Manchester • Tokyo • Cluj • Tianjin

Editors

Ricardo Branco
Department of Mechanical Engineering
University of Coimbra
Adelaide
Australia

Filippo Berto
Department of Mechanical and Industrial Engineering
NTNU
Trondheim
Norway

Andrei Kotousov
School of Mechanical Engineering
University of Adelaide
Adelaide
Australia

Editorial Office

MDPI
St. Alban-Anlage 66
4052 Basel, Switzerland

This is a reprint of articles from the Special Issue published online in the open access journal *Applied Sciences* (ISSN 2076-3417) (available at: www.mdpi.com/journal/applsci/special_issues/Advanced_High-strength_Steels).

For citation purposes, cite each article independently as indicated on the article page online and as indicated below:

LastName, A.A.; LastName, B.B.; LastName, C.C. Article Title. *Journal Name* **Year**, *Volume Number*, Page Range.

ISBN 978-3-0365-2013-1 (Hbk)
ISBN 978-3-0365-2012-4 (PDF)

Contents

About the Editors

Ricardo Branco

Ricardo Branco completed his PhD in Mechanical Engineering at the University of Coimbra. He is currently an assistant professor at the Department of Mechanical Engineering of the University of Coimbra. His research interests include the mechanical behaviour of materials, the fatigue and fracture of engineering materials, multiaxial fatigue life prediction models, the low-cycle fatigue of high-strength steels, and numerical modelling of fatigue crack growth.

Filippo Berto

Filippo Berto completed his degree in Industrial Engineering summa cum laude in 2003 at the University of Padua (Italy). After completing a PhD in Mechanical Engineering and Materials Science at the University of Florence, he worked as a researcher in the same field at the University of Padua. From 2006 to 2013, he was an assistant professor at the University of Padua, Department of Management and Engineering, Vicenza, and from October 2014 to September 2016, he was an associate professor. On 1 January 2016, he was appointed as the International Chair of Fracture Mechanics and Fatigue at the Norwegian University of Science and Technology (NTNU).

Andrei Kotousov

Andrei Kotousov has actively promoted research on Applied Mechanics in Australia and internationally through a wide international collaborative network. He has participated in the investigation of reverberated structural failures in Australia and led several large industry-funded projects. Andrei serves on the editorial board of several journals and has chaired a number of international conferences.

applied sciences

MDPI

Editorial

Special Issue on Recent Trends in Advanced High-Strength Steels

Ricardo Branco [1,*], Filippo Berto [2] and Andrei Kotousov [3]

[1] Department of Mechanical Engineering, University of Coimbra, CEMMPRE, 3030-78 Coimbra, Portugal
[2] Department of Mechanical and Industrial Engineering, Norwegian University of Science and Technology, 7491 Trondheim, Norway; filippo.berto@ntnu.no
[3] School of Mechanical Engineering, The University of Adelaide, Adelaide, SA 5005, Australia; andrei.kotousov@adelaide.edu.au
* Correspondence: ricardo.branco@dem.uc.pt

1. Introduction

Advanced high-strength steels play an essential role in many industries and engineering applications because of their excellent combination of mechanical properties important for design, e.g., strength, fatigue, fracture, wear, and manufacturability. In many challenging applications, structural components can be subjected to severe service conditions, and therefore, high-strength steels are often the only choice in these applications. Therefore, understanding the relationships between the mechanical properties and the chemical composition, microstructural features, and manufacturing methods are pivotal to develop safe and durable structures.

The goal of this Special Issue is to foster the dissemination of the latest research outcomes in the field of advanced high-strength steels from various perspectives. In this volume, a total of thirteen papers addressing several important research problems and developments have been collected, namely the relationship between microstructure and mechanical properties, the effect of heat treatment variables on the mechanical behaviour, fatigue and fracture resistance under different loading conditions, and weldability. These articles represent a significant contribution to the research field.

Citation: Branco, R.; Berto, F.; Kotousov, A. Special Issue on Recent Trends in Advanced High-Strength Steels. *Appl. Sci.* **2021**, *11*, 6914. https://doi.org/10.3390/app11156914

Received: 10 June 2021
Accepted: 26 July 2021
Published: 27 July 2021

2. Contributions

The interconnection between the microstructure and bulk mechanical properties has been the focus of many investigations. The development of deep knowledge regarding the above-mentioned relationship opens many cost-effective opportunities for new applications as well as for cost-optimisation of the current engineering designs. In this volume, Ali et al. [1] studied the effects of chromium content on the microstructure and mechanical properties of low-carbon bainitic steel containing niobium processed by thermomechanical hot rolling. Chen et al. [2] evaluated the mechanical properties and the phase transformation temperature in a transformation-induced plasticity (TRIP) steel subjected to room and high temperatures.

Currently, various heat treatment procedures are the most common methods to improve the bulk mechanical properties of advanced high-strength steels; this may justify the intensive research in this field. Haiko et al. [3] investigate the influence of tempering temperature on both the microstructure and monotonic stress-strain response in a new ultra-high-strength steel with low carbon content. Honma et al. [4] analysed the effect of intercritical quenching temperature on the strength and toughness in a low-alloy steel developed for offshore applications. Gu et al. [5] developed a numerical approach to evaluate the effect of different inclusions on the residual stress profiles during the cooling process for martensitic steels. Grajcar et al. [6] addressed the influences of isothermal holding time and temperature on the stability of retained austenite in medium manganese bainitic steels with and without niobium addition.

Mechanical behaviour under cyclic loading is crucial for designing structural components against fatigue. This maintains the need to understand the fatigue failure and fracture mechanisms of advanced high-strength steels under various loading histories. Wang et al. [7] studied the collapse of an adjustable ballast tank used in deep-sea submersibles made of an ultra-high-strength maraging steel. The fatigue crack growth in 18Ni300 maraging steel was also studied by Antunes et al. [8], using various fracture mechanic parameters, namely the classic stress intensity factor range and the newly proposed plastic crack-tip opening displacement. Dealing with the same maraging steel, Mooney et al. [9] developed a constitutive elastic-plastic model to simulate the cyclic response of the material under strain-controlled loading conditions. Ottersböck et al. [10] presented a new procedure based on the digital image correlation technique to track both the fatigue crack initiation and the fatigue crack propagation regimes in butt joints made of ultra-high-strength steel.

Establishing the link between the resultant mechanical properties of the fabricated materials and parameters of advanced manufacturing methods is always challenging and very important for potential applications. Santos et al. [11] evaluated fracture toughness of maraging steel implants fabricated by laser-beam powder-bed fusion methods. Ndubuaku et al. [12] proposed a robust and straightforward model to characterise the deformation response of high-strength steel used in pipelines. In a comprehensive state-of-the-art review, Królicka et al. [13] addressed the welding methods, process parameters, and weldability aspects of different high-strength steels; in particular, high-carbon nanobainitic steels.

Author Contributions: Conceptualization, R.B., F.B. and A.K.; Writing—original draft preparation, R.B., F.B. and A.K.; writing—review and editing, R.B., F.B. and A.K. All authors have read and agreed to the published version of the manuscript.

Funding: This research was funded by FEDER funds through the program COMPETE—Programa Operacional Factores de Competitividade—and by national funds through FCT—Fundação para a Ciência e a Tecnologia—under the project UIDB/00285/2020.

Conflicts of Interest: The authors declare no conflict of interest.

References

1. Ali, M.; Nyo, T.; Kaijalainen, A.; Hannula, J.; Porter, D.; Kömi, J. Influence of Chromium Content on the Microstructure and Mechanical Properties of Thermomechanically Hot-Rolled Low-Carbon Bainitic Steels Containing Niobium. *Appl. Sci.* **2020**, *10*, 344. [CrossRef]
2. Chen, D.; Cui, H.; Wang, R. High-Temperature Mechanical Properties of 4.5%Al δ-TRIP Steel. *Appl. Sci.* **2019**, *9*, 5094. [CrossRef]
3. Haiko, O.; Kaijalainen, A.; Pallaspuro, S.; Hannula, J.; Porter, D.; Liimatainen, T.; Kömi, J. The Effect of Tempering on the Microstructure and Mechanical Properties of a Novel 0.4C Press-Hardening Steel. *Appl. Sci.* **2019**, *9*, 4231. [CrossRef]
4. Honma, Y.; Sasaki, G.; Hashi, K. Effect of Intercritical Quenching Temperature of Cu-Containing Low Alloy Steel of Long Part Forging for Offshore Applications. *Appl. Sci.* **2019**, *9*, 1705. [CrossRef]
5. Gu, C.; Lian, J.; Bao, Y.; Xiao, W.; Münstermann, S. Numerical Study of the Effect of Inclusions on the Residual Stress Distribution in High-Strength Martensitic Steels During Cooling. *Appl. Sci.* **2019**, *9*, 455. [CrossRef]
6. Grajcar, A.; Skrzypczyk, P.; Kozłowska, A. Effects of Temperature and Time of Isothermal Holding on Retained Austenite Stability in Medium-Mn Steels. *Appl. Sci.* **2018**, *8*, 2156. [CrossRef]
7. Wang, F.; Wu, M.; Tian, G.; Jiang, Z.; Zhang, S.; Zhang, J.; Cui, W. Failure Analysis on a Collapsed Flat Cover of an Adjustable Ballast Tank Used in Deep-Sea Submersibles. *Appl. Sci.* **2019**, *9*, 5258. [CrossRef]
8. Antunes, F.; Santos, L.; Capela, C.; Ferreira, J.; Costa, J.; Jesus, J.A.D.S.D.; Prates, P. Fatigue Crack Growth in Maraging Steel Obtained by Selective Laser Melting. *Appl. Sci.* **2019**, *9*, 4412. [CrossRef]
9. Mooney, B.; Agius, D.; Kourousis, K.I. Cyclic Plasticity of the As-Built EOS Maraging Steel: Preliminary Experimental and Computational Results. *Appl. Sci.* **2020**, *10*, 1232. [CrossRef]
10. Ottersböck, M.J.; Leitner, M.; Stoschka, M.; Maurer, W. Crack Initiation and Propagation Fatigue Life of Ultra High-Strength Steel Butt Joints. *Appl. Sci.* **2019**, *9*, 4590. [CrossRef]
11. Santos, L.M.S.; De Jesus, J.; Ferreira, J.M.; Costa, J.; Capela, C. Fracture Toughness of Hybrid Components with Selective Laser Melting 18Ni300 Steel Parts. *Appl. Sci.* **2018**, *8*, 1879. [CrossRef]
12. Ndubuaku, O.; Martens, M.; Cheng, J.J.R.; Adeeb, S. Integrating the Shape Constants of a Novel Material Stress-Strain Characterization Model for Parametric Numerical Analysis of the Deformational Capacity of High-Strength X80-Grade Steel Pipelines. *Appl. Sci.* **2019**, *9*, 322. [CrossRef]
13. Królicka, A.; Ambroziak, A.; Żak, A. Welding Capabilities of Nanostructured Carbide-Free Bainite: Review of Welding Methods, Materials, Problems, and Perspectives. *Appl. Sci.* **2019**, *9*, 3798. [CrossRef]

applied sciences

Article

Influence of Chromium Content on the Microstructure and Mechanical Properties of Thermomechanically Hot-Rolled Low-Carbon Bainitic Steels Containing Niobium

Mohammed Ali [1,2,*], Tun Nyo [1], Antti Kaijalainen [1], Jaakko Hannula [1], David Porter [1] and Jukka Kömi [1]

[1] Materials and Mechanical Engineering, Centre for Advanced Steels Research, University of Oulu, P.O. Box 4200, 90014 Oulu, Finland; tun.nyo@oulu.fi (T.N.); antti.kaijalainen@oulu.fi (A.K.); jaakko.hannula@oulu.fi (J.H.); david.porter@oulu.fi (D.P.); jukka.komi@oulu.fi (J.K.)

[2] Steel Technology Department, Central Metallurgical Research and Development Institute (CMRDI), Helwan, Cairo 11421, Egypt

* Correspondence: mohammed.ali@oulu.fi or mohammedsalah2020@gmail.com

Received: 5 December 2019; Accepted: 30 December 2019; Published: 2 January 2020

Featured Application: The current results can be used as a guideline for the production of high-strength low-carbon bainitic steels with high impact toughness.

Abstract: The effect of chromium content in the range of 1 wt.%–4 wt.% on the microstructure and mechanical properties of controlled-rolled and direct-quenched 12 mm thick low-carbon (0.04 wt.%) steel plates containing 0.06 wt.% Nb has been studied. In these microalloyed 700 MPa grade steels, the aim was to achieve a robust bainitic microstructure with a yield strength of 700 MPa combined with good tensile ductility and impact toughness. Continuous cooling transformation diagrams of deformed and non-deformed austenite were recorded to study the effect of Cr and hot deformation on the transformation behavior of the investigated steels. Depending on the cooling rate, the microstructures consist of one or more of the following microstructural constituents: bainitic ferrite, granular bainite, polygonal ferrite, and pearlite. The fraction of bainitic ferrite decreases with decreasing cooling rate, giving an increasing fraction of granular bainite and polygonal ferrite and a reduction in the hardness of the transformation products. Polygonal ferrite formation depends mainly on the Cr content and the cooling rate. In both deformed and non-deformed austenite, increasing the Cr content enhances the hardenability and refines the final microstructure, shifting the ferrite start curve to lower cooling rates. Preceding austenite deformation promotes the formation of polygonal ferrite at lower cooling rates, which leads to a decrease in hardness. In hot-rolled and direct-quenched plates, decreasing the Cr content promotes the formation of polygonal ferrite leading to an increase in the impact toughness and elongation but also a loss of yield strength.

Keywords: bainite; direct quenching; CCT and DCCT diagrams; microstructure; tensile properties; impact toughness

1. Introduction

Thermomechanically controlled processing (TMCP) followed by direct quenching is considered the most powerful and effective processing route to obtain the best combination of the yield strength, good ductility, low-temperature toughness with a good hardenability in the low-carbon microalloyed steels [1]. The final microstructure of the hot-rolled steel is considered the key factor for getting the targeted combination of the mechanical properties and it is controlled by the quenching process following the TMCP. Hence, the microstructures and mechanical properties of low-carbon microalloyed steel can be improved significantly by designing a suitable accelerated cooling/direct quenching process, following the hot deformation schedule [2]. Depending on the chemical composition of the steel and the cooling path, the accelerated cooling rate may promote the formation of bainite with or without the formation of ferrite.

Low-carbon bainitic steels are used in many structural applications as a result of their good combinations of strength and toughness. Use of structural steels for weight critical applications such as mobile frames and crane booms is desirable to minimize the carbon footprint and save energy. Different alloying elements like Cr, Nb, Mo, and B are used to improve the strength and toughness through promoting the formation of bainite and refining the microstructure [3]. Bainitic ferrite and granular bainite represent the two main microstructure constituents in low-carbon bainitic steels and they are associated with distinct mechanical properties [4–6]. Cr is widely used to improve the hardenability and thereby promote the formation of bainite by suppressing the formation of high temperature transformation products [7–14].

The main objective of the current work is to develop a low-carbon steel with 700 MPa yield strength in combination with good ductility and toughness. To help reach this goal, the effect of Cr and preceding hot deformation of austenite on the continuous cooling transformation (CCT) diagrams of a candidate composition was studied. Following this, the effect of Cr on the microstructures, tensile properties, and impact toughness of hot-rolled and direct-quenched low-carbon bainitic steel plates was investigated.

2. Materials and Methods

2.1. Materials

Three 70 kg vacuum-cast ingots with the chemical composition given in Table 1 were produced at Outokumpu, Tornio, Finland. From the ingots, 200 × 80 × 55 mm (length, width, thickness) pieces were cut and soaked at 1250 °C for 2 h and then thermomechanically rolled with six passes to 12 mm thick plates from 55 mm thick slab. Final plate dimensions were approximately 1000 × 90 × 12 mm. The experimental thermomechanical process consists of several hot rolling passes in the recrystallized controlled regime followed by controlled rolling passes in the non-recrystallization regime (TNR). The temperature of the specimens was monitored during the rolling and direct quenching with the aid of thermocouples fixed in holes drilled to the centers of the pieces before hot rolling. The finish rolling temperature (FRT) was controlled to be about 880 °C, which was followed by direct quenching in a water tank with a cooling rate (CR) of about 40–50 °C/s as measured in the centers of the specimens.

Table 1. Chemical composition of the investigated steels.

Steel Code	C	Si	Mn	Cr	Nb
Steel I (1 Cr)	0.041	0.211	1.02	1.0	0.065
Steel II (2.5 Cr)	0.041	0.168	0.95	2.5	0.066
Steel III (4 Cr)	0.036	0.154	0.92	4.0	0.064

2.2. CCT Diagrams

A Gleeble 3800 thermomechanical simulator (Dynamic Systems Inc., Poestenkill, NY, USA) was used to get the dilatation curves required to construct continuous cooling transformation (CCT) diagrams and deformation continuous cooling transformation (DCCT) diagrams. Small cylindrical samples were machined with a diameter of 5 mm and a length of 7.5 mm for the Gleeble simulation tests. In order to get the dilatation curves to determine the critical transformation temperatures A_{c1} and A_{c3}, the samples were heated at 10 °C/s to 1100 °C, held for 180 s, and then quenched. To get the dilatation curves for CCT diagrams, specimens were heated at 10 °C/s to 1100 °C, held for 180 s, and then quenched at different cooling rates (CRs) ranging from 2 to 80 °C/s as shown in Figure 1a. To obtain the dilatation curves for DCCT diagrams, the specimens were heated at 10 °C/s to 1100 °C, held for 180 s, cooled at 2 °C/s to 880 °C, held there for 15 s, given a compressive deformation with a total strain of 0.6 and a strain rate of 1 s^{-1}, held for 20 s, and finally cooled at different rates in the range of 2–80 °C/s to room temperature (see Figure 1b).

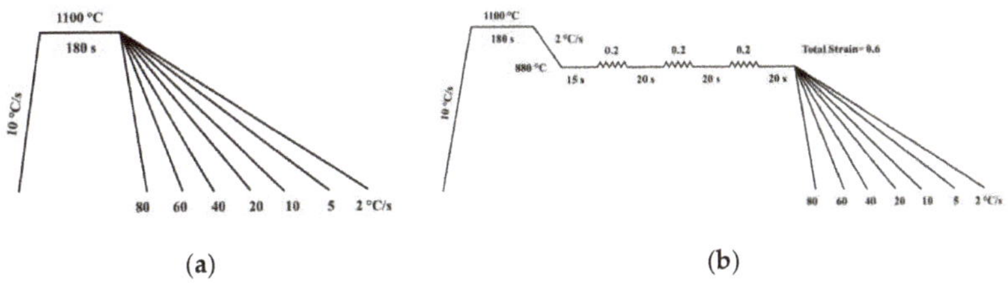

Figure 1. Gleeble thermomechanical simulator procedures for obtaining continuous cooling transformation (CCT) diagrams (**a**) and deformation continuous cooling transformation (DCCT) diagrams (**b**).

2.3. Microstructure Characterization

For the thermomechanical simulation specimens, Figure 2 shows the selected section for all further investigations. All specimens were ground to the selected section and polished following standard procedures. Then, depending on the CR, samples were etched for 15–50 s in fresh 10% nital.

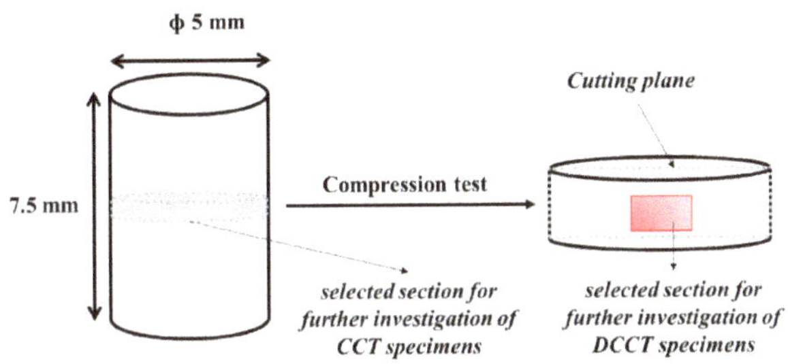

Figure 2. Selected section of investigation for the samples after thermomechanical simulation.

Specimens from hot-rolled plates were ground, polished, and etched in 2% nital. An additional polishing step was done using 0.04 µm silica for EBSD analysis. Microstructures were examined using laser scanning confocal microscopy (LSCM, VK-X200 Keyence Ltd.) and Zeiss Sigma field emission scanning electron microscopy (Carl Zeiss AG, Oberkochen, Germany) with EDAX-OIM acquisition and analysis software (7.1.0, Amatek Inc., Berwyn, PA, USA). Electron backscatter diffraction (EBSD) analyses using Zeiss Sigma FESEM, were carried out on hot-rolled specimens using 15 kV accelerating voltage with a 0.2 µm step size.

2.4. Mechanical Properties Characterization

For the tensile properties, longitudinal flat tensile specimens, 11.7 mm thick, 20 mm wide, and with a parallel length of 120 mm, were tested at room temperature according to the standard SFS-EN 10002-1 using an MTS 810 mechanical testing machine fitted with a 100 kN load cell (MTS systems, Eden Prairie, MN, USA). The results are the means of three specimens for each steel composition.

The 28 J Charpy V transition temperatures (T28J) of the rolled plates were determined using two standard (10 × 10 × 55 mm) Charpy V-notch impact test specimens per test temperature for both the longitudinal and transverse directions. Tests were made at temperatures in the range of −120 to +20 °C using a 300 J Charpy testing machine. The fracture surfaces of Charpy V impact samples broken at room temperature (20 °C) were examined using a Zeiss Sigma FESEM at 8 kV accelerating voltage at a working distance of 8.5 mm.

Hardness was measured using a Struers Duramin-A300 (Struers A/S, Ballerup, Denmark) macro-hardness tester at five random positions across the thermomechanical simulation specimens, while for the hot rolled specimens, a representative hardness value was obtained from the mean of five sets of seven suitably spaced hardness impressions through the thickness.

3. Results and Discussion

3.1. CCT and DCCT Diagrams

Figure 3 shows CCT and DCCT diagrams constructed on the basis of the results of dilatation curves, microstructures, and macrohardness values. Increasing the Cr content from 2.5 wt.% to 4 wt.% led to an increase in A_{c1} by about 35 °C, otherwise it has no effect on the critical transformation temperatures. LSCM and SEM micrographs of the final transformation products formed from non-deformed and deformed austenite after cooling at 2 and 80 °C/s are shown in Figures 4–7.

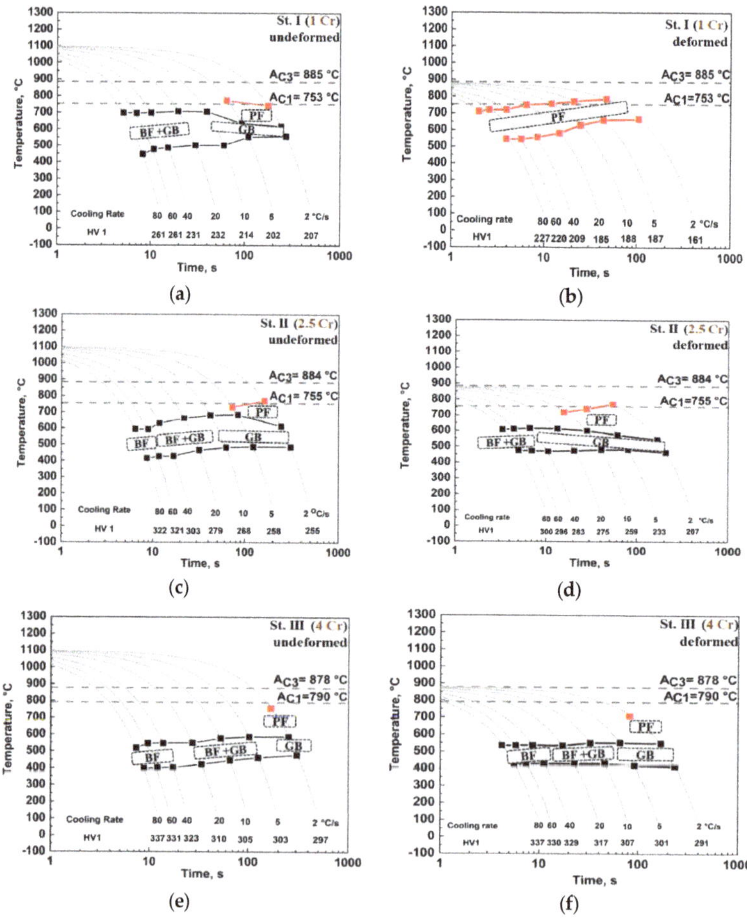

Figure 3. CCT (**a**,**c**,**e**) and DCCT (**b**,**d**,**f**) diagrams for Steels I (**a**,**b**), II (**c**,**d**), and III (**e**,**f**). Abbreviations: BF = bainitic ferrite, GB = granular bainite, and PF = polygonal ferrite.

Figure 4. Laser scanning confocal microscopy (LSCM) micrographs of Steel I (**a**,**d**), II (**b**,**e**), and III (**c**,**f**) without austenite deformation after cooling at 80 and 2 °C/s, respectively. Abbreviations: BF = bainitic ferrite, GB = granular bainite, and PF = polygonal ferrite.

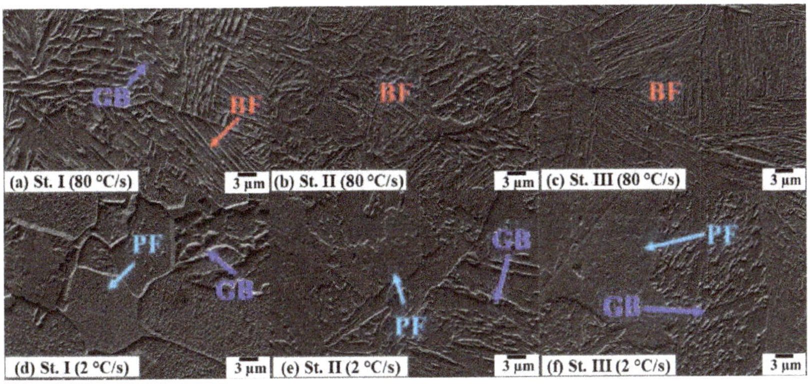

Figure 5. Scanning electron microscopy (SEM) micrographs of Steel I (**a**,**d**), II (**b**,**e**), and III (**c**,**f**) without austenite deformation after cooling at 80 and 2 °C/s, respectively. Abbreviations: BF = bainitic ferrite, GB = granular bainite, and PF = polygonal ferrite.

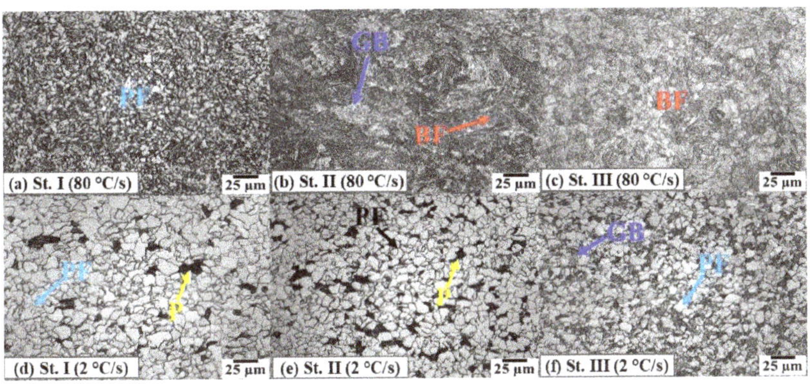

Figure 6. Laser scanning confocal microscopy (LSCM) micrographs of Steel I (**a**,**d**), II (**b**,**e**), and III (**c**,**f**) with austenite deformation after cooling at 80 and 2 °C/s, respectively. Abbreviations: BF = bainitic ferrite, GB = granular bainite, PF = polygonal ferrite, and P = pearlite.

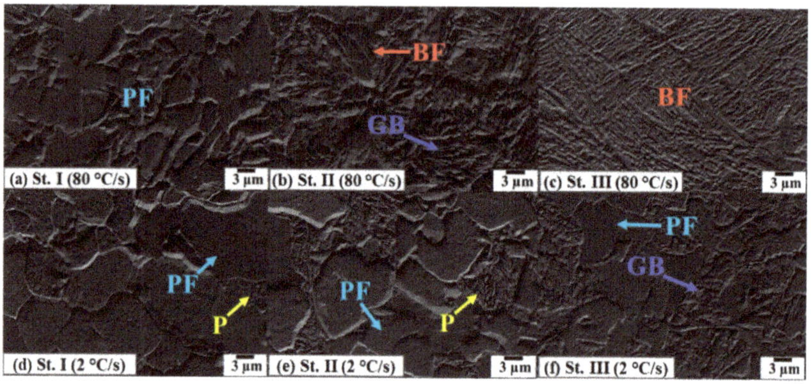

Figure 7. Scanning electron microscopy (SEM) micrographs of Steel I (**a**,**d**), II (**b**,**e**), and III (**c**,**f**) with austenite deformation after cooling at 80 and 2 °C/s, respectively. Abbreviations: BF = bainitic ferrite, GB = granular bainite, PF = polygonal ferrite, and P = pearlite.

For Steel I, with the lowest Cr content, the preceding hot deformation had a great effect on the microstructural constituents (see Figure 3a,b and Figure 4a,d, Figure 5a,d, Figure 6a,d and Figure 7a,d). In the case of the non-deformed condition, the microstructure was a mixture of bainitic ferrite (BF) and granular bainite (GB) over a wide range of CRs. Deforming the austenite changed the microstructure to fully polygonal ferrite (PF) at all the investigated CRs, apart from the appearance of a very small fraction of pearlite at CRs in the range of 2–20 °C/s (Figure 6d) which could not be detected from the dilatation curves. These changes had a clear impact in the hardness of the specimens.

For Steel II (Figure 3c,d, Figure 4b,e, Figure 5b,e, Figure 6b,e and Figure 7b,e), depending on the CR, BF, GB, and PF were observed in the non-deformed and deformed conditions. Again, a small fraction of pearlite which could not be detected in the dilatation curve was found for CRs in the range of 2–10 °C/s. Here too, preceeding hot deformation caused a drop in hardness.

For Steel III (Figure 3e,f, Figure 4c,f, Figure 5c,f, Figure 6c,f and Figure 7c,f), increasing the CR refined the microstructure, promoted the formation of BF and suppressed the formation of the high-temperature transformation products, i.e., GB and PF, in the case of both the non-deformed and deformed conditions. However, in the case of deformed austenite, the formation of BF requires a higher CR than is needed for undeformed austenite. For both non-deformed and deformed austenite, irrespective of the microstructural constituents, CR has little effect on the transformation temperatures and hardness.

3.2. *Influence of Cr Content on the CCT and DCCT Diagrams and Hardness*

As commonly observed [13–17], the hardenability of the present steels increases with increasing Cr content. This shows as a lowering of transformation temperatures in both CCT and DCCT diagrams as the Cr content increases, see Figure 8, which promotes the formation of the low-temperature transformation product BF and reduces the formation of the transformation products that form at relatively high temperatures, i.e., PF and GB.

As can be seen from Figure 9, the effect of the preceding hot deformation on the transformation products is decreased with increasing Cr content. In the case of Steel I, with the lowest Cr content (1 wt.%), prior hot deformation had a great effect on the transformation curves which shifted to significantly higher temperatures as a result of ferrite formation. This caused a significant drop in hardness as can be seen from Figure 10. This behavior results from the dislocation substructure formed in the austenite after deformation in the non-recrystallization region, which provides nucleation sites and diffusional paths for the diffusional transformation products [18,19]. In the case of Steel II (2.5 wt.% Cr), the preceding hot deformation shifted the transformation curves of ferrite to higher CRs, while B_s

shifted to lower temperatures. These changes led to an increase in the formation of PF and GB, while the formation of BF was reduced even at high CRs. However, in the case of Steel III with the highest Cr content (4.0 wt.%), the effect of the preceding hot deformation on the transformation temperatures and products became very small. The transformation curves were shifted to slightly lower temperatures, but this effect decreased with increasing cooling rate. Austenite deformation increased the fraction of GB somewhat at the lower CRs. The insensitivity to deformation also reflected in the final hardness as a function of cooling rate as shown in Figure 10.

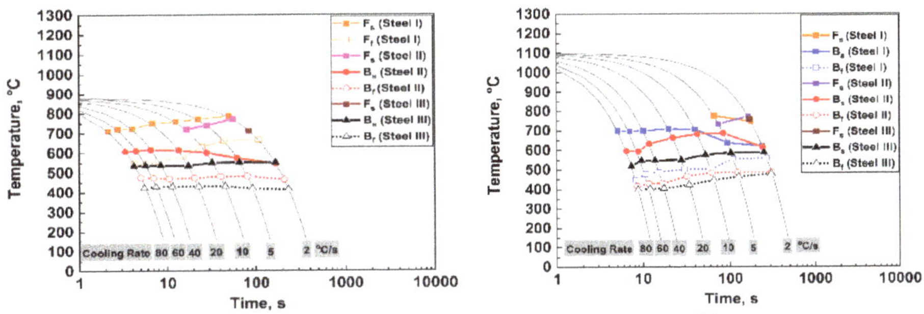

(a) (b)

Figure 8. Influence of Cr content on the transformation temperatures for CCT diagrams (**a**) and DCCT diagrams (**b**). Abbreviations: F_s = ferrite start temperature, F_f = ferrite finish temperature, B_s = bainite start temperature, and B_f = bainite finish temperature.

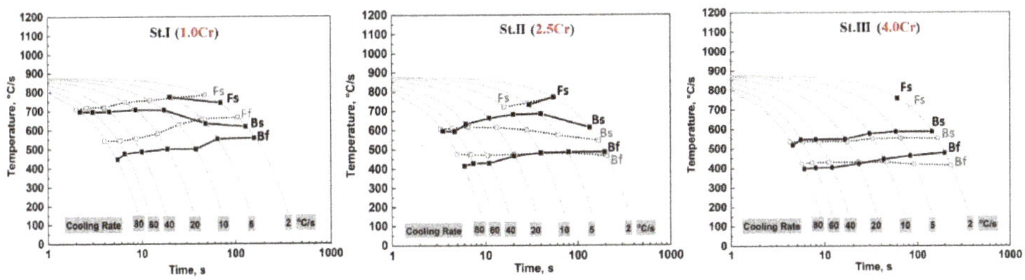

(a) (b) (c)

Figure 9. Influence of austenite deformation on CCT curves for a Cr content of 1 wt.% (**a**), 2.5 wt.% (**b**), and 4 wt.% (**c**). CCT curves bold and DCCT curves dotted. Abbreviations: F_s = ferrite start temperature, F_f = ferrite finish temperature, B_s = bainite start temperature, and B_f = bainite finish temperature.

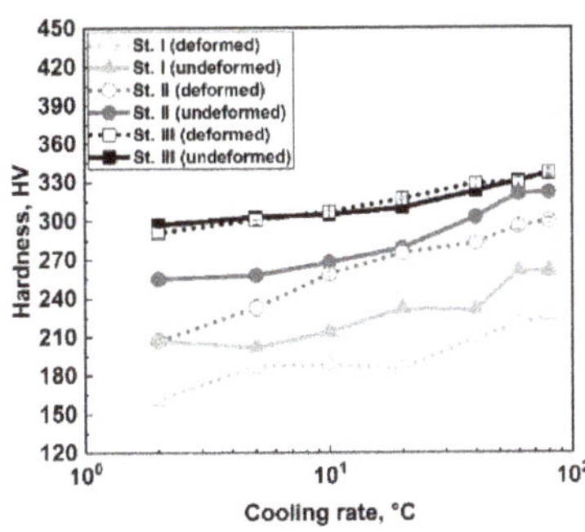

Figure 10. Influence of CR, Cr content, and prior hot deformation on hardness.

At all CRs, the hardness values increased with increasing the Cr content (Figure 10). This is due to the higher solid solution strengthening combined with the lower transformation temperatures which refine the microstructure significantly (see Figure 6a–c). Overall, it can be concluded that austenite deformation with a high Cr content lowers the transformation temperatures, while transformation temperatures are increased in the case of low Cr contents.

3.3. Microstructure of the Hot Rolled and Direct-Quenched Steel Plates

The microstructures of the hot rolled and direct-quenched steel plates have been examined in order to evaluate the effect of Cr content in the case of laboratory scale rolling. As a result of the accelerated cooling, the microstructure of the three steel plates consists mainly of one or more of the following microstructural components: BF, GB, PF, and coalesced bainite (CB) (see Figure 11). Figure 12 shows inverse pole figures (IPF) with superimposed image quality (IQ) maps, grain boundary misorientation distributions, and grain sizes. The effective grain was determined as the equivalent circle diameters (ECD) of grains surrounded by boundaries having misorientation angles >15°. The effective high-angle grain sizes at the 90th percentile in the cumulative grain size distribution ($D_{90\%}$) were also determined. The grain boundary misorientation distribution of Steel I, with the lowest Cr content, supports the LSCM and SEM observations that Steel I consisted mainly of GB with a large fraction of PF. Steel I also had a large mean effective grain size and $D_{90\%}$. Increasing the Cr content to 2.5 wt.% in case of Steel II led to an increase in the hardenability, an increase in the fraction of BF, and a reduction in the fraction of GB and PF, which are associated with reductions in the mean effective grain size from 2.66 to 2.12 μm and $D_{90\%}$ from 30.85 to 24.35 μm. The frequency of misorientation at about 58° increased compared to that of Steel I. With the further increase in the Cr content in the case of Steel III, the hardenability increased significantly, which promoted the formation of BF and suppressed the formation of PF and GB. This was associated with further reductions in the mean effective grain size from 2.12 to 1.86 μm and $D_{90\%}$ from 24.35 to 17.12 μm, together with a further increase in the frequency of misorientation at about 58° compared to that of Steel II.

In summary, increasing the Cr content was associated with a significant change in the microstructural components, an increasing grain boundary misorientation frequency at about 58°, and a reduction in the mean effective grain size and $D_{90\%}$.

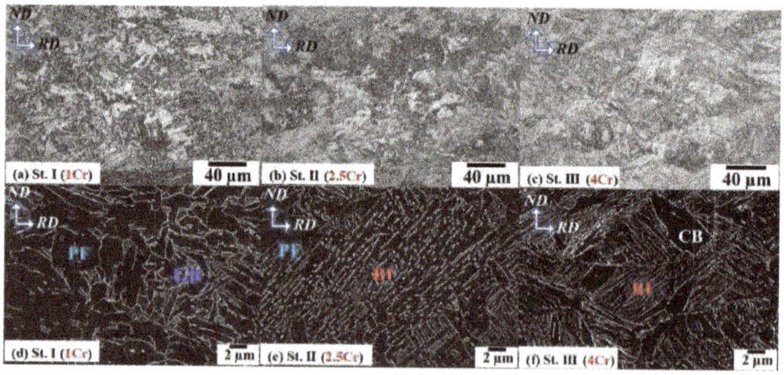

Figure 11. Typical LSCM and inlens SEM micrographs of the investigated hot-rolled and direct-quenched plates. PF and GB in Steel I (**a**,**d**), BF and PF in Steel II (**b**,**e**), and BF and CB in Steel III (**c**,**f**). Abbreviations: BF = bainitic ferrite, GB = granular bainite, PF = polygonal ferrite, and CB = coalesced bainite.

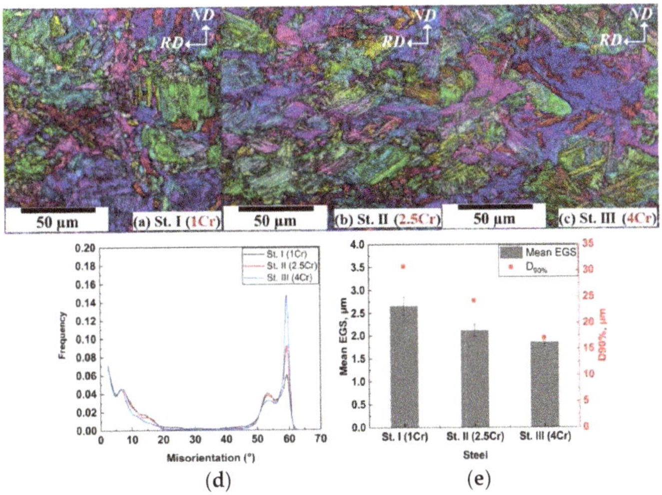

Figure 12. Inverse pole figures (IPF) with superimposed image quality (IQ) maps of hot-rolled and direct-quenched steel plates for Steel I (**a**), Steel II (**b**), Steel III (**c**). Grain boundary misorientations are given in (**d**) and mean effective grain sizes and $D_{90\%}$ in (**e**).

3.4. Mechanical Properties

The variation in the tensile properties, mean hardness values, and 28J transition temperatures (T28J) estimated from the impact tests are given in Table 2 and Figure 13. In addition, the through-thickness hardness profiles and impact transition curves are shown in Figure 14.

Table 2. Mechanical properties of the investigated steels.

Steel Code	YS [1], MPa	UTS [1], MPa	El [1], %	Hardness [2], HV	T28J (°C)	USE J/cm^2	YS/UTS	UTS*El (MPa *%)
Steel I (1 Cr)	560 ± 7	756 ± 2	17 ± 0	252 ± 6	−118	318	0.74	13034
Steel II (2.5 Cr)	699 ± 3	909 ± 4	14 ± 0	292 ± 7	−84	270	0.77	12852
Steel III (4 Cr)	780 ± 9	982 ± 8	12 ± 0	312 ± 7	−78	265	0.79	12238

[1] Error bars are standard deviations of three measurements. [2] Error bars are standard deviations of 35 measurements through the thickness. Definitions: YS = yield strength, UTS = ultimate tensile strength, El = total elongation, T28J = 28J transition temperatures, USE = upper shelf energy.

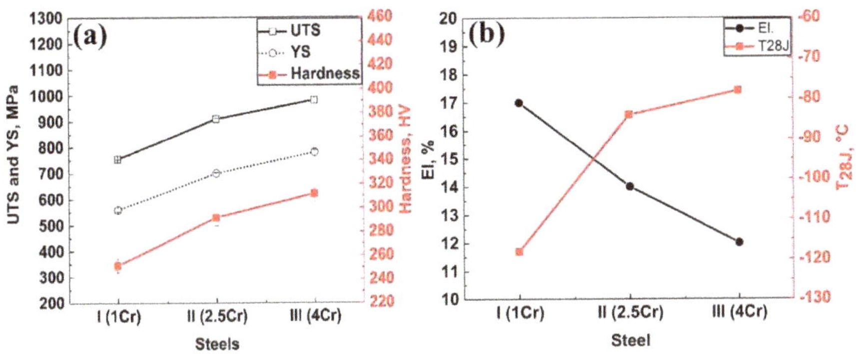

Figure 13. Effect of Cr content on the hardness, yield and ultimate tensile strength (**a**) and elongation and 28J transition temperature (**b**).

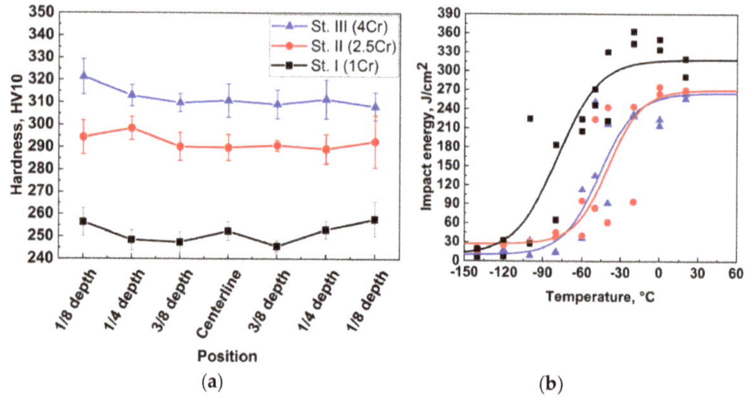

Figure 14. Through-thickness profile (**a**) and Charpy V impact transition curves (**b**).

Both Steels II (2.5 Cr) and III (4 Cr) were able to provide the targeted 700 MPa level of yield strength, which is a widely used strength class among high-strength structural steels [20]. Steel I (1 Cr) provided a lower yield strength of 560 MPa with a high elongation to fracture as a result of the formation of PF. The higher impact toughness of Steel I (1 Cr), in respect to higher upper shelf energy and lower T28J temperature compared to Steels II (2.5 Cr) and III (4 Cr) resulted from the presence of PF and GB, which contains martensite-austenite (M/A) islands (see Table 2 and Figure 14). It is well known that the presence of PF enhances the toughness properties [4]. On the other hand, the presence of M/A constituents in the lath-like granular bainite is detrimental to impact toughness [21]. Figure 15 shows that despite a refinement of the effective mean and 90th percentile grain size, increasing the Cr content was associated with poorer impact toughness, i.e., higher T28J. Presumably this is due to the combined detrimental effects caused by an increase in the fraction of BF and higher yield and tensile strength.

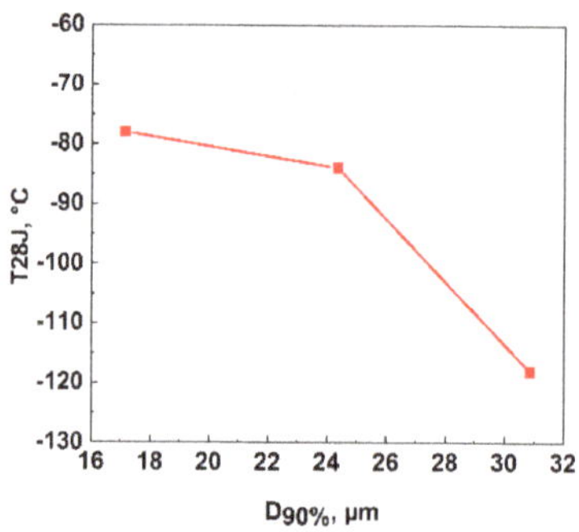

Figure 15. Relationship between effective high-angle grain sizes at the 90th percentile in the cumulative size distribution ($D_{90\%}$) and the 28-joule transition temperature (T28J).

The adverse effect of the high fraction of PF (see Figure 11a) on the hardness in Steel I can be seen clearly in Figure 14a. As mentioned above, increasing the Cr content enhanced the hardenability, promoted the formation of the low-temperature transformation product BF, and hindered the formation of the high-temperature transformation products PF and GB, thereby enhancing the hardness in the case of Steels II and III (see Table 2) [22]. Moreover, due to the higher hardenability of the high-Cr steels, the hardness was fairly uniform throughout the plate thickness as shown in Figure 14a.

As can be seen in Figure 16, there is a correlation between the yield strength and T28J. It can be seen that the targeted yield strength could still be achieved with good impact toughness in the hot-rolled and direct-quenched plates when the Cr content was in the range of 2.5 wt.%–4 wt.%, but not in the case of the lowest Cr content (1 wt.%) due to the formation of a significant fraction of PF which improved the impact toughness and tensile elongation to fracture, but at the expense of the yield strength.

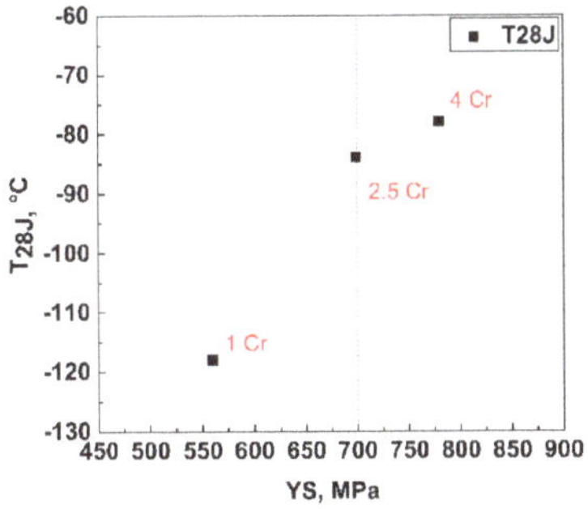

Figure 16. Relationship between yield strength and 28 joule transition temperature.

3.5. Fractography

Figure 17 shows fractographs of selected areas from the Charpy V impact samples at room temperature (20 °C). An effect of the Cr content on the fractography of the impact samples can be clearly observed. In the case of Steel I (1 Cr) (Figure 17a,b), the fracture surface consisted mainly of elongated dimples due to the microvoid coalescence associated with ductile fracture. In Steels II and III with the higher Cr content of 2.5 wt.% and 4 wt.%, the fracture surface consisted mainly of many small dimples with less ductility compared to that of Steel I (see Figure 17c–f, respectively). These results are in line with the value of upper shelf energy (USE) given in Table 2 which illustrates the closeness of the USE of Steels II and III in contrast to Steel I.

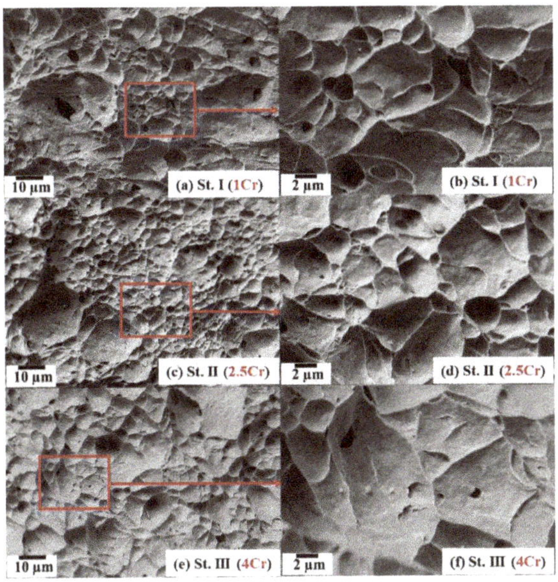

Figure 17. SEM images showing fractographs of the impact fracture surface after testing at room temperature (20 °C) for Steel I (**a**,**b**), Steel II (**c**,**d**), and Steel III (**e**,**f**).

4. Conclusions

Thee experimental steels were designed based on (in wt.%) 0.04C-0.2Si-1.0Mn-0.06Nb with 1 wt.%, 2.5 wt.%, and 4 wt.% Cr and processed through laboratory hot rolling followed by direct quenching. The effect of the Cr content on the continuous cooling transformation (CCT) diagram and deformation continuous cooling transformation (DCCT) diagram was studied. A Gleeble 3800 thermomechanical simulator was employed to determine the dilatation curves for the CCT and DCCT diagrams. In addition, the effect of the Cr content on the final microstructure and mechanical properties of the hot-rolled and direct-quenched materials was studied. The following conclusions can be drawn.

1. In the case of non-strained austenite, the microstructure consists mainly of bainitic ferrite at high CRs, a mixture of bainitic ferrite with granular bainite at intermediate CRs, and polygonal ferrite with granular bainite at the lowest CRs. Increasing the Cr content and CRs refined the microstructure significantly. In addition, increasing the Cr content led to an increase in the hardenability, i.e., a decrease in transformation temperatures, as well as the promotion of bainite at the expense of the high-temperature transformation products like polygonal ferrite and granular bainite.
2. In the low-Cr steel, deformation of austenite below its recrystallization temperature has a large effect on the final microstructural constituents, changing the microstructure completely to ferrite at all the investigated CRs and causing a large drop in hardness relative to the non-deformed case. However, this effect decreased with increasing the Cr content.
3. To get a fully bainitic microstructure in the case of deformed austenite, the Cr content should be in the range of 2.5 wt.%–4 wt.%, and the CR should be higher than 20 °C/s.
4. In the case of the 12 mm thick hot-rolled and direct-quenched steel plates, the microstructure consisted mainly of different types of bainite, i.e., bainitic ferrite, granular bainite, and coalesced bainite. A significant amount of polygonal ferrite only appears at the lowest Cr level, i.e., 1 wt.%. Increasing the Cr content led to an increase in the hardness owing to an enhanced hardenability.
5. The targeted 700 MPa yield strength with good ductility and toughness were achieved in the hot-rolled and direct-quenched plates with Cr content in the range of 2.5 wt.%–4 wt.%, but not achieved with the lowest Cr content (1 wt.%) due to the presence of a significant fraction of polygonal ferrite, which increased the impact toughness and elongation at the expense of the yield strength.

Author Contributions: Writing—review and editing, M.A., T.N., A.K., J.H. and D.P.; supervision, D.P. and J.K. All authors have read and agreed to the published version of the manuscript.

Funding: This research was funded by CBMM (Companhia Brasileira de Metalurgia e Mineracao).

Conflicts of Interest: The authors declare no conflict of interest.

References

1. Xie, H.; Du, L.-X.; Hu, J.; Misra, R.D.K. Microstructure and Mechanical Properties of a Novel 1000MPa Grade TMCP Low Carbon Microalloyed Steel with Combination of High Strength and Excellent Toughness. *Mater. Sci. Eng. A* **2014**, *612*, 123–130. [CrossRef]
2. Bandyopadhyay, P.S.; Ghosh, S.K.; Kundu, S.; Chatterjee, S. Evolution of Microstructure and Mechanical Properties of Thermomechanically Processed Ultrahigh-Strength Steel. *Metall. Mater. Trans. A* **2011**, *42*, 2742–2752. [CrossRef]
3. Hu, H.; Xu, G.; Zhou, M.; Yuan, Q. Effect of Mo Content on Microstructure and Property of Low-Carbon Bainitic Steels. *Metals* **2016**, *6*, 173. [CrossRef]
4. Krauss, G.; Thompson, S.W. Ferritic Microstructures in Continuously Cooled Low- and Ultralow-carbon Steels. *ISIJ Int.* **1995**, *35*, 937–945. [CrossRef]
5. Lan, H.; Du, L.; Zhou, N.; Liu, X. Effect of Austempering Route on Microstructural Characterization of Nanobainitic Steel. *Acta Metall. Sin. Engl. Lett.* **2014**, *27*, 19–26. [CrossRef]

6. Yang, X.L.; Xu, Y.B.; Tan, X.D.; Yu, Y.M.; Wu, D. Microstructures and Mechanical Properties of High Strength Low Carbon Bainitic Steel. *Mater. Sci. Forum* **2015**, *817*, 257–262. [CrossRef]
7. Caballero, F.G.; Bhadeshia, H.K.D.H.; Mawella, K.J.A.; Jones, D.G.; Brown, P. Design of Novel High Strength Bainitic Steels: Part 1. *Mater. Sci. Technol.* **2001**, *17*, 512–516. [CrossRef]
8. Caballero, F.G.; Bhadeshia, H.K.D.H.; Mawella, K.J.A.; Jones, D.G.; Brown, P. Design of Novel High Strength Bainitic Steels: Part 2. *Mater. Sci. Technol.* **2001**, *17*, 517–522. [CrossRef]
9. Caballero, F.G.; Bhadeshia, H.K.D.H. Very Strong Bainite. *Curr. Opin. Solid State Mater. Sci.* **2004**, *8*, 251–257. [CrossRef]
10. Caballero, F.G.; Santofimia, M.J.; García-Mateo, C.; Chao, J.; de Andrés, C.G. Theoretical Design and Advanced Microstructure in Super High Strength Steels. *Mater. Des.* **2009**, *30*, 2077–2083. [CrossRef]
11. Hasan, H.S.; Peet, M.J.; Avettand-Fènoël, M.-N.; Bhadeshia, H.K.D.H. Effect of Tempering upon the Tensile Properties of a Nanostructured Bainitic Steel. *Mater. Sci. Eng. A* **2014**, *615*, 340–347. [CrossRef]
12. Yao, Z.; Xu, G.; Hu, H.; Yuan, Q.; Tian, J.; Zhou, M. Effect of Ni and Cr Addition on Transformation and Properties of Low-Carbon Bainitic Steels. *Trans. Indian Inst. Met.* **2019**, *72*, 1167–1174. [CrossRef]
13. Zhou, M.; Xu, G.; Tian, J.; Hu, H.; Yuan, Q. Bainitic Transformation and Properties of Low Carbon Carbide-Free Bainitic Steels with Cr Addition. *Metals* **2017**, *7*, 263. [CrossRef]
14. Tian, J.; Xu, G.; Zhou, M.; Hu, H.; Wan, X. The Effects of Cr and Al Addition on Transformation and Properties in Low-Carbon Bainitic Steels. *Metals* **2017**, *7*, 40. [CrossRef]
15. Kong, L.; Liu, Y.; Liu, J.; Song, Y.; Li, S.; Zhang, R.; Li, T.; Liang, Y. The Influence of Chromium on the Pearlite-Austenite Transformation Kinetics of the Fe–Cr–C Ternary Steels. *J. Alloys Compd.* **2015**, *648*, 494–499. [CrossRef]
16. Zhang, G.-H.; Chae, J.-Y.; Kim, K.-H.; Suh, D.W. Effects of Mn, Si and Cr addition on the Dissolution and Coarsening of Pearlitic Cementite during Intercritical Austenitization in Fe-1mass%C alloy. *Mater. Charact.* **2013**, *81*, 56–67. [CrossRef]
17. Siwecki, T.; Eliasson, J.; Lagneborg, R.; Hutchinson, B. Vanadium Microalloyed Bainitic Hot Strip Steels. *ISIJ Int.* **2010**, *50*, 760–767. [CrossRef]
18. Opiela, M.; Zalecki, W.; Grajcar, A. Influence of Plastic Deformation on CCT-Diagrams of New-Developed Microalloyed Steel. *J. Achiev. Mater. Manuf. Eng.* **2012**, *51*, 78–89.
19. Zhao, M.-C.; Yang, K.; Xiao, F.-R.; Shan, Y.-Y. Continuous Cooling Transformation of Undeformed and Deformed Low Carbon Pipeline Steels. *Mater. Sci. Eng. A* **2003**, *355*, 126–136. [CrossRef]
20. The High-Strength Structural Steel at 700 MPa. Available online: https://www.ssab.com/products/brands/strenx/products/strenx-700-mc (accessed on 2 September 2019).
21. Bhadeshia, H.K.D.H. Local Brittle Zones and the Role of Niobium. *Mater. Sci. Forum* **2014**, *783–786*, 2129–2135. [CrossRef]
22. Lee, H.; Lee, H. Effect of Cr Content on Microstructure and Mechanical Properties of Low Carbon Steel Welds. *Int. J. Electrochem. Sci.* **2015**, *10*, 8028–8040.

applied sciences

Article

High-Temperature Mechanical Properties of 4.5%Al δ-TRIP Steel

Dayu Chen, Heng Cui * and Rudong Wang

Collaborative Innovation Center of Steel Technology, University of Science and Technology Beijing, Beijing 100083, China; g20179075@xs.ustb.edu.cn (D.C.); rdwang87@126.com (R.W.)
* Correspondence: cuiheng@ustb.edu.cn; Tel.: +86-1367-123-9796

Received: 28 October 2019; Accepted: 18 November 2019; Published: 25 November 2019

Featured Application: δ-TRIP steel, a product of ultimate tensile strength (UTS) and total elongation (TE) similar to the third generation of advanced high-strength steels (AHSS), is expected to become the materials of automobile safety parts and structural parts.

Abstract: The high-temperature mechanical properties of a 4.5% Al-containing δ-transformation-induced plasticity (TRIP) steel were studied by using the Gleeble 3500 thermomechanical simulator. The zero ductility temperature (ZDT) and the zero strength temperature (ZST) were measured, and the brittle zones were divided. The phase transformation zone was determined by differential scanning calorimetry (DSC). The temperature of the phase transformation and the proportion of the phase were calculated by the Thermo-Calc software. The ZDT and the ZST of the 4.5% Al-containing δ-TRIP steel are 1355 and 1405 °C, respectively. The first brittle zone and the third brittle zone of the steel are 1300–1350 °C and 800–975 °C, respectively. The reason for the embrittlement of the third brittle zone of the 4.5% Al-containing δ-TRIP steel is that the α-ferrite formed at the austenite grain boundary causes the sample to crack along the grain boundary under stress. The ductility of the 4.5% Al-containing δ-TRIP steel decreases first and then increases with the increase of the α-ferrite. When the proportion of the α-ferrite reaches 37%, the reduction of area (RA) of the 4.5% Al-containing δ-TRIP steel is reduced to 44%. The 4.5% Al-containing δ-TRIP steel has good resistance to the high-temperature cracking.

Keywords: high-temperature mechanical properties; crack-sensitive zone; brittle zones; phase transformation; equilibrium phase diagram

1. Introduction

Transformation-induced plasticity (TRIP) steel refers to ultrahigh-strength steel that is ductility-elevated due to gradual martensitic transformation processes in steel structures. When there is a high content of Al in alloying elements (the Al content is higher than 3%), it is found that, during the solidification process, the δ-ferrite does not participate in the reaction and remains in the whole temperature range, so it is named δ-TRIP steel. Recently, δ-TRIP steel, due to its low density, a product of ultimate tensile strength (UTS) and total elongation (TE) (up to 23 GPa·%) similar to the third generation of advanced high-strength steels (AHSS), and excellent welding properties, is expected to become the materials of automobile safety parts and structural parts [1,2]. Yi et al. [3] found an alloy (its precise composition is Fe–0.4C–0.5Mn–0.2Si–0.5Cr–3.5Al wt %), having a density of 7.455×10^3 kg/m^3, is approximately 5% lighter than that without the aluminum addition, which has a density of 7.8145×10^3 kg/m^3. Abedi et al. [4] found a duplex low-density steel with an exact chemical composition of Fe–17.5Mn–8.3Al-0.74C–0.14Si (wt %), having a density of 6.8×10^3 kg/m^3. The measurement shows that the density of a 4.5% Al-containing δ-TRIP steel is 6.8325×10^3 kg/m^3, which is approximately 8% lighter than that with a 3.5 wt % aluminium addition. It is approximately 12.6% lighter than that without the aluminum addition, which has a density of 7.8145 × 10^3 kg/m^3. Many studies have been reported on mechanical deformation properties [5,6], microstructures [7,8], and properties of δ-TRIP steel [9,10]. However, there is an insufficient understanding of the high-temperature crack sensitivity of the δ-TRIP steel. The influence of high Al content on it is still lacking, so it is necessary to understand its high-temperature mechanical properties further.

There are many researches, which are regarding high-temperature mechanical properties of Nb-containing TRIP steel [11], P-containing TRIP steel [12], twinning-induced plasticity (TWIP) steel [13–15], dual-phase (DP) steel [16], and FeCrAl [17,18]. A Gleeble thermal-mechanical testing machine is a common method for studying high-temperature mechanical properties [19]. He et al. [18] studied the hot ductility of FeCrAl alloys by using a Gleeble thermal-mechanical testing machine and thought the hot ductility of the alloys mostly depends on the temperature rather than the content of Al. The ferritic grain coarsening and the precipitation of carbides result in decreased ductility between 800 and 900 °C. He et al. [18] found high-Cr-content ferrite stainless steel suffers from embrittlement when subjected to temperatures up to around 475 °C, due to the separation of the α (Fe-rich) ferrite and the α′ (Cr-rich) ferrite and the transformation of the Fe-rich α phase and the Cr-rich α′ phase to the α-(Fe,Cr) ferrite phase by using differential scanning calorimetry (DSC). Hechu et al. [20] found the solid-state transformation of a δ dendrite core to a single γ phase without a peritectic reaction at 1358 °C. Some researches [15,21,22] show that the increase of Al content leads to a worse hot ductility and high Al content in steel can easily lead to deterioration of surface quality of continuous casting slabs [23]. However, Su et al. [12] found that the increase of Al content in TRIP steel (0.03–0.87% Al) led to an overall increase of a reduction of area (RA) and better hot ductility. It was indicated that the effect of high Al content on the high-temperature mechanical properties of steel is a complicated change. There are many research results on mechanical properties of δ-TRIP steel, but research on high-temperature mechanical properties of high-aluminum δ-TRIP steel is relatively limited.

In this paper, a high-temperature tensile test was carried out on the Gleeble 3500 thermomechanical simulator, and the characteristic curves were obtained. Brittle zones and a crack-sensitive zone of δ-TRIP steel were calculated by the characteristic curves. The fracture morphology of the specimen after the high-temperature tensile fracture was observed by scanning electron microscopy (SEM) to determine the fracture type. The equilibrium phase diagram of the 4.5% Al-containing δ-TRIP steel was calculated by the Thermo-Calc software. The phase transformation temperatures of the 4.5% Al-containing δ-TRIP steel were measured by the DSC.

2. Materials and Methods

A δ-TRIP steel ingot was prepared by vacuum induction melting using commercial-purity materials with a target composition of 0.4C–0.5Si–2.0Mn–4.5%Al–Fe (wt %). The dimensions and the shape of the δ-TRIP steel ingot are shown in Figure 1. The concentrations of C and S were measured using the infrared absorptiometric method after combustion in current oxygen. Besides, the inert-gas carrier melting–thermoconductometric method was employed for the measurement of N concentration. The contents of the other alloying elements (Si, Mn, and Al) were measured by the inductively coupled plasma optical emission spectrometric (ICP-OES) methodology.

Figure 1 shows the dimensions and the shape of the δ-TRIP steel ingot as well as the dimensions and the locations of the specimens. The shape of δ-TRIP steel ingot, like an inverted cone, had an upper diameter (Φ) of 120 mm and a total length of 240 mm. Gleeble tensile specimens were cut from the columnar grain region of the ingots. The Gleeble tensile specimens were standard cylindrical tensile specimens, of which the diameter was 10 mm and the length was 121.5 mm. The sampling position of the DSC specimen was near the sampling position of the Gleeble tensile specimens. The DSC specimen was required to have be a round flake with a diameter of 5 mm or a rectangular flake with a diagonal length no more than 5 mm, and its thickness was generally less than 1 mm. Due to the large contents of Al and Mn in the specimen, to minimize the contamination of the instrument support during the melting process, the DSC specimen size (diameter × thickness) in this test was 3 mm × 1 mm. Drill 6 equidistant holes were in the axis of the δ-TRIP steel ingot and had a small amount of steel scrap. A cylindrical specimen with a diameter of 33 mm and a height of 10 mm at the bottom of the ingot was used, and this cylindrical specimen and the steel scrap were analyzed to determine the chemical composition of the ingot.

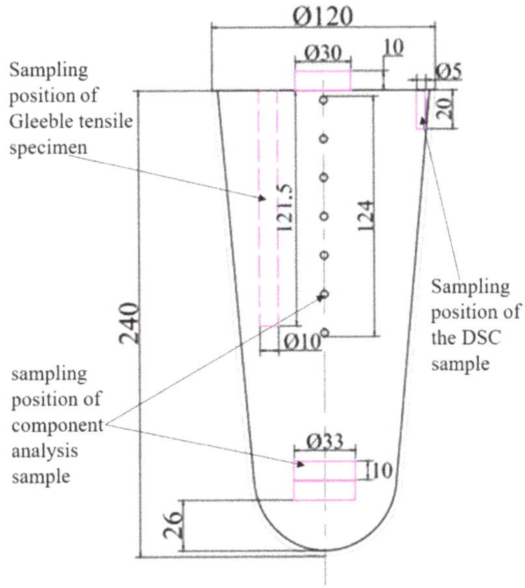

Figure 1. Sampling positions and dimensions of the δ-transformation-induced plasticity (TRIP) steel ingot and the specimens.

The high-temperature mechanical properties of δ-TRIP steel were studied by using the Gleeble 3500 thermomechanical simulator. With the electrical discharge machining (EDM) method to shape the ingot-carved cylindrical tensile steel and the area of the columnar crystal being placed at a temperature range of 700 to 1350 °C, the specimen was tensioned at a constant temperature and at a constant strain rate of $1 \times 10^{-3}\ s^{-1}$. All the tests were carried out in an atmosphere of argon to avoid steel oxidation. The temperature control diagram of the tensile test is shown in Figure 2. The steel samples were first heated to 1300 °C with a heating rate of 10 °C/s, and then they were subjected to insulation during the solid solution treatment for 180 s to homogenize the microstructure and eliminate the internal stresses. The steel was cooled at a rate of 3 °C/s for 3 min until its temperature reached the test temperature, which was maintained. At temperatures above 1300 °C, the steel was heated to the test temperature at a rate of 3 °C/s and incubated for another 3 min. Finally, the sample was tensioned until it fractured and cooled rapidly in argon. The measured area reduction was used as a function of the deformation temperature to evaluate the hot ductility behavior. The morphology of the fracture was observed under SEM. The fracture type was analyzed. The RA and the σ_{max} of the steel were obtained in the high-temperature tensile test. The temperatures, at which both the RA and the σ_{max} value of the sample were zero, were measured corresponding to the zero ductility temperature (ZDT) and the zero strength temperature (ZST), respectively [24,25].

The solidification path with the equilibrium phase of the δ-TRIP steel was studied according to the phase equilibrium fraction diagram, which was calculated by using the Thermo-Calc software and the TCFE9 database.

The rough DSC specimens were finely ground for a smooth surface and a uniform thickness. The phase transformation of the high-aluminium-content δ-TRIP steel in a range of 40–1400 °C was analyzed by DSC. The measurement error of the instrument was within 2.5%, the temperature error was ±1.5 °C, and the temperature rise was rated at 10 °C/min during the test. The test was carried out in an inert gas (Ar) using a corundum crucible at a temperature range from 40 to 1400 °C.

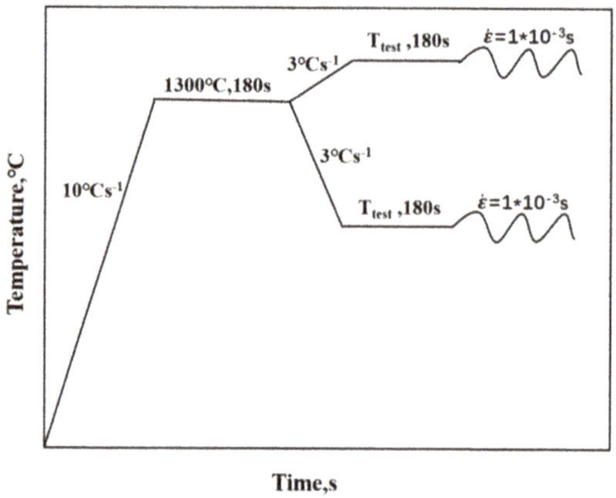

Figure 2. Thermal schedule for the hot tensile tests.

3. Results and Discussion

3.1. Hot Ductility

Figure 3 shows the measured tensile strength relative to the temperature. σ_{max} was decreased with the increasing temperature. Between 700 and 950 °C, σ_{max} decreased with temperature, while the decreasing speed of the σ_{max} at the temperatures above 950 °C was significantly slower. The σ_{max} value was the smallest at 1350 °C, and its value was 1.54 MPa. The measured tensile strength curve was extended by the slope of the final two data points. The ZST of the sample was around 1405 °C.

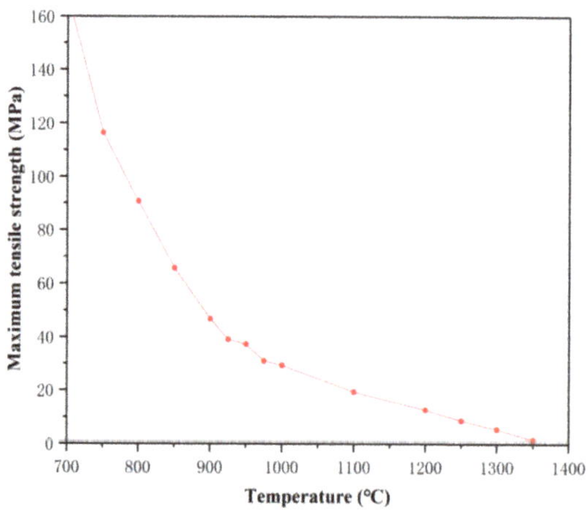

Figure 3. Maximum tensile strength as a function of temperature.

The trend of the RA of the specimen as a function of temperature is shown in Figure 4. In order to better determine the RA of the third brittle zone, the test temperature interval was changed from 50 to 25 °C, and several specimens tested at small temperature intervals were added in the first brittle zone. At a temperature range between 700 and 1300 °C, the RA was basically higher than 40%. In this temperature range, the minimum value of RA reached approximately 44% at 925 °C. The RA reached its maximum value of 78% at 1250 °C. The RA dropped rapidly between 1250 and 1350 °C. The RA was the smallest at 1350 °C, and its value was 2%. It was indicated that the sample had good resistance to the high-temperature cracking. The RA–temperature curve can be extrapolated. The ZDT of the sample was around 1355 °C.

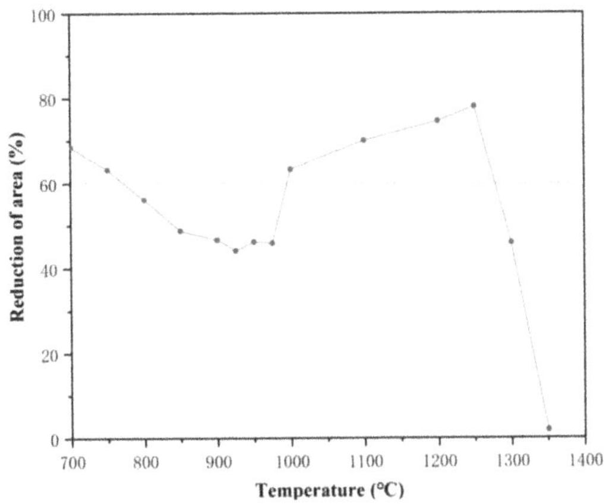

Figure 4. Reduction of area of the specimen as a function of temperature.

ΔT is the difference between the ZST and the ZDT, which is the temperature range of the crack-sensitive zone. Suzuki et al. [26] thought the embrittlement in zone I of a titanium alloy is related to the solidification cracking and the sensitivity of the solidification cracking may be negligibly small if ΔT is less than 50 °C. Li et al. [27] found the ΔT of TWIP steel with 0.05% C, 25% Mn, 3% Al, and 3% Si is about 50 °C, which is wider compared with those of other steels. The result also indicates that this TWIP steel has stronger high-temperature crack sensitivity. In this research, the ΔT of the sample was 50 °C. The high-temperature crack sensitivity of the 4.5% Al-containing δ-TRIP steel is excellent.

The research of Suzuki et al. [28] shows the steel is not easy to crack for RA > 60% and the sensitivity of the crack increases for RA < 60%. Based on the RA of 60% as a critical value for dividing ductility, we found that the sample had three temperature zones: the first brittle zone (1300–1350 °C), the second ductility zone (1000–1250 °C), and the third brittle zone (800–975 °C).

3.2. Fracture Morphology

Figure 5 shows the typical tensile fracture morphologies of the steel at four temperatures (700 °C, 925 °C, 1250 °C, and 1300 °C) under SEM, which corresponds to the tensile fracture morphologies of the typical points in the four regions bounded by an RA of 60, respectively.

Figure 5c shows the tensile fracture morphology of the steel at 1250 °C. It can be seen that the fracture has visible dimple characteristics and the ductility is great. Figure 5c corresponds to the fracture morphology of the highest RA points in the hot ductility curve of the steel.

Due to the microscopic holes generated by the plastic deformation of the steel in the microregion, the nucleation, the growth, and the aggregation were finally connected to each other to cause a fracture, and the diameter and the depth of the dimple were large. The larger the size of the dimples, the better the ductility of the material was. However, the steel with poor ductility had a little change in diameter, the size of the dimple was small and shallow, the shape of the section was flat, the fracture was characterized by the brittle fracture, and the RA was small. Figure 5b shows the fracture morphology of the lowest RA points in the hot ductility curve at 700–1300 °C. The RA was 44% at 925 °C, the dimple had a smaller diameter and a shallow depth, the cross-section was relatively flat, and "rock sugar" particles appeared. The fracture had brittle fracture characteristics, belonging to the brittle intergranular fracture. Take Figure 5c for example. The crack propagation process before the temperature reached 1250 °C was slow, and it consumed high plastic deformation energy. Therefore, the diameter and the depth of the fracture dimple were large, and there was no flat plane. The fracture was a ductile fracture. At this time, the RA was as high as 78%.

Figure 5d shows the fracture morphology of the steel at 1300 °C. At this time, since the temperature was high, the fracture was composed of a plurality of facets, and the grain boundary was in a molten state.

By comparing the SEM observations, the rock-sugar-shape fracture temperature was found to be above 1300 °C or between 800 and 975 °C. The results are consistent with those of the brittle zones, and the temperature of the dimple fracture was mostly in the range of 1000–1250 °C or 700–800 °C. The observation result is basically consistent with that of the second ductility zone.

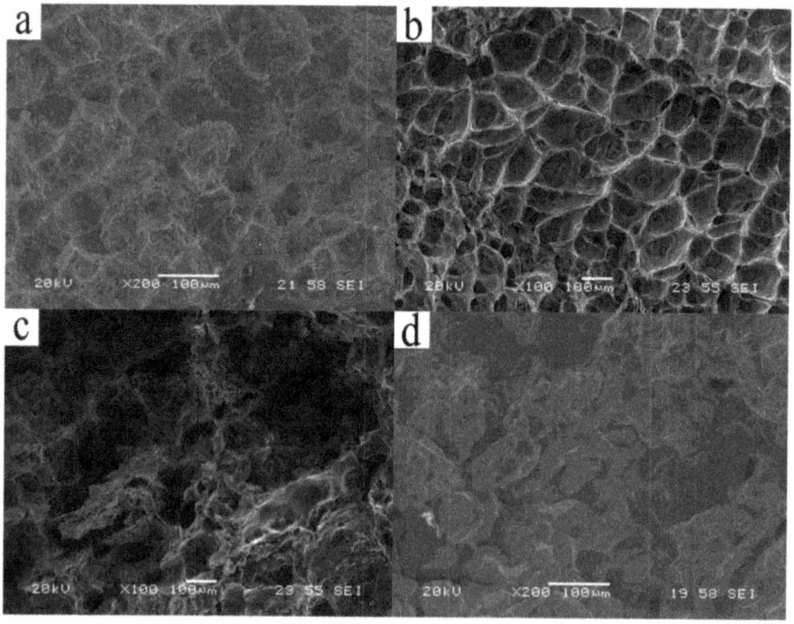

Figure 5. Steel tensile fracture morphologies at different temperatures: (**a**) 700 °C; (**b**) 925 °C; (**c**) 1250 °C; (**d**) 1300 °C.

3.3. Brittleness Analysis

Figure 6 is a comparison diagram of the high-temperature phase transformation of the sample calculated by the Thermo-Calc software and the TCFE9 database. In Figure 6, the sample did not go through the single-phase (γ) region, but through (δ + γ) two-phase region, in the temperature range of 705–1406 °C. However, the ductility of the sample was excellent due to the fact that most of the RAs were higher than 40%. This is because the amount of the δ-ferrite of the sample was high and the proportion of the film-like α-ferrite was small.

Figure 6 shows that, at the temperatures between 705 and 1406 °C, the δ-ferrite was continuously transformed into the γ phase. The proportion of γ increased to 76% at 1100 °C and then began to decrease, and the ferrite content began to rise from 24%. It is speculated that the transformation of the γ phase to the α-ferrite occurred at this time.

The austenite transformed into the α-ferrite at the temperature range between 705 and 1100 °C. This indicated that the reason why the ductility of the sample was relatively poor in the third brittle zone is that the massive transforms from the austenite to the α-ferrite and the primary film-like ferrite grains generated along the austenite grain boundary produced slips. The ductility of the sample decreased first and then increased with the increase of the α-ferrite. In the second ductility zone, the sample was in the two-phase regions of the ferrite phase and the γ phase. The ductility of the sample was excellent in the second ductility zone. Al is an element that inhibited the formation of the austenite grains, so that the austenite grains were small and the ductility was excellent.

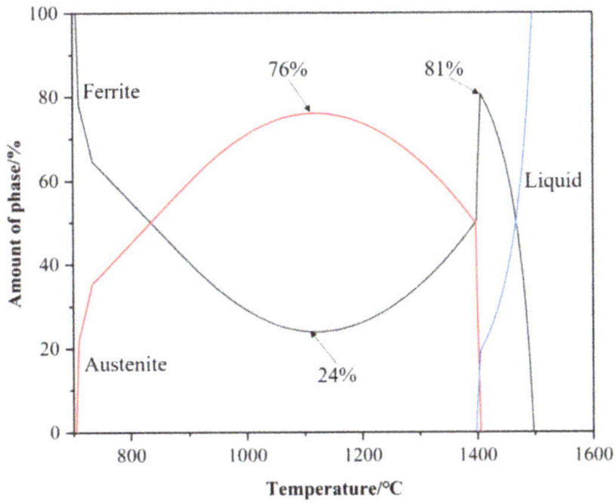

Figure 6. Amount of phase in the 4.5% Al-containing δ-TRIP steel.

In the temperature range of 1110–1406 °C, the δ-ferrite transformed into austenite, and the crystal structure was transformed from a body-centerd cubic to a face-centerd cubic. Moreover, within the temperature range of 705–1100 °C, the austenite transformed into the α-ferrite, the crystal structure was transformed from a face-centerd cubic to a body-centerd cubic, and the strength of the face-centerd cubic crystal was greater than that of the body-centerd cubic crystal. With the transformation going on and the α-ferrite content increasing, this will lead to lowering the substrate ductility.

3.4. Phase Transformation Comparison

In order to compare the actual phase transformation temperature of the sample during heating with the phase transformation temperature by using the Thermo-Calc calculation, the sample was tested by the DSC. Figure 7 shows the DSC analysis curve of the 4.5% Al-containing δ-TRIP steel heated from 40 to 1400 °C at a heating rate of 10 °C/min to identify the phase transformation temperature.

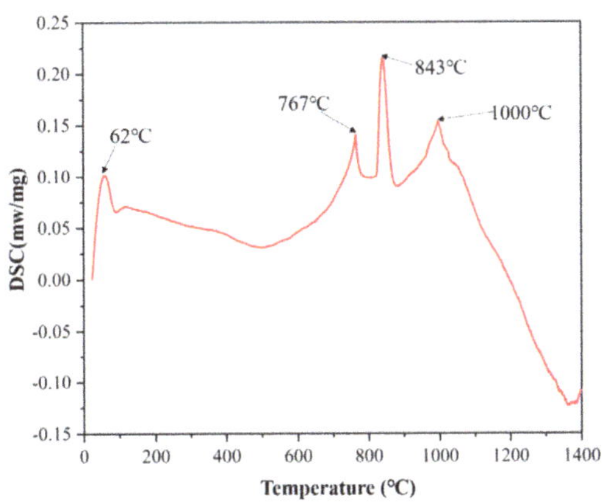

Figure 7. Differential scanning calorimetry (DSC) analysis curve.

It can be seen from the figure that there is a significant endothermic peak of the sample at 767, 843, and 1000 °C. It indicates that the matrix transformed from the γ phase to the α-ferrite at 1000 °C and the phase transformation ended at 767 °C. Ignoring the test error and the instrument temperature fluctuation, the third brittle zone of the sample is consistent with its phase transformation zone. It shows that the reason for the embrittlement of sample in the third brittle zone is the transformation of the austenite to the α-ferrite. The primary film-like α-ferrite has low strength and will lead to lowering substrate ductility. The peak temperature of 62 °C is the result of heat-up.

4. Conclusions

1. The ZDT of the sample was 1355 °C, and the ZST was 1405 °C. The ΔT of the sample was small. The 4.5% Al-containing δ-TRIP steel had good resistance to the high-temperature cracking.
2. The sample had the first brittle zone at 1300–1350 °C and the third brittle zone at 800–975 °C at a constant strain rate of 1×10^{-3} s^{-1}. During the continuous casting process, the surface temperature of the slab should be kept away from 800 to 975 °C to reduce the possibility of surface cracking on the slab.
3. The reason for the embrittlement of the third brittle zone of the 4.5% Al-containing δ-TRIP steel is that the α-ferrite formed at the austenite grain boundary caused the sample to crack along the grain boundary under stress.
4. The ductility of the 4.5% Al-containing δ-TRIP steel decreased first and then increased with the increase of the α-ferrite. When the proportion of the α-ferrite reached 37%, the RA of the 4.5% Al-containing δ-TRIP steel was reduced to 44%.

Author Contributions: Investigation, D.C.; Methodology, D.C. and H.C.; Data curation, D.C.; Writing of original draft preparation, D.C.; Writing of review and editing, D.C., H.C., and R.W.; Supervision, H.C.; Funding acquisition, H.C.

Funding: The research was financially supported by the National Natural Science Foundation of China (grant number: U1860106) and the China Scholarship Council (grant number: 201806465050).

Acknowledgments: The authors would like to thank C. Bernhard and P. Presoly from Montanuniversität Leoben for providing access to the Thermo-Calc software and the TCFE9 database.

Conflicts of Interest: The authors declare no conflicts of interest.

References

1. Yi, H.L.; Sun, L.; Xiong, X.C. Challenges in the formability of the next generation of automotive steel sheets. *Mater. Sci. Technol.* **2018**, *34*, 1112–1117. [CrossRef]
2. Chatterjee, S.; Murugananth, M.; Bhadeshia, H. δ-TRIP steel. *Mater. Sci. Technol.* **2007**, *23*, 819–827. [CrossRef]
3. Yi, H.L.; Chen, P.; Hou, Z.Y.; Hong, N.; Cai, H.L.; Xu, Y.B.; Wu, D.; Wang, G.D. A novel design: Partitioning achieved by quenching and tempering (Q–T & P) in an aluminium-added low-density steel. *Scr. Mater.* **2013**, *68*, 370–374.
4. Abedi, H.R.; Hanzaki, A.Z.; Liu, Z.; Xin, R.; Haghdadi, N.; Hodgson, P.D. Continuous dynamic recrystallization in low density steel. *Mater. Des.* **2017**, *114*, 55–64. [CrossRef]
5. Shiri, S.G.; Jahromi, S.A.J.; Palizdar, Y.; Belbasi, M. Unexpected Effect of Nb Addition as a Microalloying Element on Mechanical Properties of δ-TRIP Steels. *J. Iron Steel Res. Int.* **2016**, *23*, 988–996. [CrossRef]
6. Jiao, Z.B.; Luan, J.H.; Miller, M.K.; Yu, C.Y.; Liu, C.T. Effects of Mn partitioning on nanoscale precipitation and mechanical properties of ferritic steels strengthened by NiAl nanoparticles. *Acta Mater.* **2015**, *84*, 283–291. [CrossRef]
7. Choi, Y.J.; Suh, D.W.; Bhadeshia, H. Retention of δ-ferrite in aluminium-alloyed TRIP-assisted steels. *Proc. R. Soc. A Math. Phys. Eng. Sci.* **2012**, *468*, 2904–2914. [CrossRef]
8. Hwang, S.W.; Ji, J.H.; Lee, E.G.; Park, K.-T. Tensile deformation of a duplex Fe-20Mn-9Al-0.6C steel having the reduced specific weight. *Mater. Sci. Eng. A* **2011**, *528*, 5196–5203. [CrossRef]
9. Kaar, S.; Krizan, D.; Schwabe, J.; Hofmann, H.; Hebesberger, T.; Commenda, C.; Samek, L. Influence of the Al and Mn content on the structure-property relationship in density reduced TRIP-assisted sheet steels. *Mater. Sci. Eng. A* **2018**, *735*, 475–486. [CrossRef]

10. Xiong, X.C.; Sun, L.; Wang, J.F.; Jin, X.Y.; Wang, L.; Xu, B.Y.; Chen, P.; Wang, G.D.; Yi, H.L. Properties assessment of the first industrial coils of low-density duplex δ-TRIP steel. *Mater. Sci. Technol.* **2016**, *32*, 1403–1408. [CrossRef]
11. Tuling, A.; Banerjee, J.R.; Mintz, B. Influence of peritectic phase transformation on hot ductility of high aluminium TRIP steels containing Nb. *Mater. Sci. Technol.* **2011**, *27*, 1724–1731. [CrossRef]
12. Su, H.; Gunawadarna, W.D.; Tuling, A.; Mintz, B. Influence of Al and P additions on hot ductility of steels. *Mater. Sci. Technol.* **2007**, *23*, 1357–1366. [CrossRef]
13. Qaban, A.; Mintz, B.; Kang, S.E.; Naher, S. Hot ductility of high Al TWIP steels containing Nb and Nb-V. *Mater. Sci. Technol.* **2017**, *33*, 1645–1656. [CrossRef]
14. Liu, H.; Liu, J.; Wu, B.; Shen, Y.; He, Y.; Ding, H.; Su, X. Effect of Mn and Al contents on hot ductility of high alloy Fe-xMn-C-yAl austenite TWIP steels. *Mater. Sci. Eng. A* **2017**, *08*, 60–374. [CrossRef]
15. Kang, S.E.; Tuling, A.; Banerjee, J.R.; Gunawardana, W.D.; Mintz, B. Hot ductility of TWIP steels. *Mater. Sci. Technol.* **2011**, *27*, 95–100. [CrossRef]
16. Xiong, Z.P.; Kostryzhev, A.G.; Stanford, N.E.; Pereloma, E.V. Effect of deformation on microstructure and mechanical properties of dual phase steel produced via strip casting simulation. *Mater. Sci. Eng. A* **2016**, *651*, 291–305. [CrossRef]
17. He, Y.; Liu, J.; Qiu, S.; Deng, Z.; Yang, Y.; McLean, A. Microstructure and high temperature mechanical properties of as-cast FeCrAl alloys. *Mater. Sci. Eng. A* **2018**, *726*, 56–63. [CrossRef]
18. He, Y.; Liu, J.; Han, Z.; Deng, Z.; Su, X.; Ji, Y. Phase transformation and precipitation during solidification of FeCrAl alloy for automobile exhaust gas purifying systems. *J. Alloy. Compd.* **2017**, *714*, 251–257. [CrossRef]
19. Grajcar, A.; Kwaśny, W. Microstructural study on retained austenite in advanced high-strength multiphase 3Mn-1.5 Al and 5Mn-1.5 Al steels. *J. Exp. Bot.* **2012**, *54*, 168–177.
20. Hechu, K.; Slater, C.; Santillana, B.; Clark, S.; Sridhar, S. A novel approach for interpreting the solidification behaviour of peritectic steels by combining CSLM and DSC. *Mater. Charact.* **2017**, *133*, 25–32. [CrossRef]
21. Li, Z.C.; Misra, R.D.K.; Cai, Z.H.; Li, H.X.; Ding, H. Mechanical properties and deformation behavior in hot-rolled 0.2 C-1.5/3Al-8.5 Mn-Fe TRIP steel: The discontinuous TRIP effect. *Mater. Sci. Eng. A* **2016**, *673*, 63–72. [CrossRef]
22. Li, Z.C.; Ding, H.; Cai, Z.H. Mechanical properties and austenite stability in hot-rolled 0.2C-1.6/3.2Al-6Mn-Fe TRIP steel. *Mater. Sci. Eng. A* **2015**, *639*, 559–566. [CrossRef]
23. Cui, H.; Zhang, K.; Wang, Z.; Chen, B.; Liu, B.; Qing, J.; Li, Z. Formation of surface depression during continuous casting of high-Al TRIP steel. *Metals* **2019**, *9*, 204. [CrossRef]
24. Nakagawa, T.; Umeda, T.; Murata, J.; Kamimura, Y.; Niwa, N. Deformation behavior during solidification of steels. *ISIJ Int.* **1995**, *35*, 723–729. [CrossRef]
25. Yu, C.H.; Suzuki, M.; Shibata, H.; Emi, T. Simulation of crack formation on solidifying steel shell in continuous casting mold. *ISIJ Int.* **1996**, *36*, S159–S162. [CrossRef]
26. Suzuki, H.G.; Eylon, D. Hot ductility of titanium alloy: A challenge for continuous casting process. *Mater. Sci. Eng. A* **1998**, *243*, 126–133. [CrossRef]
27. Li, S.Q.; Liu, J.H.; Liu, H.B.; Zhuang, C.L.; Liu, J.; Han, Z.B. Study on high-temperature mechanical properties of low-carbon Fe-Mn-Si-Al TWIP steel. *High Temp. Mater. Process.* **2017**, *36*, 505–513. [CrossRef]
28. Suzuki, H.G.; Nishimura, S.; Imamura, J.; Nakamura, Y. Hot Ductility in Steels in the Temperature Range between 900 and 600 C. *Trans. Iron Steel Inst. Jpn.* **1981**, *67*, 1180–1189. [CrossRef]

applied sciences

MDPI

Article

The Effect of Tempering on the Microstructure and Mechanical Properties of a Novel 0.4C Press-Hardening Steel

Oskari Haiko [1,*], Antti Kaijalainen [1], Sakari Pallaspuro [1], Jaakko Hannula [1], David Porter [1], Tommi Liimatainen [2] and Jukka Kömi [1]

1 Materials and Mechanical Engineering, Centre for Advanced Steels Research, University of Oulu, 90014 Oulu, Finland; antti.kaijalainen@oulu.fi (A.K.); sakari.pallaspuro@oulu.fi (S.P.); jaakko.hannula@oulu.fi (J.H.); david.porter@oulu.fi (D.P.); jukka.komi@oulu.fi (J.K.)

2 Raahe Works, SSAB Europe, 92100 Raahe, Finland; tommi.liimatainen@ssab.com

* Correspondence: oskari.haiko@oulu.fi

Received: 12 September 2019; Accepted: 4 October 2019; Published: 10 October 2019

check for updates

Featured Application: Potential wear-resistant steel for harsh environments in agricultural sector, i.e., chisel ploughs and disc harrows.

Abstract: In this paper, the effects of different tempering temperatures on a recently developed ultrahigh-strength steel with 0.4 wt.% carbon content were studied. The steel is designed to be used in press-hardening for different wear applications, which require high surface hardness (650 HV/58 HRC). Hot-rolled steel sheet from a hot strip mill was austenitized, water quenched and subjected to 2-h tempering at different temperatures ranging from 150 °C to 400 °C. Mechanical properties, microstructure, dislocation densities, and fracture surfaces of the steels were characterized. Tensile strength greater than 2200 MPa and hardness above 650 HV/58 HRC were measured for the as-quenched variant. Tempering decreased the tensile strength and hardness, but yield strength increased with low-temperature tempering (150 °C and 200 °C). Charpy-V impact toughness improved with low-temperature tempering, but tempered martensite embrittlement at 300 °C and 400 °C decreased the impact toughness at −40 °C. Dislocation densities as estimated using X-ray diffraction showed a linear decrease with increasing tempering temperature. Retained austenite was present in the water quenched and low-temperature tempered samples, but no retained austenite was found in samples subjected to tempering at 300 °C or higher. The substantial changes in the microstructure of the steels caused by the tempering are discussed.

Keywords: steel; martensite; tempering; press-hardening

1. Introduction

Steels with yield strength greater than 1300 MPa are becoming more widely used in different applications. An increase in strength provides the possibility to use less material, i.e., to design and use lighter structures. The result is increased energy efficiency and fuel savings that not only translate into financial gains but also may be required by different norms and regulations. Also, increased steel strength leads to higher hardness and abrasive wear resistance. Generally, the strongest steels used in structural, wear, and protection applications have martensitic microstructures. The hardness of quenched martensite increases with the increasing carbon content (up to 0.6–0.8 wt.%) up to more than 800 HV with tensile strength levels exceeding 2500 MPa for quenched plain carbon and low-alloy compositions [1]. Martensite may be lath or plate-like or a mixture of both. Below the 0.6 wt.% carbon content, the fine lath martensite is dominant. The substructure of prior austenite grains in martensite can be divided into packets which contain blocks and sub-blocks of individual laths [2,3]. These features comprise the fine and complex structure of the martensite phase that can be altered by chemical composition and processing. Carbon atoms segregated in the structure, prior austenite grain size, dislocation density, different precipitates, and retained austenite together contribute to the superior strength and hardness of martensite, with carbon responsible for the main contribution [1–8]. Furthermore, the highest hardness and strength levels are often achieved by expensive, heavy alloying, or by the utilization of time-consuming heat treatments in order to meet requirements related to usability and toughness. However, thermomechanical controlled processing (TMCP) is nowadays utilized to refine the grain structure for improved impact toughness, especially in the case of ultrahigh-strength steels. The rolling schedule is adjusted to include passes in the non-recrystallization regime (NRX), which results in an elongated austenite grain structure with an increased dislocation density and improved impact toughness [9–11]. However, the formability of ultrahigh-strength steels with relatively high carbon content is limited and might turn out be challenging when producing complex shapes or small components, i.e., blades for furrow ploughs or disc harrows. Therefore, these types of high-hardness steels are provided in the hot-rolled condition to the parts manufacturer for further processing: the final shape is given to the part in the reheating process while in the soft and easily formable condition, after which suitable quenching can be applied. Different press and hardening methods are utilized for processing high-hardness steels. Oil, water, or different aqueous solutions are used for the quenching, which requires careful planning to avoid quench-cracking. Often the process includes tempering to improve the impact toughness of the material. When abrasive wear resistance is required, high hardness needs to be retained after tempering. AISI 4340, 300M, and 51CrV4 with a carbon content of around 0.4–0.5 wt.% are commonly used in the aforementioned applications.

Tempering is a common method to improve the ductility and impact toughness of steels. It is widely applied for ultrahigh-strength martensitic steels due to the limited toughness properties of quenched medium- and high-carbon steels. Often, the increasing carbon content and the subsequent increase in hardness and strength impairs the toughness properties of carbon steels in their as-quenched state. Therefore, different temperatures and tempering times can be used depending on the desired mechanical properties, such as suitable hardness and impact toughness levels [1,12,13]. Tempering temperatures up to 650 °C can be utilized to adjust the mechanical properties of steels for different applications: low-alloy, low-carbon steels benefit from improved low-temperature impact toughness and bendability properties. Low-temperature tempering in the temperature range of 150–200 °C is applied to ultrahigh-strength carbon steels with a medium carbon content, as it usually provides a significant improvement of impact toughness without drastically decreasing hardness and tensile strength. Generally, the tempering temperature affects the mechanical properties more than the tempering time when tempering temperatures are 500 °C or less [14]. Apart from the tempering applied after the quenching process, auto-tempering may also take place during the cooling process. The higher the martensite start temperature (M_s), the more time there is for the carbon atoms to diffuse and form carbides during quenching [12]. The martensite that is formed near the M_s temperature has the highest probability for auto-tempering with the remaining cooling process [13]. After auto-tempering, the fraction of less-tempered hard martensite decreases, which in turn may improve ductility [15] and impact toughness [16].

In order to study the effect of tempering on an ultrahigh-strength quenched steel, a newly developed quench and press-hardening steel, was investigated. Different tempering temperatures were applied to understand the microstructural changes and the evolution of mechanical properties. The steel has been designed to be used in its as-quenched or tempered condition in different applications that require high abrasive wear resistance.

2. Materials and Methods

A hot-rolled steel sheet with 10 mm thickness was obtained from the strip mill production line from SSAB, Raahe, Finland. The large sheet was cut to smaller test plates with dimensions of 300 mm × 250 mm. The plates were then reheated and held at 980 °C for 40 min before submerging into a water tank for a rapid quenching (cooling rate > 80 °C/s). The quenched samples were then tempered for 2 h at selected temperatures (150 °C, 200 °C, 300 °C, and 400 °C). A hot-rolled sample and water quenched sample without any tempering were also included for comparison (abbreviated HR and WQ in the article, respectively).

A Gleeble 3800 thermomechanical simulator was used in order to create a continuous cooling transformation (CCT) diagram for the steel. Samples were cut from the received hot-rolled material in the longitudinal direction. A total of six cooling rates in the range 3–96 °C/s were applied. Samples were machined to a cylindrical shape with dimensions of 6 mm × 9 mm (cooling rates from 3 to 48 °C/s) or 5 mm × 7.5 mm (cooling rate 96 °C/s). The heating rate was 10 °C/s and austenitization temperature was 920 °C. The holding time was 10 min for each sample. No strain was applied during the Gleeble testing. Phase transformation temperatures were identified from the Gleeble dilatation data and phase fractions were calculated from the images taken with an optical microscope. In order to simulate the industrial reheating and quenching process, the samples were not homogenized prior to the Gleeble testing.

The chemical composition for the steel is presented in Table 1. The composition has been designed to give full hardening for a plate thickness of approximately 12 mm. Carbon content was 0.4 wt.%, which should provide hardness exceeding 650 HV10 (Vickers hardness number) or 58 HRC (Rockwell C hardness). The avoidance of expensive alloying elements, such as nickel, was a major consideration in the alloy design. Instead of nickel and chromium, molybdenum has been selected as the main hardenability element along with the medium carbon content. Earlier trials with laboratory casts showed that molybdenum had a strong effect on the hardenability of the steel and could be utilized to achieve the desired hardness levels. Manganese has been kept relatively low to avoid the possible deterioration of impact toughness inflicted by segregation. Microalloying with Nb and Ti has been used to prevent excessive grain growth at higher austenitization temperatures. The studied steel has a lower manganese, chromium, and nickel content in comparison to the well-known commercial steel grades, such as the AISI 4340, 300M, and 51CrV4. Furthermore, commercial boron steels used in similar applications often have notably higher manganese levels.

Table 1. Chemical composition of the steel (wt.%, balance Fe).

C	Si	Mn	Al	Cr	Mo	Ni	Ti	Nb	V	N
0.43	0.19	0.25	0.035	0.05	0.7	0.05	0.013	0.013	0.04	0.004

The test plates were water jet cut and machined to suitable dimensions for mechanical testing and microstructural inspection. Three longitudinal flat tensile test specimens of each material variant were tested using an MTS 810 servohydraulic universal testing machine. The dimensions for the reduced section of the tensile test specimens were 10 mm × 20 mm × 120 mm. Charpy-V impact toughness testing was conducted at room temperature and at −40 °C with three specimens for each material condition at both temperatures. The transverse impact tests were made according to ISO 148-1 standard. The specimen size was 7.5 mm × 10 mm × 55 mm for the impact tests for all samples. A Struers Duramin A-300 tester was used to measure hardness with a minimum of seven indentations (HV10) through the thickness of the material.

Microstructural characterization was done with a field emission scanning electron microscope (FESEM, Zeiss Sigma) and with a laser scanning confocal microscope (Keyence VK-X200). Samples were polished and etched with Nital (2%) prior to the inspection. Prior austenite grain size (PAGS) measurements were done from planar sections of samples etched with picric acid.

X-Ray diffraction (XRD) studies were carried out on specimens with their surfaces at the quarter-thickness depth from the surface using a Rigaku SmartLab 9 kW X-ray diffractometer with cobalt Kα radiation. PDXL2 analysis software (version 2.6.1.2, Rikagu Corporation, Tokyo, Japan) was used to estimate the lattice parameters, microstrains, and crystallite sizes of the studied steels. Rietveld's whole powder profile fitting method was used also for the determination of retained austenite content. Furthermore, dislocation densities were calculated using the Williamson–Hall method (Equation (1)) [17,18]:

$$\rho = \sqrt{\rho_s \rho_p} \tag{1}$$

where ρ_s is the dislocation density calculated from strain broadening, and ρ_p is the dislocation density calculated from crystallite size. According to Williamson et al. ρ_s and ρ_p can be calculated using the following Equations (2) and (3) [17,18]:

$$\rho_s = \frac{k\,\varepsilon^2}{F\,b^2} \tag{2}$$

and

$$\rho_p = \frac{3n}{D^2} \tag{3}$$

where ε is the microstrain, b is the Burgers vector, F is an interaction factor assumed to be 1, factor k is assumed as 14.4 for body-centred cubic metals, and D is crystallite size. In the equation, n is dislocations per block face, assumed as 1 [18].

3. Results

3.1. Dilatometry and CCT Diagram

The continuous cooling transformation (CCT) diagram determined from the Gleeble dilatometer traces is presented in Figure 1. The hardness was nearly 650 HV10 with the highest cooling rate (96 °C/s), but 600 HV10 was still achieved with 24 °C/s. Since the hardness level of 600 HV was still obtained with lower cooling rates, more time for auto-tempering could be provided with the use of a slower cooling media, such as oil or a polymer suspension. The slower cooling rates will not result in a fully martensitic microstructure, as shown in the CCT diagram of Figure 1 and the micrographs in Figure 2. The martensite start temperature was estimated as approximately 350–380 °C based on the CCT diagram.

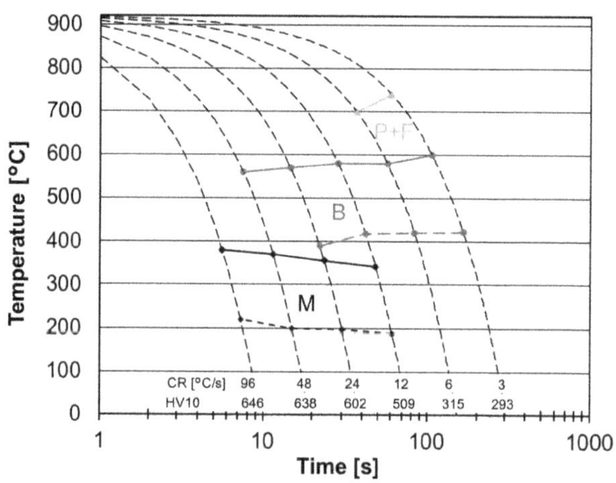

Figure 1. Continuous cooling transformation (CCT) diagram for the steel. Abbreviations: P + F is pearlite and ferrite, B is bainite and M is martensite.

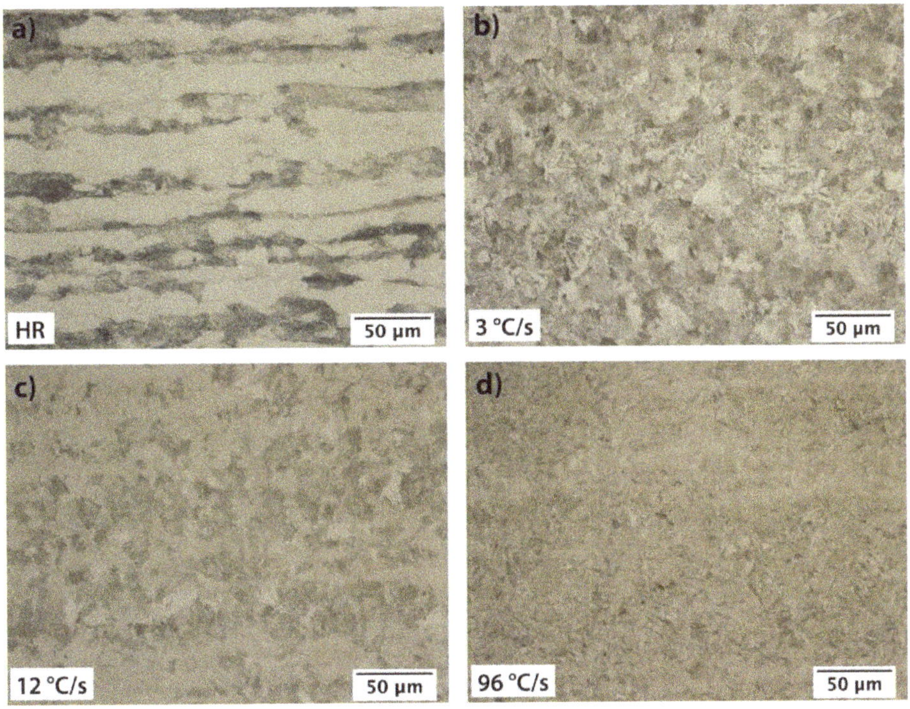

Figure 2. Optical micrographs of the hot-rolled (HR) microstructure and Gleeble samples with different cooling rates: (**a**) hot-rolled, (**b**) 3 °C/s, (**c**) 12 °C/s, (**d**) 96 °C/s.

The optical microscope images show that low cooling rates produced multiphase microstructures. The hot-rolled (HR) sample represents the microstructure prior to Gleeble testing or reheating and quenching. The lowest cooling rate (3 °C/s) tested with the Gleeble had a highly mixed microstructure consisting of bainite (approximately 50–55%), pearlite (30–40%), and ferrite (5–10%). There was very little pearlite and ferrite with a cooling rate of 12 °C/s and the phase fractions were estimated at 50%–60% for martensite and 40–50% for bainite. Bainite was still present at 48 °C/s but at less than 10%. The highest cooling rate (96 °C/s) showed close to 100% martensite with only a minor fraction of bainite. The phase fractions were highly localized within the Gleeble samples with the slow and medium cooling rates (<48 °C/s). Based on the dilatometry, high quenching rates can be used to obtain hard, fully martensitic microstructures, whereas slower cooling rates with possible isothermal holding provide possibilities for obtaining different bainitic or multiphase structures.

3.2. *Microstructures*

The hot-rolled sample (HR) showed a banded ferritic-pearlitic microstructure with no martensite present (Figure 2a) and with a hardness below 200 HV10. The absence of martensite and bainite was due to the low level of alloying combined with the fact that the hot-rolled material was coiled at around 700 °C, after which it cooled very slowly, which can take up to three days to cool down to room temperature. Figure 3a shows the resulting ferritic-pearlitic microstructure. This is a typical microstructure for plain carbon steels when quenching has not been applied.

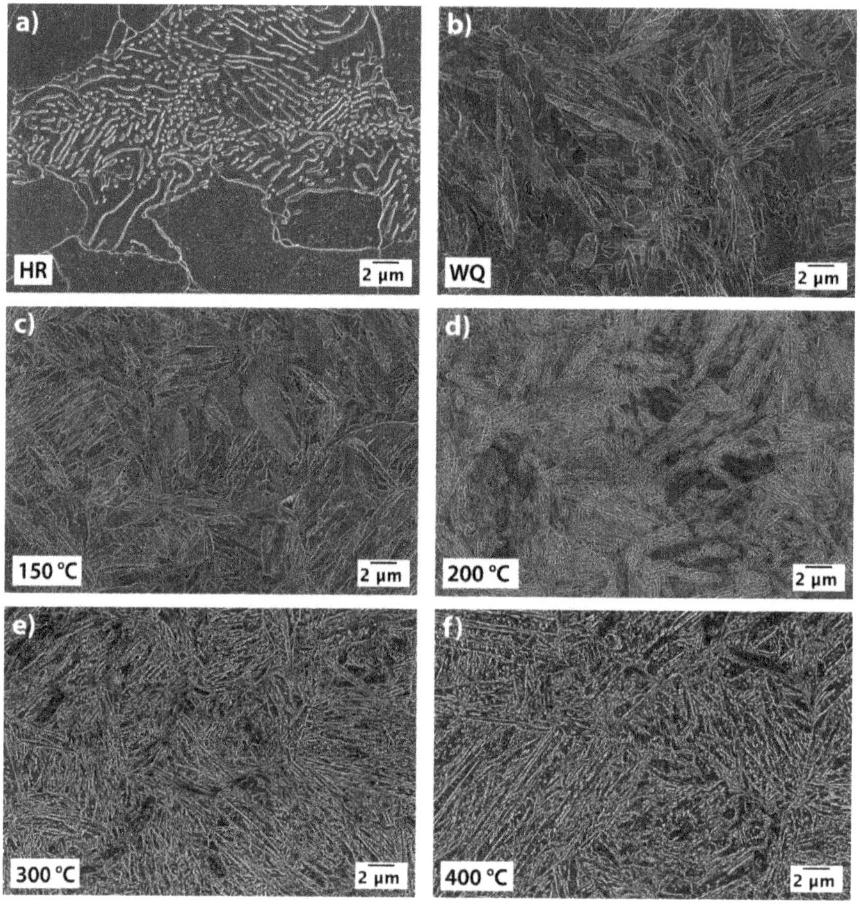

Figure 3. Field emission scanning electron microscope (FESEM) micrographs of the tested steels. Images taken from $\frac{1}{4}$ depth from the top surface in normal-to-rolling direction (longitudinal). Samples: (**a**) hot-rolled, (**b**) water quenched, (**c**) tempered at 150 °C, (**d**) tempered at 200 °C, (**e**) tempered at 300 °C, (**f**) tempered at 400 °C.

Microstructural evolution during tempering is seen in the quenched and tempered samples in the FESEM images in Figure 3b–f. Starting from the water quenched sample, the microstructure consisted of lath-like martensite comprising packets, blocks, and sub-blocks. The darkest auto-tempered martensite was the most dominant phase in the water quenched microstructure (Figure 3b), as is typical of other commercial medium carbon steels, such as some wear-resistant steels [19–23]. Auto-tempering is desirable for this type of steel in order to improve the impact toughness in the as-quenched state. Despite being subjected to rapid water cooling, the samples did not display any signs of immediate or delayed quench-cracking indicating that milder quenching media are not required for the 10 mm plate thickness concerned, at least.

The effects of tempering begin to appear more distinctly at 150 °C. The fraction of precipitate-rich, light areas increased, but the precipitate size was generally very small. The sample tempered at 200 °C showed significantly less untempered martensite, i.e., the darkest areas without visible traces of precipitates, than the water quenched (WQ) and 150 °C samples. This is consistent with the first stages of tempering, i.e., precipitation of fine, presumably transition, carbides [12,13]. Both the 150 °C and

200 °C samples contained a high number density of fine transition carbides, but in the sample tempered at 200 °C the precipitates were notably larger, and their needle-like shape could be distinguished more easily. These differences result in the microstructure of the 200 °C sample showing more light contrast in the low magnification FESEM images.

The water quenched (WQ) sample had more carbide-free areas while the auto-tempering had introduced some carbide formation (Figure 4a). In the FESEM images, this was also seen as the different depth of the areas as the regions could be assumed to have different carbon concentrations in the solid solution [24]. This supports the concept of martensite formation in rapid sequences rather than as a single massive transformation throughout the whole sample at once. The darker areas appeared as there would have been nearly a total absence of precipitates. Therefore, this martensite may be assumed to have had formed in the later stages of the phase transformation, having less time for the interstitial carbon to diffuse into precipitates.

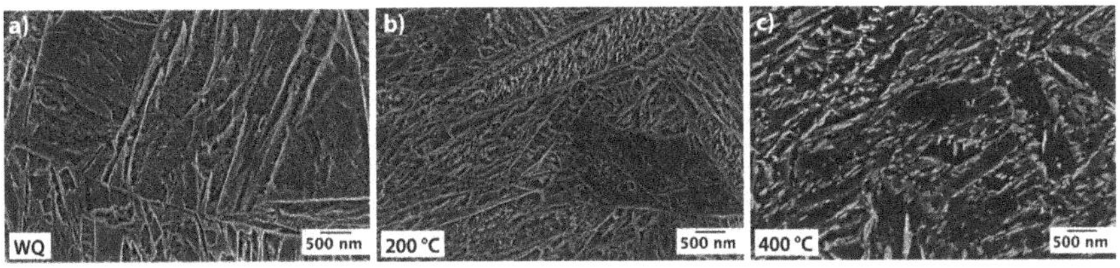

Figure 4. FESEM images of (**a**) water quenched, (**b**) tempered at 200 °C for 2 h and (**c**) tempered at 400 °C for 2 h.

At the higher tempering temperatures of 300 °C and 400 °C there is a higher contrast between the dark martensite matrix and the white carbides (Figure 4b,c), presumably cementite (θ-carbide), which have replaced the small transition carbides found at the lower tempering temperatures [12]. The various lath, block, packet, and grain boundaries are seen to act as sites for cementite nucleation [25]. The precipitates are clearly coarser after tempering at 400 °C than at 300 °C. Therefore, the microstructure of the 400 °C tempered sample was assumed closer to conventional tempered ferrite and cementite than that of martensite [12,13].

Based on the FESEM imaging, the evolution of the martensitic steel from the as-quenched state to the 400 °C tempered microstructure appeared as to have decreased homogeneity. The water quenched sample had far less precipitates and they were clearly smaller than those of the tempered steels. Then, the carbon migration slowly altered the structure into a clearer two-component structure with the dark matrix and the white carbides when the tempering temperature exceeded 300 °C. The water quenched variant and the 150 °C tempered samples had the most uniform structure, and also showed mechanical properties very close to each other, which will be discussed further in the next section.

The prior austenite grain size (PAGS) was measured for the 300 °C tempered sample using a calculation tool recently developed by Seppälä et al. [26]. The calculation is based on the mean linear intercept method of Sellars and Higginson [27]. The calculated mean prior austenite grain size was 12.3 μm with a standard deviation of 0.6 μm. Therefore, the somewhat high austenitization temperature of 980 °C had not caused excessive grain growth as seen in both the optical image (Figure 5a) and in the grain size distribution calculations (Figure 5b). Only a few large grains were detected (24–30 μm) and no abnormal grain growth was present. For comparison, some examples of PAGS for austenitization temperatures below 1000 °C of similar grade steels have been reported by Ritchie et al. [28]. For AISI 4340, austenitization at 870 °C for 1 h resulted in a grain size ranging from 24 to 32 μm. Austenite grain growth for the 300M ultrahigh-strength steel (a modified version of 4340 with higher silicon content) has also been studied in the temperature range of 850 to 1150 °C with different holding times. Zhang et al. [29] discovered that the grain growth rate increased rapidly when soaking 300M above 900 °C. Heat treatment at 950 °C resulted in an average grain size of 31 μm, while a heating temperature of 1000 °C increased the grain size to 46 μm. Chen et al. [30] have found in their work that the average grain size for 300M was greater than 50 μm in when austenized at 900 °C or above for 40 min. Horn et al. [31] measured PAGS of 20 μm for 4340 and 300M steels with an 870 °C austenitization temperature. Therefore, the mean prior austenite grain size (12.3 μm) can be considered to be small for austenitization temperatures of 950 to 1000 °C. This is probably attributable to the presence of the microalloying elements Ti and Nb, which have inhibited the grain growth through the formation of small finely dispersed carbides or carbonitrides [32]. Also, the relatively large amount of molybdenum (0.7 wt.%) can hinder the austenite grain growth by producing strong solute drag effect by segregating at the austenite grain boundaries and thus pinning the grain boundary movement [33].

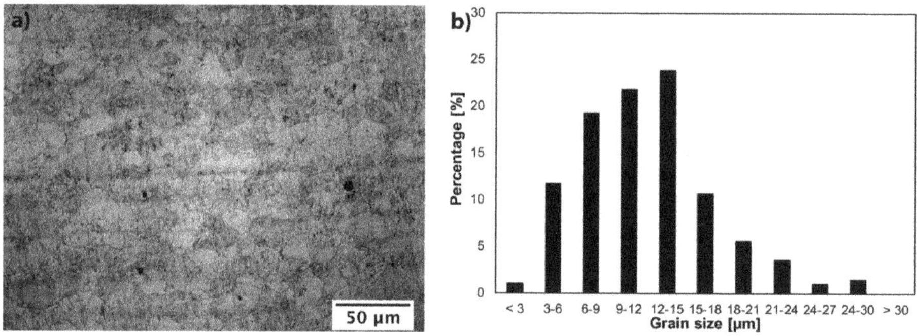

Figure 5. (**a**) Picric acid etched sample (300 °C tempered) revealing the prior austenite grain boundaries (normal-to-rolling direction) and (**b**) the prior austenite grain size distribution.

3.3. Mechanical Properties

3.3.1. General Observations

The mechanical properties are presented in Figure 6. The hot-rolled variant showed poor performance when compared to the quenched and tempered variants. The strength and hardness were inferior to the other tested samples that had a martensitic microstructure. Yield strength remained under 500 MPa at 471 ± 24 MPa while the tensile strength was 711 ± 23 MPa for the hot-rolled state (Figure 6a). The measured hardness was 195 ± 4 HV10 (Figure 6b). Also, the ferritic-pearlitic steel had very poor impact toughness as the tested Charpy-V samples showed only a few joules of impact energy. Only the uniform elongation exceeded that of the martensitic variants. Nevertheless, the steel has not been designed to be used in the hot-rolled condition, but the HR variant was included for comparison to understand the effect of both quenching and tempering on the mechanical properties of the steel.

The water quenched version without any tempering exhibited the highest tensile strength exceeding 2200 MPa (Figure 6c). However, the highest hardness (668 ± 15 HV10) was measured for the variant tempered at 150 °C. The highest yield strength (1616 ± 13 MPa) was measured for the sample tempered at 200 °C, which surpassed the WQ (1528 ± 13 MPa) and 150 °C (1606 ± 22 MPa) samples. Both yield and tensile strength decreased rapidly with tempering temperatures of 300 °C and 400 °C, and hardness decreased as well. Uniform elongation also followed the same decreasing trend with the tensile strength and hardness. Impact toughness was improved with the low-temperature tempering up to 200 °C but tempering at 300 °C or 400 °C did not produce any further improvement in the low-temperature Charpy-V energies (Figure 6d). Impact energies were measured in joules and converted directly proportional to J/cm^2.

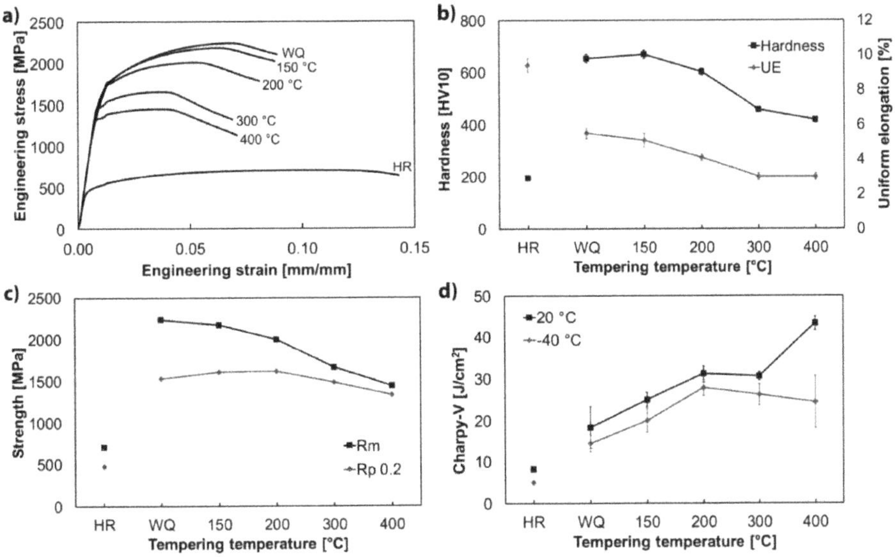

Figure 6. Mechanical properties of the tested steels: (**a**) engineering tensile test curves, (**b**) material hardness and plastic component of the uniform elongation (UE), (**c**) yield strength (0.2% offset, $R_{p0.2}$) and tensile strength (R_m), and (**d**) Charpy-V impact test energies at +20 and −40 °C temperatures. Standard deviations are included in (**b**–**d**).

3.3.2. Strength and Hardness

The general strength decrease during tempering can be attributed to the diffusion of carbon atoms, decomposition of retained austenite, and dislocation recovery effects [24] in the tempering temperatures used in the current study. As stated earlier, the precipitation of cementite was observed in the FESEM images, and the microstructures of the 300 °C and 400 °C samples differed greatly from the quenched and low-temperature tempered samples. Now, the effect on the mechanical properties can be distinguished in Figure 6. Ultimate tensile strength (UTS) began to decrease already with the 150 °C tempering temperature. The measured difference was very small (WQ: 2233 ± 9 MPa vs. 150 °C: 2168 ± 28 MPa) and almost within the limits of measuring error. The decrease of UTS was more prominent for the 200 °C tempered variant. The temperature increase of 50 °C from the 150 to 200 °C tempering temperature had a significant effect on the tensile strength, when comparing to the clearly less dramatic difference in tensile strength between the WQ and 150 °C samples. Further, the tempering at 300 °C and 400 °C caused the ultimate tensile strength to decrease even more.

The hardness for the water quenched steel was on the expected level reaching 653 ± 16 HV10. The low tempering at 150 °C also resulted in very high hardness exceeding the water quenched sample. However, the values were very close to each other, almost within the limits of error, similar to the tensile strength values. Hardness is generally considered to be related to the tensile strength, and therefore it can be stated that both the tensile strength and hardness were both almost on par with the WQ and 150 °C samples. Nevertheless, the lowest tempering temperature did not significantly reduce the strength nor the hardness. Tempering at 200 °C produced a slightly more distinct decrease of hardness as it also did for the tensile strength. Yet, the hardness was above 600 HV10 for the 200 °C tempered sample. Higher tempering temperatures caused a more severe drop of hardness to levels below 500 HV10. The ratio of tensile strength to hardness varied within 3.2 to 3.4 (R_m/HV10) and 2.3 to 3.2 for yield strength and hardness ($R_{p0.2}$/HV10). The latter showed an increase with increasing tempering temperature, following the increasing yield-to-tensile strength ratio.

Yield strength of the tested steels did not directly follow the same behavior as the tensile strength. Low-temperature tempering increased the yield strength to 1606 ± 2 MPa (150 °C) and to 1616 ± 13 MPa (200 °C), as the water quenched variant had yield strength of 1528 ± 13 MPa. The yield-to-tensile strength ratio increased steadily from 0.66 to 0.68 (HR and WQ) to 0.92 for the 400 °C tempered variant. Uniform elongation decreased linearly for the quenched and tempered samples. A plateau for the elongation was reached at 400 °C as the martensite strength decrease was already substantial. Hutchinson et al. [5,6] have discussed the yielding phenomena in martensitic steels, noting that uniform elongation may increase with increasing strength level, i.e., tempered samples have decreased uniform elongation as was also seen in the current study. They argued that the relaxation of internal stresses was the main cause for earlier instability that subsequently led to the decrease of uniform elongation.

Also, Hutchinson et al. [5,6] studied the increase of yield strength caused by low-temperature tempering, which could be also explained by the relaxation of internal stresses. The quenched martensite has very high internal stresses, which are decreased by the tempering. This does not affect the flow stress as drastically, hence the ultimate tensile strength does not necessarily decrease. Their model predicted a large increase of yield strength for tempered steel when internal stresses were removed [6]. There was a very smooth curvature to the stress–strain relation of the quenched and low-temperature tempered samples, but a sharp yield point appeared at higher tempering temperatures, as can be seen in Figure 7. Due to the different shapes and nature of the stress–strain curves, attention has to be paid to the shape of the curve and not just the numerical value of the 0.2% proof stress.

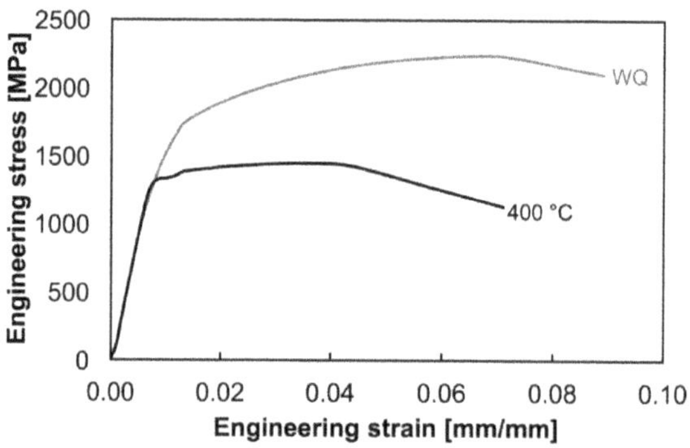

Figure 7. Engineering stress–strain curves for the water quenched (WQ) and 400 °C tempered samples.

3.3.3. Impact Toughness

For the studied steels, the tempering up to 200 °C improved the impact toughness, but eventually the higher temperatures in the temper martensite embrittlement (TME) range (300 °C and 400 °C) resulted in even or slightly decreasing Charpy-V impact energies when tested at −40 °C (Figure 6d). Interestingly, the hot-rolled sample also fitted this curve quite well, though other mechanical properties did not seem to have any direct linear relationship with the quenched and tempered variants. Only the yield-to-tensile strength ratio was very close for the HR and WQ steels. The room temperature testing (+20 °C) showed that the 400 °C tempered sample had better impact toughness over the 300 °C sample. The main reason for the higher room temperature impact toughness energy was presumably a shift of the transition temperature region towards lower temperatures. Nevertheless, impact energies are all relatively low when tested at subzero or room temperatures. The improved Charpy-V result for the 400 °C might have been due to the tempering temperature surpassing the temper embrittlement temperature range.

Temper embrittlement in martensitic steels is an extensively studied phenomenon [12–14,31,34], and a reduction of impact toughness is often found in steels tempered at these temperatures. It is related to the formation of cementite in the temperature range of approximately 300 °C to 400 °C, depending on the alloying content. Interlath retained austenite decomposes and creates brittle cementite films, which form pathways for cracks to propagate. Concurrently, impurities may also segregate to the grain boundaries inflicting brittle behavior by providing weak interfaces for crack propagation. The severity of the tempered martensite embrittlement is affected by chemical composition. For example, silicon shifts cementite formation to higher temperatures [31]. In the current tests, it was found that samples tempered in the TME region showed a slight decrease of Charpy-V impact energies in the subzero temperature testing (−40 °C) when compared to the low-temperature tempered samples. The microstructure of these steels (300 °C and 400 °C) already revealed the formation of cementite (Figure 3e,f), which causes the temper embrittlement. In addition, the XRD results showed that retained austenite was present in the quenched and low-temperature tempered samples but was diminished with the higher tempering temperatures.

Fracture surfaces of the Charpy-V-notch impact toughness testing specimens are presented in Figures 8 and 9. Fracture occurred completely in the brittle mode in HR samples without any significant formation of shear lips. In the samples tempered at 200 °C or higher, the formation of shear lips is pronounced, which partially explains higher energy absorption of these samples. The shear lips cover already about 30–40% of the fracture surface area with the samples tempered at 400 °C. Similar behavior was observed in the fracture toughness specimens in the study of Li et al. [35], where AISI 4340 steel was tempered at various temperatures. In their study, shear lips on the edges of the samples increased with the increasing tempering temperature, which also occurred in the present study.

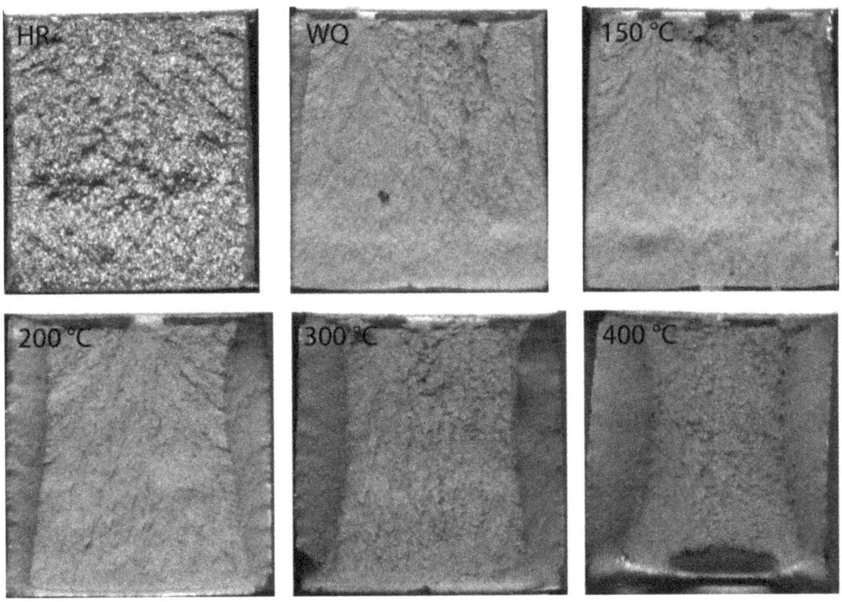

Figure 8. Macroscopic fracture surfaces of the Charpy-V samples (width 7.5 mm) tested at +20 °C. Notch is in the upper part of the specimens.

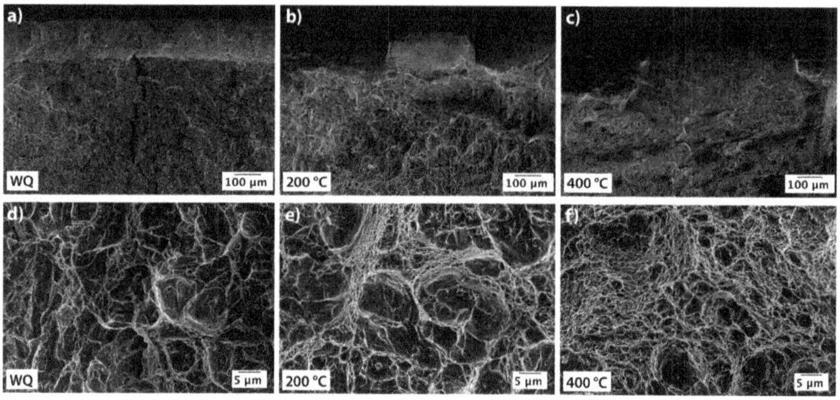

Figure 9. FESEM images of fracture surfaces taken from the upper part (**a**) water quenched, (**b**) tempered at 200 °C, (**c**) tempered at 400 °C and the middle part (**d**) water quenched, (**e**) tempered at 200 °C, (**f**) tempered at 400 °C of selected impact toughness samples tested at +20 °C.

FESEM images taken from near the notch tip (a–b) and the middle of the fracture surface area (d–f) of chosen samples are presented in Figure 9. Surprisingly, even the fracture surface of the WQ sample consisted of a significant amount of small and scattered dimples with most of the surface area consisting of separate cleavage facets and locally sheared features between the narrow multi-void-coalesced strips (Figure 9a). As the tempering temperature increases, the fraction of the dimpled area increases steadily from WQ to the samples tempered at 400 °C. The respective impact energies for the given samples in Figure 9 were 18 ± 5 J/cm^2 (WQ), 31 ± 2 J/cm^2 (200 °C), and 43 ± 2 J/cm^2 (400 °C), i.e., the absorbed energy almost doubled from the WQ sample to 400 °C tempered sample. This is credited to the increased dimple fraction, and the increase of shear lips, of which the latter consumes about half the energy of ductile fracture [36]. These observations are also in line with the ones made by Li et al. [35] in case of fracture toughness specimens of tempered AISI 4340. Though the absolute values for the impact toughness energies were low throughout all the samples tested due to the nature of the given steel, the evolution of the fracture surfaces and moreover the increasing impact energy could be clearly seen in the fracture surface images.

3.4. XRD Results

Table 2 presents the microstrains, crystallite sizes, dislocation densities, and retained austenite contents of tested materials based on the XRD measurements. Tempering increased the crystallite size and reduced the amount of internal strains, which can be assumed to be attributed from the reduced amount of dislocations after tempering. Calculated dislocation densities correlated well with the condition of the steels giving the highest value of 8.31 × 10^{15} m^{-2} in the case of the untempered, martensitic water quenched steel, and decreasing with higher tempering temperature as seen also from Figure 10. The strengthening effect of dislocation density on martensitic steel can be calculated using Equation (4) presented by Bhadeshia et al. [37].

$$\sigma_{\text{disl.}} = K_D \rho^{1/2} \quad (4)$$

where K_D is 0.38 μb for bcc metals, μ is the shear modulus (used here 82 GPa for Fe), and b is the Burgers vector, which is 0.2485 nm in the present case. Based on the calculations using Equation (1), the strengthening effect of measured dislocation densities starting from WQ steel moving towards the higher tempering temperatures were 668 MPa (WQ), 620 MPa (150 °C), 522 MPa (200 °C), 444 MPa

(300 °C), and 373 MPa (400 °C). Therefore, part of the softening, especially at tempering temperatures of 300 °C and 400 °C, can be attributed to clearly lower dislocation density.

Table 2. Microstrain, crystallite size, dislocation density, and retained austenite content based on the X-Ray diffraction (XRD) measurements.

Material	Crystallite Size [Å]	Microstrain [%]	Dislocation Density [$\times 10^{15}$ m^{-2}]	Retained Austenite [%]
WQ	231	0.590	8.3	3.6
150 °C	277	0.611	7.1	2.8
200 °C	369	0.577	5.1	2.8
300 °C	338	0.382	3.7	-
400 °C	352	0.281	2.6	-

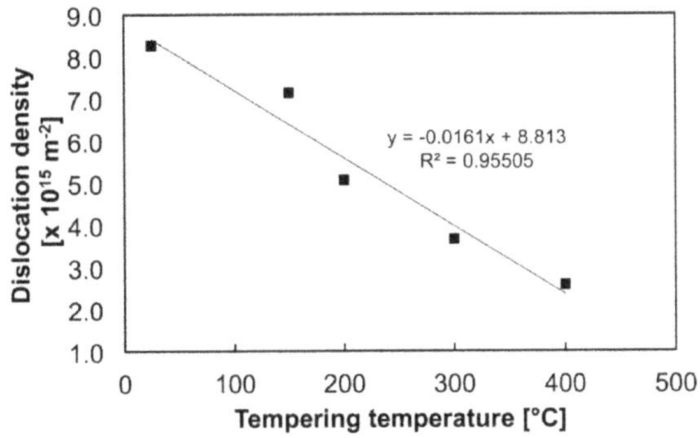

Figure 10. Dislocation densities of the tested steels (WQ tempering temperature set as 25 °C).

The decrease of dislocation density with increasing tempering temperature followed the results found in the literature [22,25,38–40]. Saastamoinen et al. [38] reported dislocation density for both direct quenched (DQ) and reheated and quenched (RAQ) around 4.0×10^{15} m^{-2} for martensite containing 0.1% C, which explains some of the difference with the current results. Using the same Williamson–Hall method as used here, in the case of 0.3% C martensite, Kennett et al. [39] found dislocation densities in the as-quenched state in the range 8–10 $\times 10^{15}$ m^{-2}, depending on PAGS, while Takebayashi et al. [40] found a dislocation density of 2.0×10^{16} m^{-2}.

Furthermore, retained austenite contents were determined from the XRD data showing ~3% austenite content in the case of WQ, 150 °C and 200 °C steels. Tempering at 300 °C and 400 °C decomposed the retained austenite, as seen from retained austenite values in Table 2. The previously discussed transformation of retained austenite during tempering was therefore confirmed by the XRD measurements. This also affects the mechanical properties of the steel along with precipitation and carbon redistribution during tempering.

4. Conclusions

The recently developed ultrahigh-strength press-hardening steel was studied in this paper. The steel was tempered at different temperatures in the range of 150 °C to 400 °C, and subsequently tested for mechanical properties. Microstructural characterization along with XRD measurements and fracture surface inspection were conducted. The following findings were made:

1. A hardness greater than 650 HV/58 HRC was obtained with a tensile strength exceeding 2200 MPa. Reheating at 980 °C prior to quenching resulted in a prior austenite grain size of 12.3 μm and no quench-cracking was detected after the water quenching. Therefore, milder quenching media are not necessary to avoid quench cracking in the case of the 10 mm thick plate studied.

2. Tempering had a significant effect on both the mechanical properties and microstructure of the steel. The most notable changes in the microstructure were the amount and size of the precipitates formed during tempering. The water quenched variant showed only minor traces of auto-tempering, but transition carbides began to form when tempering at 150 °C and 200 °C. Tempering at 300 °C and 400 °C altered the microstructure from martensite to more ferritic with cementite.
3. The hardness and tensile strength reductions were not substantial for the low-temperature tempered variants (150 °C and 200 °C). Further increases in tempering temperature resulted in a decrease of hardness, strength, and uniform elongation. The yield-to-tensile strength ratio increased with increasing tempering temperatures. Low-temperature Charpy-V impact toughness energies increased with tempering until the tempered martensite embrittlement (TME) range was reached.
4. Fracture surface inspection showed that the shear lip formation increased with increasing tempering temperature for the impact toughness samples tested at room temperature, with the highest fraction of ductile fracture in the specimens tempered at 400 °C. However, tempering at 300 °C or above resulted in decrease in impact toughness when tested at −40 °C due to martensite temper embrittlement.
5. X-ray diffraction (XRD) analysis revealed a decrease of dislocation density with the increasing tempering temperature, as expected. The retained austenite content was approximately 3% for the quenched and low-temperature tempered steels but after tempering at 300 °C and 400 °C there were no traces of retained austenite, i.e., the decomposition of austenite had occurred.

Author Contributions: Supervision, D.P., J.K.; writing–original draft, O.H.; writing–review and editing, D.P., A.K., J.K., T.L.; XRD analyses, J.H.; fracture surface analysis, S.P.

Funding: This research has been done within the program Steel Ecosystem for Focused Applications (StEFA). We gratefully acknowledge financial support from Business Finland and the companies participating in the program.

Acknowledgments: The corresponding author would like to express his gratitude for the support provided by the University of Oulu Graduate School through the Advanced Materials Doctoral Program (ADMA-DP).

Conflicts of Interest: The authors declare no conflict of interest.

References

1. Krauss, G. Martensite in steel: Strength and structure. *Mater. Sci. Eng. A* **1999**, *273–275*, 40–57. [CrossRef]
2. Morito, S.; Yoshida, H.; Maki, T.; Huang, X. Effect of block size on the strength of lath martensite in low carbon steels. *Mater. Sci. Eng. A* **2006**, *438–440*, 237–240. [CrossRef]
3. Morito, S.; Adachi, Y.; Ohba, T. Morphology and Crystallography of Sub-Blocks in Ultra-Low Carbon Lath Martensite Steel. *Mater. Trans.* **2009**, *50*, 1919–1923. [CrossRef]
4. Hutchinson, B.; Hagström, J.; Karlsson, O.; Lindell, D.; Tornberg, M.; Lindberg, F.; Thuvander, M. Microstructures and hardness of as-quenched martensites (0.1–0.5%C). *Acta Mater.* **2011**, *59*, 5845–5858. [CrossRef]
5. Hutchinson, B.; Lindell, D.; Barnett, M. Yielding Behaviour of Martensite in Steel. *ISIJ Int.* **2015**, *55*, 1114–1122. [CrossRef]
6. Hutchinson, B.; Bate, P.; Lindell, D.; Malik, A.; Barnett, M.; Lynch, P. Plastic yielding in lath martensites—An alternative viewpoint. *Acta Mater.* **2018**, *152*, 239–247. [CrossRef]
7. Galindo-Nava, E.I.; Rivera-Díaz-Del-Castillo, P.E.J. Understanding the factors controlling the hardness in martensitic steels. *Scr. Mater.* **2016**, *110*, 96–100. [CrossRef]
8. Galindo-Nava, E.I.; Rivera-Díaz-Del-Castillo, P.E.J. A model for the microstructure behaviour and strength evolution in lath martensite. *Acta Mater.* **2015**, *98*, 81–93. [CrossRef]
9. Ouchi, C. Development of Steel Plates by Intensive Use of TMCP and Direct. *ISIJ Int.* **2001**, *41*, 542–553. [CrossRef]

10. Kaijalainen, A.J.; Suikkanen, P.P.; Limnell, T.J.; Karjalainen, L.P.; Kömi, J.I.; Porter, D.A. Effect of austenite grain structure on the strength and toughness of direct-quenched martensite. *J. Alloys Compd.* **2013**, *577*, S642–S648. [CrossRef]
11. Kaijalainen, A.J.; Suikkanen, P.; Karjalainen, L.P.; Jonas, J.J. Effect of Austenite pancaking on the microstructure, texture, and bendability of an ultrahigh-strength strip steel. *Metall. Mater. Trans. A Phys. Metall. Mater. Sci.* **2014**, *45*, 1273–1283. [CrossRef]
12. Krauss, G. Tempering of martensite in carbon steels. In *Phase Transformations in Steels: Volume 2: Diffusionless Transformations, High Strength Steels, Modelling and Advanced Analytical Techniques*; Woodhead Publishing Limited: Cambridge, UK, 2012; pp. 126–150.
13. Bhadeshia, H.K.D.H.; Honeycombe, R. The Tempering of Martensite. In *Steels: Microstructure and Properties*; Butterworth-Heinemann: Oxford, UK, 2006; pp. 183–208.
14. Lee, W.S.; Su, T.T. Mechanical properties and microstructural features of AISI 4340 high-strength alloy steel under quenched and tempered conditions. *J. Mater. Process. Technol.* **1999**, *87*, 198–206. [CrossRef]
15. Matsuda, H.; Mizuno, R.; Funakawa, Y.; Seto, K.; Matsuoka, S.; Tanaka, Y. Effects of auto-tempering behaviour of martensite on mechanical properties of ultra high strength steel sheets. *J. Alloys Compd.* **2013**, *577*, S661–S667. [CrossRef]
16. Li, C.N.; Yuan, G.; Ji, F.Q.; Ren, D.S.; Wang, G.D. Effects of auto-tempering on microstructure and mechanical properties in hot rolled plain C-Mn dual phase steels. *Mater. Sci. Eng. A* **2016**, *665*, 98–107. [CrossRef]
17. Williamson, G.K.; Hall, W.H. X-Ray broadening from filed aluminium and tungsten. *Acta Metall.* **1953**, *1*, 22–31. [CrossRef]
18. Williamson, G.K.; Smallman, R.E., III. Dislocation densities in some annealed and cold-worked metals from measurements on the X-ray Debye-Scherrer spectrum. *Philos. Mag.* **1956**, *1*, 34–46. [CrossRef]
19. Haiko, O.; Miettunen, I.; Porter, D.; Ojala, N.; Ratia, V.; Heino, V.; Kemppainen, A. Effect of Finish Rolling and Quench Stop Temperatures on Impact-Abrasive Wear Properties of 0.35% Carbon Direct-Quenched Steel. *Tribol. Finnish J. Tribol.* **2017**, *35*, 5–21.
20. Haiko, O.; Somani, M.; Porter, D.; Kantanen, P.; Kömi, J.; Ojala, N.; Heino, V. Comparison of impact-abrasive wear characteristics and performance of direct quenched (DQ) and direct quenched and partitioned (DQ&P) steels. *Wear* **2018**, *400–401*, 21–30.
21. Vuorinen, E.; Heino, V.; Ojala, N.; Haiko, O.; Hedayati, A. Erosive-abrasive wear behavior of carbide-free bainitic and boron steels compared in simulated field conditions. *Proc. Inst. Mech. Eng. Part J J. Eng. Tribol.* **2018**, *232*, 3–13. [CrossRef]
22. Saha, D.C.; Biro, E.; Gerlich, A.P.; Zhou, Y. Effects of tempering mode on the structural changes of martensite. *Mater. Sci. Eng. A* **2016**, *673*, 467–475. [CrossRef]
23. Khani Sanij, M.H.; Ghasemi Banadkouki, S.S.; Mashreghi, A.R.; Moshrefifar, M. The effect of single and double quenching and tempering heat treatments on the microstructure and mechanical properties of AISI 4140 steel. *Mater. Des.* **2012**, *42*, 339–346. [CrossRef]
24. Ohlund, C.E.I.C.; Schlangen, E.; Erik Offerman, S. The kinetics of softening and microstructure evolution of martensite in Fe-C-Mn steel during tempering at 300 °C. *Mater. Sci. Eng. A* **2013**, *560*, 351–357. [CrossRef]
25. Furuhara, T.; Kobayashi, K.; Maki, T. Control of Cementite Precipitation in Lath Martensite by Rapid Heating and Tempering. *ISIJ Int.* **2004**, *44*, 1937–1944. [CrossRef]
26. Seppälä, O.; Uusikallio, S.; Larkiola, J. A tool for computer-aided calculation of grain size. In Proceedings of the Abstracts of the XXVI International Conference on Computer Methods in Materials Technology KomPlasTech 2019, Zakopane, Poland, 13–16 January 2019; pp. 128–130, ISBN 978-83-947091-4-3.
27. Higginson, R.; Sellars, M. *Worked Examples in Quantitative Metallography*, 1st ed.; CRC Press: Boca Ralton, FL, USA, 2003; ISBN 9781902653808.
28. Ritchie, R.O.; Francis, B.; Server, W.L. Evaluation of toughness in AISI 4340 alloy steel austenitized at low and high temperatures. *Metall. Trans. A* **1976**, *7*, 831–838. [CrossRef]
29. Zhang, S.S.; Li, M.Q.; Liu, Y.G.; Luo, J.; Liu, T.Q. The growth behavior of austenite grain in the heating process of 300 M steel. *Mater. Sci. Eng. A* **2011**, *528*, 4967–4972. [CrossRef]
30. Chen, R.C.; Hong, C.; Li, J.J.; Zheng, Z.Z.; Li, P.C. Austenite grain growth and grain size distribution in isothermal heat-treatment of 300M steel. *Procedia Eng.* **2017**, *207*, 663–668. [CrossRef]
31. Horn, R.M.; Ritchie, R.O. Mechanisms of tempered martensite embrittlement in medium-carbon steels. *Metall. Trans. A* **1978**, *9*, 1039–1053. [CrossRef]

32. Baker, T.N. Microalloyed steels. *Ironmak. Steelmak.* **2016**, *43*, 264–307. [CrossRef]
33. Lee, S.J.; Lee, Y.K. Prediction of austenite grain growth during austenitization of low alloy steels. *Mater. Des.* **2008**, *29*, 1840–1844. [CrossRef]
34. Krauss, G. Low toughness and embrittlement phenomena in steels. In *Steels: Processing, Structure, and Performance*; ASM International: Cleveland, OH, USA, 2015; pp. 439–485. ISBN 9781627080842.
35. Li, H.F.; Wang, S.G.; Zhang, P.; Qu, R.T.; Zhang, Z.F. Crack propagation mechanisms of AISI 4340 steels with different strength and toughness. *Mater. Sci. Eng. A* **2018**, *729*, 130–140. [CrossRef]
36. Wallin, K. *Fracture Toughness of Engineering Materials—Estimation and Application*; FESI Publishing: Warrington, UK, 2011; ISBN 978-0-9552994-6-9.
37. Bhadeshia, H.K.D.H.; Young, C.H. Strength of mixtures of bainite and martensite. *Mater. Sci. Technol.* **1994**, *10*, 209–214.
38. Saastamoinen, A.; Kaijalainen, A.; Heikkala, J.; Porter, D.; Suikkanen, P. The effect of tempering temperature on microstructure, mechanical properties and bendability of direct-quenched low-alloy strip steel. *Mater. Sci. Eng. A* **2018**, *730*, 284–294. [CrossRef]
39. Kennett, S.C.; Krauss, G.; Findley, K.O. Prior austenite grain size and tempering effects on the dislocation density of low-C Nb-Ti microalloyed lath martensite. *Scr. Mater.* **2015**, *107*, 123–126. [CrossRef]
40. Takebayashi, S.; Kunieda, T.; Yoshinaga, N.; Ushioda, K.; Ogata, S. Comparison of the Dislocation Density in Martensitic Steels Evaluated by Some X-ray Diffraction Methods. *ISIJ Int.* **2010**, *50*, 875–882. [CrossRef]

applied sciences

MDPI

Article

Effect of Intercritical Quenching Temperature of Cu-Containing Low Alloy Steel of Long Part Forging for Offshore Applications†

Yuta Honma *, Gen Sasaki and Kunihiko Hashi

The Japan Steel Works. Ltd., 4 Chatsucho, Muroran-shi, Hokkaido 051-8505, Japan; gen_sasaki@jsw.co.jp (G.S.); kunihiko_hashi@jsw.co.jp (K.H.)

* Correspondence: yuta_honma@jsw.co.jp; Tel.: +81-143-22-0337

† This paper is an extended version of paper published in ASME 2017 36th International Conference on Ocean, Offshore and Arctic Engineering, Trondheim, Norway, 25–30 June 2017.

Received: 26 February 2019; Accepted: 18 April 2019; Published: 25 April 2019

Abstract: In our previous study, intercritical quenching from the dual-phase region of ferrite and austenite regions, which is called lamellarizing (L) treatment, brought a clear improvement of balance between the strength and toughness of Cu-containing low alloy steel based on American Society for Testing and Materials (ASTM) A707 5L grade. Moreover, the results imply that the reverse transformation behavior during L treatment is very important in order to optimize the L treatment temperature. Hence, the purpose of this paper is to clarify the mechanism by which L treatment improves the mechanical properties in terms of reverse transformation behavior. Additionally, offshore structures require good weldability, because the structures generally have a lot of weld joints. Therefore, weldability was also investigated in this study. The investigation revealed that coarse Cu precipitates are observed in the not-transformed α phase, so the strength tendency in relation to the L treatment temperature is relevant to the area ratio of the not-transformed α phase and the transformed γ phase during L treatment. From the in situ electron back scatter diffraction (EBSD) results, it is believed to be possible to enhance the mechanical properties of Cu-containing low alloy steel by controlling the area ratio of the reverse-transformed gamma phase and selecting the appropriate L treatment temperature. Furthermore, the long part forging of Cu-containing low alloy steel has a good weldability, since the maximum hardness of the heat-affected zone (HAZ) is less than 300 HV, and the HAZ of steel has a good crack tip opening displacement (CTOD) property with less than 2.3 kJ/mm of heat input of GTAW.

Keywords: ASTM A707; Cu-containing low alloy steel; intercritical quenching; long part forging; offshore application

1. Introduction

Oil and natural gas are major energy sources, accounting for about 60% of the primary energy in the world. Ultra-deepwater development is expected to increase in the future, as demands for oil and natural gas increase [1]. Copper-containing low alloy steel based on the American Society for Testing and Materials (ASTM) A707 5L grade is widely used for structural parts of offshore wells because of its low carbon equivalent (C_{eq}) and weld crack sensitivity composition (P_{cm}) [2–4]. From the viewpoint of preventing failure and accidents, steel with high strength and excellent low-temperature toughness is required. Generally, the forging steel is subjected to quenching (Q) and tempering (T). The steel is reverse-transformed from α to γ by holding the quenching temperature; after that, γ is transformed to bainite or martensite by water cooling. However, manufacturing large forged parts that have homogeneous and excellent properties over the whole length and thickness is difficult, since there are various reasons such as the segregation of elements, mass effects, and so on.

Therefore, in our previous research, we investigated the effect of intercritical quenching—which is called lamellarizing (L) treatment—from the dual-phase region of ferrite (α) and austenite (γ) on the mechanical properties and microstructures of A707 modified steel using 50-kg test ingots [5]. L-treated steels such as structural steel and steel for very low temperatures have been studied [6–13]. On the other hand, the effects of L treatment on Cu-containing low alloy steel have not been reported. The strength of Cu-containing low alloy steel is improved by Cu deposited during heat treatment [2–14]. Therefore, for the A707 modified steel, the Cu precipitation state is very important to obtain the strength. On the other hand, it has been reported that the precipitation behavior of Cu-bearing steel differs depending on each stage of the refining heat treatment, especially intercritical quenching or tempering [15]. So, it is extremely important to evaluate the mechanical properties of the material whose L treatment conditions are changed and also show the optimum treatment conditions in A707 modified steel as well in order to apply for offshore structures. In our previous research, L treatment brought clear improvement in the balance between the strength and toughness of A707 modified steel [5]. Therefore, the purpose of this research is to clarify the mechanism of improvement of mechanical properties by L treatment in terms of reverse transformation behavior, because the behavior during L treatment is considered to affect the mechanical properties of heat-treated steel due to grain refinement and the hardness of each phase. Based on our previous results, we applied L treatment to a long forged product about 20,000 mm long, and studied the mechanical properties, including crack tip opening displacement (CTOD) and nil ductility transition temperature (NDTT) over the trial production to check whether or not L treatment has a good effect on the mechanical properties.

In the manufacturing of offshore structures, the weldability of steel is very important because the structures have a lot of weld joints. In general, the equivalent amount of C is kept at a low level, and the component design is in consideration of the weldability of the A707 steel. However, since the element, such as C, Ni, and Mn, diffuses to the reverse-transformed γ phase during the L treatment and becomes concentrated locally [16], there is also a concern that the local non-uniformity of the element might reduce the weldability. Therefore, the weldability of the forging that applied L treatment was evaluated from the CTOD property of the welded part in this study.

2. Materials and Methods

2.1. Material

The specimens used in this study were cut from a 50-kg steel ingot made by vacuum induction melting (steel A) and a full-scale large forging that weighed approximately 10 metric tonnes (steel B). Table 1 shows the chemical composition and transformation temperatures of the steels. The steels are characterized by low C content to obtain good weldability, and high Ni and Cu content to guarantee adequate hardenability and strength. Moreover, Al and Nb are added to refine the prior austenite grain size.

Table 1. Chemical composition and transformation temperature of the steels investigated.

	Chemical Composition (mass%)						Transformation Temp. (K)	
	C	Si	Mn	Ni	Cu	Other	A_{C1}	A_{C3}
Steel A (50 kg test ingot)	0.03	0.35	1.40	2.15	1.27	Cr, Mo, Al, Nb	927	1083
Steel B (Full-size product)	0.02	0.33	1.30	2.11	1.23	Cr, Mo, Al, Nb	927	1081

Steel A was forged in the temperature range of 1523 K to 1173 K to a 45-mm thickness at a forging ratio of approximately 4.0 followed by air-cooling to room temperature. After that, normalizing and quality heat treatment with intercritical quenching were carried out. The heat treatment conditions are shown in Figure 1. The samples were quenched at 1173 K after normalizing at 1233 K, and L treatment was subsequently conducted at temperatures from 953 K to 1068 K in order to clarify the effects of L treatment temperature on the mechanical properties and its mechanism in terms of reverse transformation behavior. After L treatment, these were tempered at 873 K for 4 h. Since steel A is a forging plate of 45-mm thickness, the cooling rate of a center part of a water-quenched 300-mm thick section of the long part forging can't be simulated by ordinary water cooling during quenching and L treatment. Since the difference in the cooling rate greatly affects the mechanical properties, it is necessary to control the cooling rate of quenching and L treatment for various investigations using steel A, which is a screening test for examining the applicability of L processing. Hence, in this research, the cooling rate of the test of steel A was simulated to be the central part of the water-quenched 300-mm thick section by the heat treatment furnace, which can control the rate.

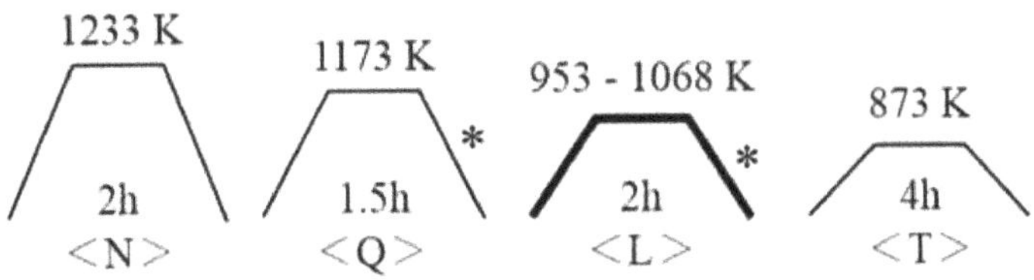

* The cooling rate of quenching and L treatment was simulated to be the central part of the water-quenched 300-mm thick section by the heat treatment furnace, which can control the rate.

Figure 1. Heat treatment condition producing Q–L–T samples. Q: quenching; L: lamellarizing; T: temperature.

Figure 2 shows schematic illustration of the full-size product of steel B—which was 20,000 mm long, with a maximum diameter of 1200 mm and weight of 10 metric tons—and sampling locations. The production process is as follows: ingot making, hot working, and preliminary and quality heat treatment with L treatment at 1068 K. The casting ingot, which was 90 metric tons, was hot-forged to near the final product shape, as shown in Figure 3a. After forging, a quality heat treatment was conducted using a vertical furnace, as shown in Figure 3b.

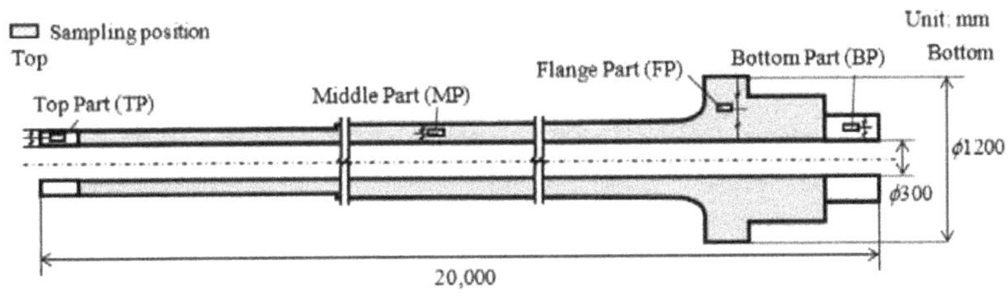

Figure 2. Schematic illustration of the long part product and sampling positions.

(a) Hot working

(b) Heat treatment

Figure 3. Photographs of the manufacturing process.

2.2. Experimental Procedure of Mechanical Properties of the Steel with L Treatment

In this study, mechanical properties such as tensile strength, toughness, and fracture toughness were evaluated by a tensile test, Charpy impact test, drop weight test (DWT), and CTOD test. Following the ASTM E8 specification, cylindrical tensile test specimens 50 mm in gauge length and 12.5 mm in diameter were prepared such that the length direction of the specimen was perpendicular to the main forging direction. The Charpy impact test was conducted using 2-mm V-notched ASTM A370 specimens. The fracture appearance transition temperature (FATT) was also evaluated. The grain sizes (hereafter called the EBSD grain size, in which EBSD stands for electron back scatter diffraction) of the samples were estimated from the boundaries with misorientations of 15 degrees or larger determined by the analysis of EBSD patterns. The EBSD grain sizes and toughness have a correlation [3]. Thus, the EBSD grain size was measured from a cross-section of the Charpy impact test specimen in order to clarify the relationship between the L treatment temperature and the EBSD grain size.

The NDTT was measured by DWT in accordance with ASTM E208. The specimen type was P-3 with a crack starter bead. In addition, the CTOD test was carried out using the B × 2B and B × B types per ISO 15653:2010, as shown in Figure 4. The value of B was 16.7 mm for the B × 2B type specimen, and 100 mm for the B × B type specimen. The specimens were taken from the center of the forging. The CTOD tests for the B × 2B type thin series and B × B type thick series were conducted at 253 K and 273 K, respectively. The loading rate (dK/dt) was 0.5–3.0 MPa $m^{1/2}$/s. CTOD values were calculated using Equation (1) [17]:

$$\delta = \frac{K^2(1-\nu^2)}{2\sigma_Y E} + \frac{r_p(W-a_0)V_p}{r_p(W-a_0)+a_0+z} \tag{1}$$

where δ (mm) is the CTOD value, K (MPa $m^{1/2}$) is the stress intensity factor, E is Young's modulus, r_p is the rotation factor, W (mm) is the specimen width, a_0 is the initial crack length, Vp (mm) is the plastic component of the notch-opening displacement, and z (mm) is the distance of the notch-opening gauge location on the surface of the specimen.

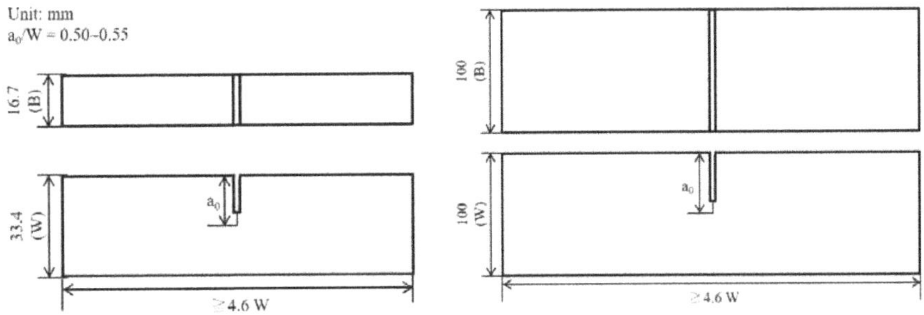

(**a**) Thin series (B × 2B) (**b**) Thick series (B × B)

Figure 4. Crack tip opening displacement (CTOD) specimen (three-point bending type) sizes of thin series (**a**) and thick series (**b**).

Moreover, Cu precipitation and the behavior of the reverse transformation was observed in this study to clarify the mechanism of the L treatment on the steel. Cu precipitation in the Q (hereafter called as-Q) and the Q-L (hereafter called as-L) material was observed using a transmission electron microscope (TEM). Moreover, reverse transformation behavior was checked using in situ EBSD, which was able to measure at high temperatures. The in situ EBSD test condition is shown in Figure 5. The reverse transformation behavior was researched at temperatures from 823 K (below A_{C1}) to 1103 K (above A_{C3}), as shown in Figure 5a, using the as-quenched sample of steel A. Moreover, the test was carried out at 1053 K and 1068 K for 2 h to clarify the relationship between the area ratio of these phases and the holding time of the L treatment (see Figure 5b).

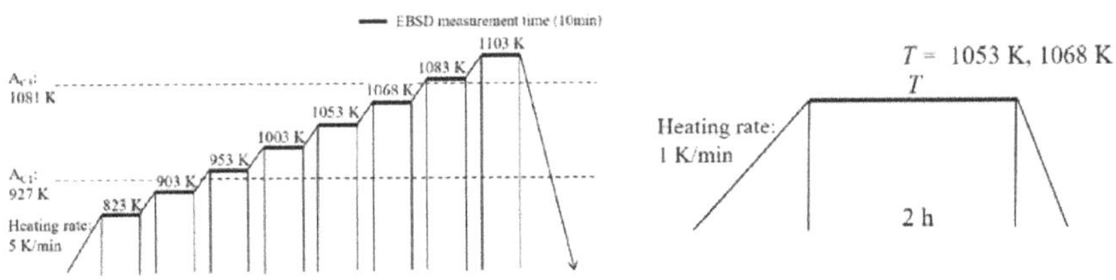

(**a**) Reverse transformation behavior (**b**) change with time

Figure 5. In situ EBSD test conditions of the evaluation of reverse transformation behavior (**a**) and change over time (**b**).

2.3. Experimental Procedure of Weldability of the Steel

The weldability was evaluated using welding test plates 250 × 400 × 20 mm in size, which were cut from the long part of the full-sized production forging. The test plates were welded using gas tungsten arc welding (GTAW) or submerged arc welding (SAW). Figure 6 and Table 2 show schematic illustrations of the welding test plate and welding conditions, respectively. The welding heat input was changed from 1.5 kJ/mm to 3.5 kJ/mm in order to investigate the relationship between heat input and mechanical properties, especially the toughness of the heat-affected zone (HAZ) for the weld joint. After welding, each specimen was cut from the weld test plate without preheating or post-weld heat treatment.

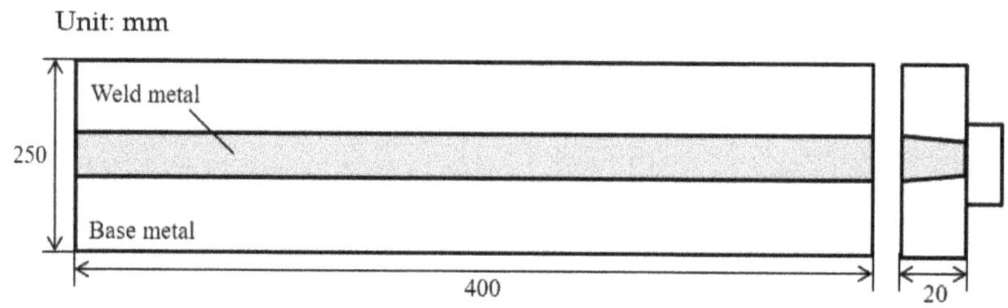

Figure 6. Schematic illustration of welding test plate.

Table 2. Welding conditions.

Welding Process	Abbreviated Expression	Consumable	Maximum Heat Input (KJ/mm)
Submerged arc welding	SAW	A5.23 F9A8-EG-G	3.5 2.5 1.8
Gas tungsten arc welding	GTAW	A5.28 ER100S-G	2.7 2.3 1.6

The macrostructures of the weld joints were observed using cross-sections of the joints. Specimens was cut perpendicularly to the welding direction and etched in a copper chloride ammonia and hydrochloric acid solution. The Vickers hardness profile was measured at 1-mm depth from the surface of the weld joints. In addition, the CTOD test was conducted in this study. The specimens for the CTOD test were B × 2B (15 × 30 mm) in accordance with BS 7448 Part 2, and the notch of the CTOD specimen was located at the center of the weld metal, which is called the fusion line (F.L.), as well as at F.L. + 1 mm, F.L. + 2 mm, and F.L. + 5 mm. After preparation of the specimen, local compression treatment was conducted according to BS 7448 Part 2 to introduce straight initial fatigue pre-cracking. Then, the test was carried out at 273 K.

3. Results and Discussion

3.1. *Effect of L Treatment Temperature on Mechanical Properties*

In our previous study [5], we investigated the mechanical properties of a Q–T sample and Q–L–T sample and the effect of L treatment temperature on the mechanical properties of the steel. The Table 3 shows a summary of the mechanical test data of the Q–T and Q–L–T samples for various L temperatures. Based on those results, the effects of L treatment on mechanical properties are as follows:

1. The tensile strength (T.S.) of the Q–L–T sample is comparable to that of the Q–T sample. Also, 0.2% yield strength (Y.S.) is slightly decreased by L treatment.
2. The toughness is dramatically improved by L treatment.
3. The strength–toughness balance was obviously improved by L treatment at 1068 K (near the A_{C3} point of the steel).

Table 3. Summary of the mechanical test data of the Q–T sample and Q–L–T samples obtained at various L temperatures.

Heat Treatment Process	L Treatment Temp. (K)	0.2% Y.S. (MPa)		T.S. (MPa)		El. (%)		R.A. (%)		YR (-)	FATT (K)
		Each	Ave.	Each	Ave.	Each	Ave.	Each	Ave.		
Q-T	-	642 638	640	729 729	729	28 28	28	75 76	76	0.88	233
Q-L-T	1053	569 570	569	725 727	726	27 27	27	76 75	75	0.78	193
Q-L-T	953	564 575	570	675 684	680	29 28	29	80 80	80	0.84	198
	1003	525 525	525	661 661	661	31 31	31	81 82	82	0.79	190
	1038	532 527	530	665 663	664	30 30	30	80 79	80	0.80	185
	1053	531 532	532	663 663	663	31 30	31	81 81	81	0.80	190
	1068	568 566	567	684 683	684	29 31	30	80 82	81	0.83	178

Moreover, based on the above results, we investigated whether or not it is possible to obtain a good effect of L treatment on mechanical properties using a full-scale long forging product manufactured from A707 Grade L5 modified steel. The mechanical properties at each location are summarized in Table 4. From the above results, it was confirmed that the full-scale forging product was chemically homogeneous across its whole length, and had excellent tensile properties and fracture toughness using L treatment.

Table 4. Typical production test results in a trial product.

Sampling Location		0.2% Y.S. (MPa)		T.S. (MPa)		El. (%)		R.A. (%)		YR (-)	FATT (K)	NDTT (K)	CTOD Value δ (mm)
		Each	Ave.	Each	Ave.	Each	Ave.	Each	Ave.				
Thin part*	TP	618 616	617	692 691	692	28 29	29	82 82	82	89	163	208	1.52 (δ_m) 1.27 (δ_m) 1.17 (δ_m)
	MP	606 607	607	681 681	681	29 29	29	82 82	82	89	159	208	1.27 (δ_m) 1.46 (δ_m) 1.38 (δ_m)
Thick part*	FP	552 542	547	669 663	666	30 30	30	81 81	81	82	159	228	2.89 (δ_m) 2.98 (δ_m) 2.35 (δ_u)
	BP	557 566	562	676 678	677	30 30	30	81 81	81	83	160	228	2.96 (δ_m) 2.86 (δ_m) 3.23 (δ_m)

* The shapes of the specimens for the CTOD test were B × 2B at the thin part and B × B at the thick part. The value of B was 16.7 mm for the B × 2B type specimen and 100 mm for the B × B type specimen. The CTOD tests for the B × 2B type thin series were conducted at 253 K, and the B × B type thick series was tested at 273 K.

These investigations clarified that L treatment is the better method to improve the mechanical properties of full-scale long forging products, but it is necessary to consider the mechanism. We considered that the effect of the L treatment temperature on the balance of strength and toughness depends on the ratio of the reverse transformed γ phase and the not-transformed α phases during L treatment. Therefore, the reverse transformation was investigated over a temperature range from below A_{C1} to above A_{C3} using in situ EBSD in order to clarify the microstructure variation during the treatment. The phase maps are shown in Figure 7. The gray region shown in Figure 7 indicates the not-transformed α phase, and the light yellow region indicates the reverse transformed γ phase at the measurement temperature. Figure 7a shows that all of the regions are in the α phase, and reverse transformation has not occurred, since 823 K—which is the measurement temperature—is about 100 K lower than the A_{C1} point of the steel. However, small amounts of reverse transformed γ phases were generated from the grain boundaries of the α phase at 903 K, and the reverse transformed γ phase region spread with increasing temperature. Furthermore, the insides of grains were observed at 1003 K. When the temperature was raised to 1053 K or 1083 K, there was a complicated structure consisting of not-transformed α phases and transformed γ phases. Over 1103 K, the microstructure consisted entirely of transformed γ phases. The in situ EBSD measurement results indicate that a fine and complicated microstructure is formed by L treatment at higher temperatures between A_{C1} and A_{C3}. The complicated structure seems to indicate that the crystal grain after the L treatment become extremely fine.

The area ratio of the transformed γ phase depends not only on the temperature of the L treatment but also on the holding time [8]. Hence, it is necessary to grasp the relationship between the γ phase and holding time to develop the mechanical property-improving effect due to the complication of the microstructure caused by the L treatment. Figure 8 shows changes with time in the area ratio of the not-transformed α phase and the transformed γ phase measured by in situ EBSD. The area ratios of the transformed γ phase when the temperature reached 1053 K and 1068 K were approximately 40% and 60%, respectively. At both temperatures, the proportion of the reverse transformation γ phase increased with the increase of the holding time, and the area ratio of the transformed γ phase reached approximately 60% at 1053 K and 90% at 1068 K after holding for 120 min. The area ratio of the γ phase at 1068 K also became almost constant beyond 100 min. Therefore, the necessary L treatment time is thought to be at least 100 min in order to stably obtain the mechanical property-improving effect.

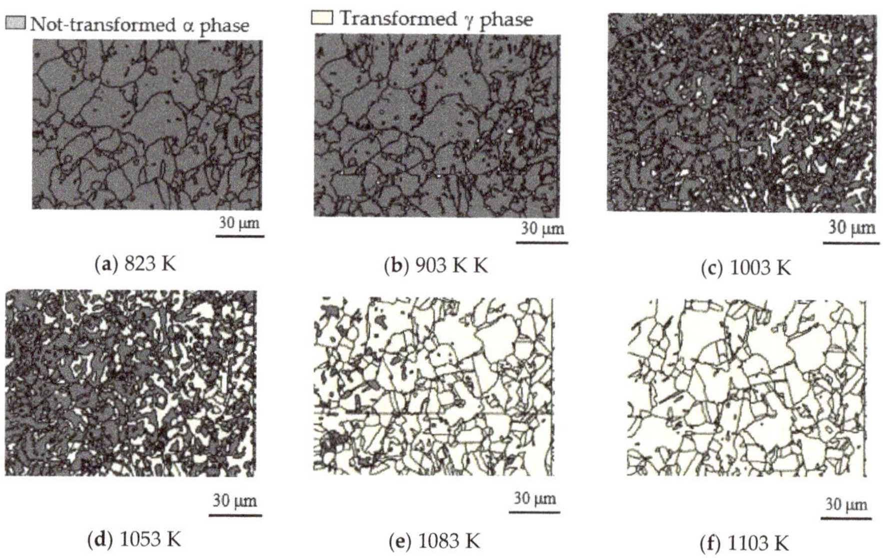

Figure 7. Phase maps of not-transformed α phase and transformed γ phase at 823 K (**a**), 903 K (**b**), 1003 K (**c**), 1053 K (**d**), 1083 K (**e**), and 1103 K (**f**).

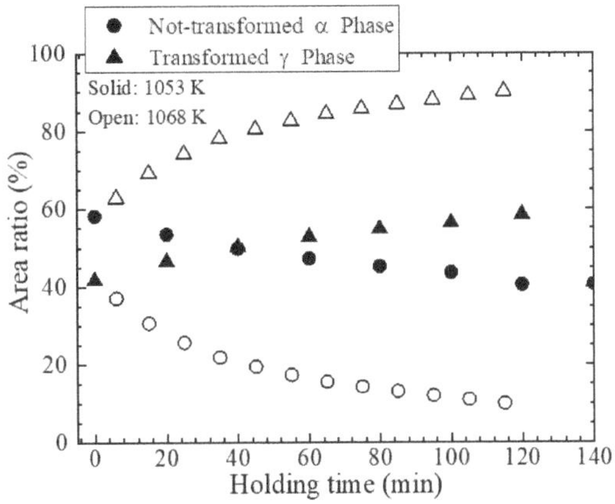

Figure 8. Changes with time in the area ratio of the not-transformed α phase and the transformed γ phase during L treatment at 1053 K and 1068 K.

As mentioned above, the material is Cu precipitation-strengthened steel. Hence, it is important to elucidate the Cu precipitation behavior in the not-transformed α phase and in the transformed γ phase during L treatment. Figures 9 and 10 show TEM images and element (Ni, Cu, and Mn) maps using TEM-Energy Dispersive X-ray Spectroscopy (EDS), respectively. No Cu precipitates were observed in the as-Q sample, while Cu precipitated significantly in the as-L sample. It should be noted that there are Cu precipitated region and non-precipitated regions. The region containing coarse Cu precipitates seems to be in the not-transformed α phase, because the α phase tempered at high temperatures during L treatment. On the other hand, the region containing no Cu precipitates is considered to be the transformed γ phase, and Cu may be dissolved in the matrix. The region at which coarse Cu precipitated is presumed to be the not-transformed α phase, because the not-transformed α phase is tempered during the L treatment, and the precipitated Cu is simply coarsened.

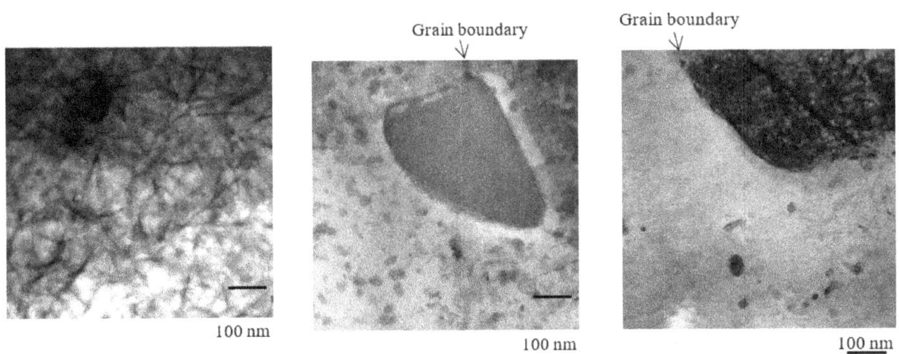

(**a**) As Q (**b**) 953 K (**c**) 1053 K

Figure 9. TEM images of as-Q sample (**a**) and as-L sample heated to 953 K (**b**) and 1053 K (**c**).

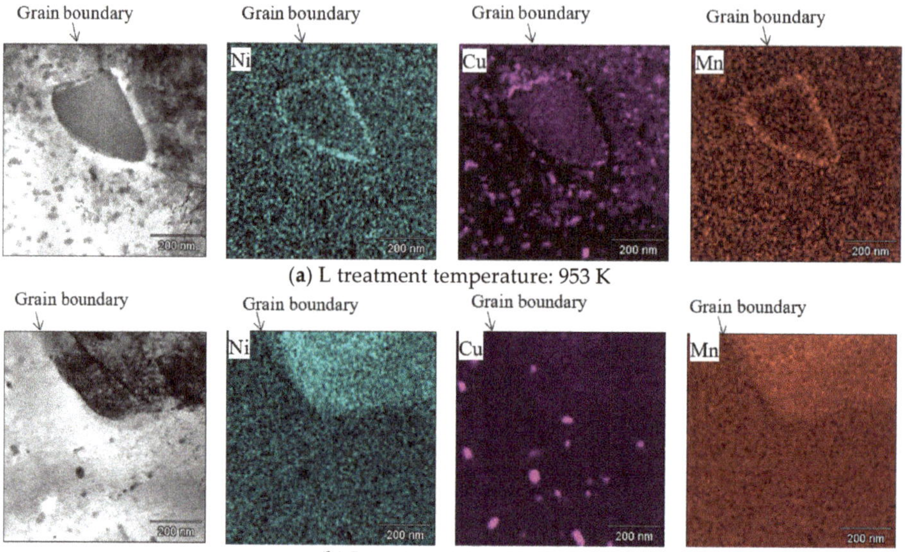

(a) L treatment temperature: 953 K

(b) L treatment temperature: 1053 K

Figure 10. Element mapping images of TEM-EDS of Q–L samples heated to 950 K (**a**) and 1053 K (**b**).

3.2. Mechanism of Improvement of Mechanical Properties by L Treatment

As mentioned above, the influences of the L treatment temperature on mechanical properties are as follows:

1. The strength of the Q–L–T material becomes lower than that of the Q–T material in the cases where the L treatment was carried out in the temperature range from 1003 K to 1053 K (between A_{C1} and A_{C3}).
2. The toughness gradually improves as the L treatment temperature increases.

Firstly, we discuss the main reason for phenomenon (1). The strength of Cu-added steel generally depends on the dispersed state of Cu precipitation resulting from heat treatment [2,14], and the precipitated state is dependent on the L treatment temperature [15]. When the steel is heated to about 723 K to 823 K, the strength increases by age hardening, but the strength decreases in an overaging treatment at temperatures over 823 K. Additionally, L treatment is a heat treatment in the two phases (α and γ). The investigation implied that the not-transformed α phase is softened during L treatment, because it is heated up to 1068 K. Hence, the strength could be decreased in the softened not-transformed α phase. The results shown in Figures 9 and 10 also indicate that the decreased strength is consistent with coarse Cu precipitates. On the other hand, Cu precipitates dissolve into transformed γ phases again, and then the γ phases are quenched. Hence, the γ phases harden due to Cu precipitation during tempering. L treatment at 953 K (right above A_{C1}) has a trivial effect on the strength, because the not-transformed α phase was tempered at low temperatures. However, as the L temperature for treating the steel ranged between 1003–1053 K, the strength was notably decreased due to the not-transformed α phase treated with high-temperature tempering. For the L treatment at 1068 K (right below A_{C3}), the strength of the sample was higher than that treated at approximately 1003 K to 1053 K. According to Figure 8, the transformed γ phase became the main phase after holding at 1068 K for 2 h; therefore, the harder phase inherited the γ phase, leading to improve strength.

Secondly, we discuss the main reason for phenomenon (2). Previous research has suggested that the retained γ phase is stabilized by L treatment and is helpful for enhancing the low-temperature toughness of the 9% Ni steel [6]. However, the Cu-containing low alloy steel did not have a retained

γ phase after L treatment, so the improvement of toughness owing to the L treatment is not related to the retained γ phase for this steel. Therefore, it is necessary to verify the mechanism of toughness improvement other than the retained γ phase. Thus, we focused on the EBSD grain size that indicates the distribution of high angle grain boundary (HAGB), because the HAGB generally has high crack propagation resistance [18]. In other words, the refinement of EBSD particle size indicates an increase in the frequency of HAGB. The EBSD grain size of each sample underwent L treatment at a temperature range from 953 K to 1068 K in order to investigate phenomenon (2). The results are shown in Table 5 and Figure 11. The EBSD grain size became finer as the L treatment temperature increased, and also, the grain size of the sample treated at an optimized temperature, which was 1063 K, was the finest among the four samples. These results suggest that the toughness improvement with the increasing L temperature is due to the grain refining. Furthermore, in past literature [19], it is reported that the effective grain size corresponding to the destructive unit of the brittle fracture is refined, and the toughness is improved along with the refinement of the EBSD grain size. This report supports our results.

Table 5. The electron back scatter diffraction (EBSD) grain size of Q–L–T samples treated at various temperature.

L Treatment Temperature (K)	EBSD Grain Size (μm)	
	Average	Maximum
953	12.5	129
1003	10.8	96
1053	10.5	84
1068	7.2	49

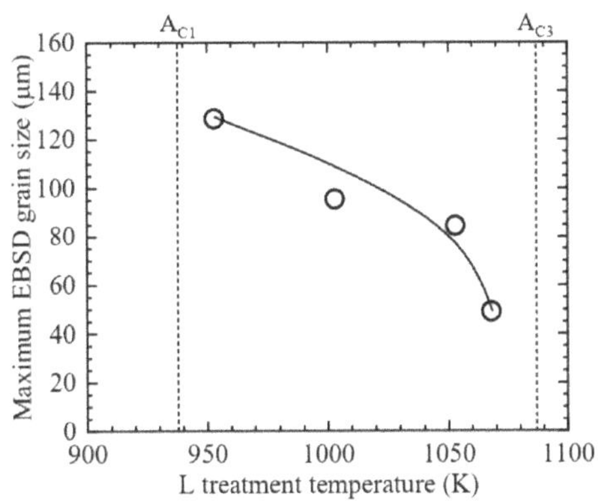

Figure 11. Relationship between L treatment temperature and maximum EBSD grain size.

Reference [8] reported that the optimum L treatment of SA508 Gr.3 steel is at a relatively low temperature between the A_{C1} and A_{C3} of the steel, and the trend is different from the research results using A707 modified steel. In A707 modified steel, when the L treatment temperature is relatively high—that is, near the A_{C3} point—a good balance of strength and toughness could be obtained. This is considered to be the difference in whether the mechanical properties after the L treatment are due to the precipitation or dispersion of carbides, or the area ratio of the reverse transformed γ phase and the Cu precipitates. In the SA508 Gr.3 steel, the C content is about 0.20 mass%, and the carbides precipitated during the L treatment have an effect on the mechanical properties. On the other hand, the carbides might be slightly precipitated during the treatment in A707 modified steel, because the steel has 0.03 mass% C. Accordingly, the refinement of the grain size by reverse transformation during the L treatment and precipitation state of the Cu become dominant with respect to the mechanical properties.

From these results, it is considered that the optimum L treatment condition differs depending on the steel type, because the influence factor affecting the mechanical properties is different.

Based on these results, schematic illustrations regarding changes in the microstructure and precipitates during the L treatment of A707 modified steel are shown in Figure 12. Reverse transformation does not proceed in the state at 953 K that is shown in Figure 12a, so the grain refining and the toughness improvement are not achieved. At 1053 K, as shown in Figure 12b, the toughness improves due to the refined grain size with the reverse transformation, but the strength is decreased by the not-transformed α phase containing coarse Cu precipitates, because the area ratio of the not-transformed α phase still remain approximately 40% after L treatment. At 1068 K, as shown in Figure 12c, the α phase almost transforms to the γ phase after 2 h. The transformed γ phase bears the strength of the steel, because it becomes fresh bainitic ferrite during the cooling process of L treatment, and furthermore, Cu precipitates in a matrix during tempering. It is important to retain a small amount of not-transformed α phase during L treatment for grain refining. From the above hypotheses, it is believed to be possible to enhance the mechanical properties of Cu-containing low alloy steel by controlling the area ratio of the reverse transformed γ phase and by selecting the appropriate L treatment temperature.

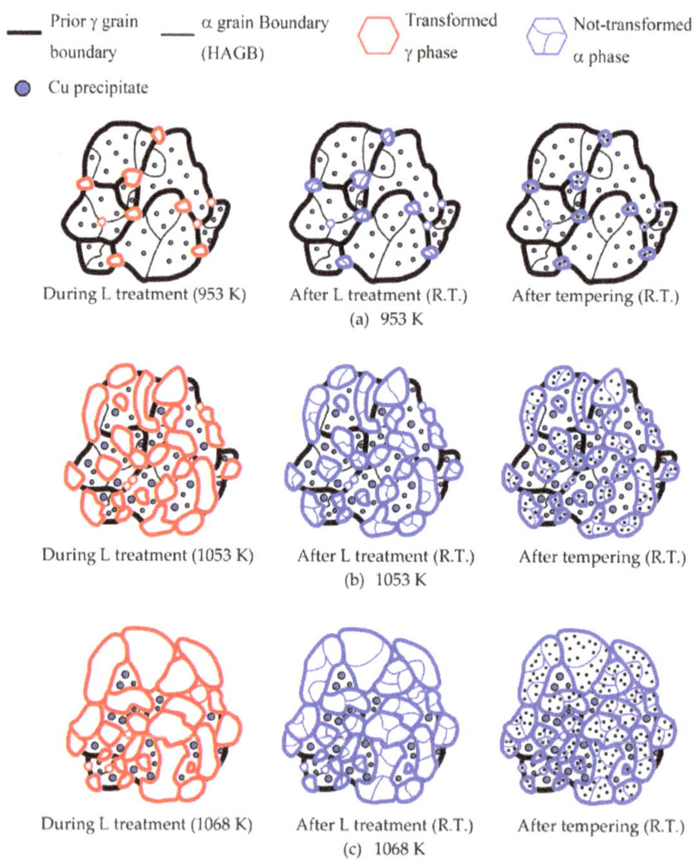

Figure 12. Schematic illustrations of changes in microstructure and precipitates during L treatment at 953 K (**a**), 1053 K (**b**), and 1068 K (**c**).

3.3. Weldability of Cu-Containing Low Alloy Steel

The weldability of the steel was evaluated using a weld plate cut from near the top part (TP) of the full-scale forged product. The macrostructures of the cross-sections of the joints welded at 2.5 kJ/mm with SAW and 2.3 kJ/mm with GTAW are shown in Figure 13. No welding defects were observed in either macrostructure photo. Figure 14 shows the Vickers hardness distribution of each weld joint at 1-mm depth from the surface. Both weld joints had over-matching profiles, and also a maximum HAZ hardness of about 290 HV, which were acceptable values for offshore applications.

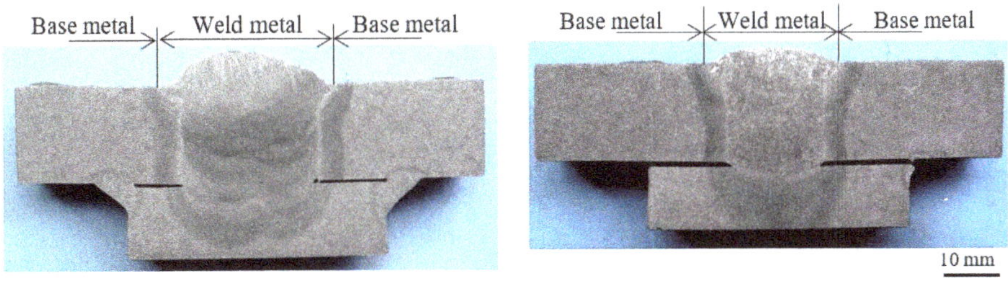

(**a**) SAW (**b**) GTAW

Figure 13. Macrostructures of cross-sections of submerged arc welding (SAW) (**a**) and gas tungsten arc welding (GTAW) (**b**) weld joints.

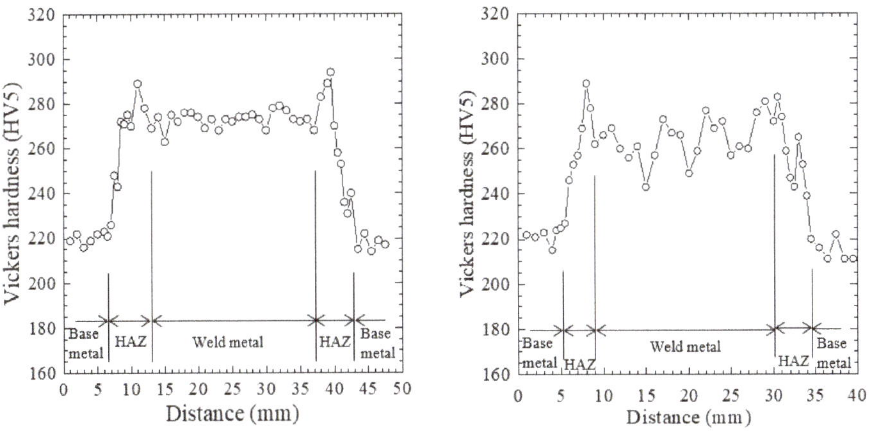

(**a**) SAW (**b**) GTAW

Figure 14. Hardness profiles of the SAW (**a**) and GTAW (**b**) weld joints.

CTOD tests were performed at 273 K on five different notch locations of the SAW weld joint made under a heat input of 3.5 kJ/mm. The results are shown in Figure 15. The test load reached a maximum at the weld metal, F.L. + 2 mm and F.L. + 5 mm, and the CTOD values were judged to be δ_m. However, the CTOD value decreased approaching the F.L., and showed the lowest value at the F.L. The critical CTOD value was 0.06 mm at the F.L. location. An inverse pole figure (IPF) map with HAGB was measured at the F.L. and F.L. + 2 mm locations using EBSD in order to estimate the low CTOD value of the F.L. The results are shown in Figure 16. Fine grains were observed at the F.L. + 2 mm location, but there were coarse grains at the F.L. The temperature rise caused by welding probably leads to significant grain coarsening near the F.L. It is well-known that the toughness of the HAZ is decreased by the formation of coarse grain HAZ. Hence, it is important to reduce the heat input to the welded part, and suppress the grain coarsening in order to improve the toughness of the welded part.

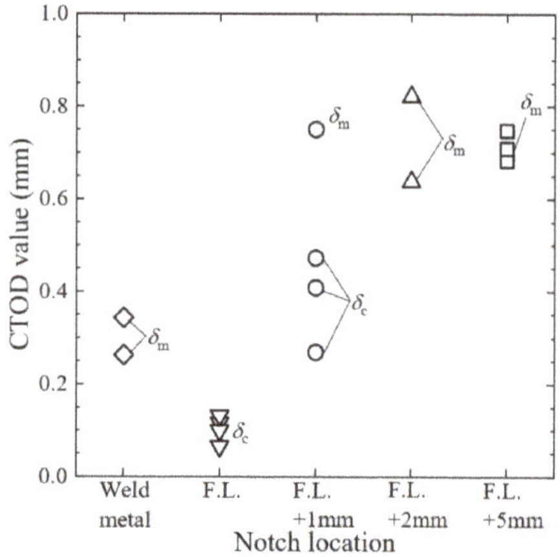

Figure 15. CTOD test results at each location welded using SAW of 3.5 kJ/mm (test temp. 273 K).

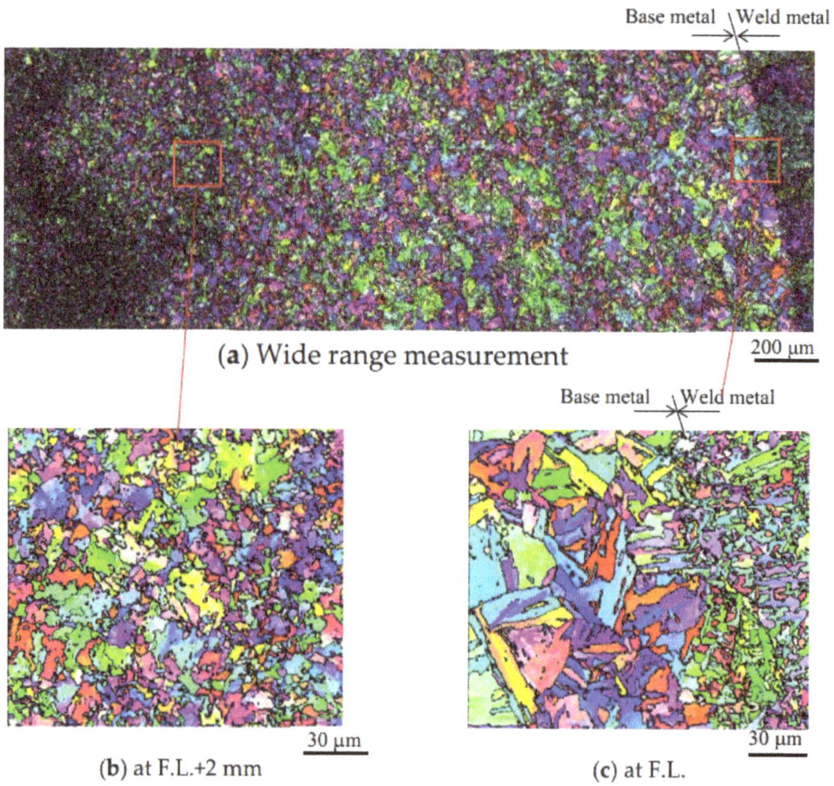

Figure 16. Inverse pole figure (IPF) maps with HAGB (black line) of weld joint (**a**) at the fusion line (F.L.) + 2 mm location (**b**) and the F.L. (**c**).

Figure 17 shows the relationship between heat input and CTOD values at the F.L. of SAW or GTAW. For SAW, the CTOD values with a heat input of 2.5 kJ/mm were almost the same as those at 3.5 kJ/mm, but those of 1.8 kJ/mm were slightly higher than the other two. On the other hand, the CTOD properties were significantly improved with low-heat input GTAW. The critical CTOD values with heat inputs of 2.7 kJ/mm and 1.6 kJ/mm were 0.17 mm and 0.62 mm, respectively. The results suggest that the CTOD property has a tendency to improve with the decreasing heat input of the weld, and the tendency in GTAW was clearer than that in SAW. The difference between the SAW and GTAW results might be due to the shape of the HAZ, and we will work to clarify this in future studies. We have concluded that the long part forgings of Cu-containing low alloy steel have a good weldability, because the maximum hardness of HAZ is less than 300 HV, and the HAZ of steel has a good CTOD property with heat input less than 2.3 kJ/mm using GTAW.

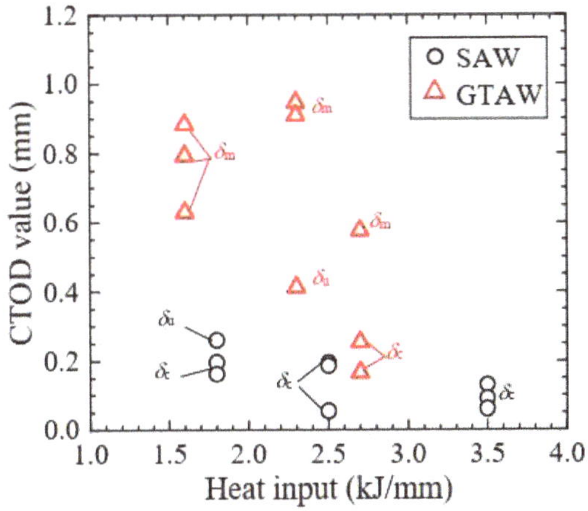

Figure 17. Relationship between CTOD value and heat input for SAW and GTAW weld joint at F.L. (test temperature: 273 K).

4. Conclusions

The effects of L treatment temperature on the mechanical properties of A 707 Grade L5 modified steel were investigated, and the mechanisms of the effects were also discussed in terms of the reverse transformation behavior and precipitates. In addition, L treatment was applied to a full-scale long part product, and the weldability of the product was evaluated. In conclusion:

- The strength–toughness balance was obviously improved by L treatment at 1068 K (near the A_{C3} point of the steel), and an investigation of the mechanical and fracture toughness properties of the overall product revealed that L treatment resulted in high quality characteristics of the forging for use in an offshore structure.
- The in situ EBSD measurement results indicate that a fine and complicated microstructure is formed by L treatment at higher temperatures between A_{C1} and A_{C3}. The complicated structure seems to indicate that the crystal grain after the L treatment become extremely fine. Moreover, it is clear that the necessary L treatment time is at least 100 min in order to stably obtain the mechanical property-improving effect.
- TEM–EDS analysis shows that coarse Cu precipitates are observed in the not-transformed α phase. Thus, the strengthening effect of the L treatment temperature is relevant to the area ratio of the not-transformed α phase and the transformed γ phase during L treatment. The strength, especially the Y.S., seems to be decreased by the not-transformed α phase acting as a softer phase.
- The EBSD results indicate that the improvement of toughness is due to the refining of the EBSD grain size by the transformed γ phase that is generated during L treatment.

- The long part forgings of Cu-containing low alloy steel have good weldability, since the maximum hardness of HAZ is less than 300 HV, and the HAZ of steel has a good CTOD property with less than 2.3 kJ/mm of heat input using GTAW.

Author Contributions: Y.H. and G.S. conceived and performed the experiments; Y.H. wrote the paper; K.H. reviewed and supervised the work.

Funding: This research received no external funding.

Conflicts of Interest: The authors declare no conflict of interest.

References

1. Saint-Marcoux, J.; Legras, J. Impact on Risers and Flowlines Design of the FPSO Mooring in Deepwater and Ultra Deepwater. In Proceedings of the Offshore Technology Conference 2014, OTC, Texas, 5–8 May 2014; 2014; pp. 946–956.
2. Martin, A.; Walsh, S.P. The Chemistry Modifications to ASTM A707 for Offshore Structural Integrity. In *Steel Forgings Second Volume*; ASTM International: Pennsylvania, 1997; pp. 196–209.
3. Venturino, P.; Otegui, L.J.; Teutónico, M. GAS Pipeline Leakage in Urban Subsurface Soil. *Procedia Mater. Sci.* **2012**, *1*, 289–296. [CrossRef]
4. Papavinasam, S. *Corrosion Control in the Oil and Gas Industry, Chapter3: Material*; Elsevier: Amsterdam, The Netherlands, 2014; pp. 133–177.
5. Honma, Y.; Sasaki, G.; Hashi, K. Improvement on Mechanical Properties of Cu-Containing Low Alloy Steel of Long Part Forging for Offshore Applications by Manufacturing Process. In *ASME 2017 36th International Conference on Ocean, Offshore and Arctic Engineering*; American Society of Mechanical Engineers: Trondheim, Norway, 2017.
6. Saitoh, N.; Yamaba, R.; Muraoka, H.; Saeki, O. Dvelopment of Heavy 9% Nickel Steel Plates with Superior Low-Temperature Toughness for LNG Storage Tanks. *Nippon Steel Tech. Rep.* **1993**, *58*, 9–16.
7. Shikanai, N.; Mitao, S.; Endo, S. Recent Development in Microstructural Control Technologies through the Thermo-Mechanical Control Process (TMCP) with JFE Steel's High-Performance Plates. *JFE Tech. Rep.* **2008**, *11*, 1–6.
8. Kim, H.; Ahn, Y.; Oh, Y.; Kim, G.; Hong, J. Effect of Intercritical Annealing on the Toughness and Strength of SA508 Gr.3 Heavy Section Forging Steel. In Proceedings of the 13th Internatinal Forgemasters Meeting, Pusan, Korea, 12–16 October 1997; pp. 197–208.
9. Xie, B.S.; Cai, Q.W.; Yun, Y.; Li, G.S.; Ning, Z. Development of high strength ultra-heavy plate processed with gradient temperature rolling, intercritical quenching and tempering. *Mater. Sci. Eng. A* **2017**, *680*, 454–468. [CrossRef]
10. Tsuchiyama, T.; Inoue, T.; Tobata, J.; Akama, D.; Takaki, S. Microstructure and mechanical properties of a medium manganese steel treated with interrupted quenching and intercritical annealing. *Scrip. Mater.* **2016**, *122*, 36–39. [CrossRef]
11. Yan, N.; Di, H.; Misra, R.D.K.; Huang, H.; Li, Y. Enhancing austenite stability in a new medium-Mn steel by combining deep cryogenic treatment and intercritical annealing: An experimental and theoretical study. *Mater. Sci. Eng. A* **2019**, *753*, 11–21. [CrossRef]
12. Thompson, S.W. Fine–scale structural features of intercritically aged HSLA-100 plate steel and their influence on yield strength and low-temperature impact toughness. *Mater. Charact.* **2018**, *136*, 425–434. [CrossRef]
13. Hwang, B.; Lee, G.G. Influence of thermomechanical processing and heat treatments on tensile and Charpy impact properties of B and Cu bearing high-strength low-alloy steels. *Mater. Sci. Eng. A* **2010**, *527*, 4341–4346. [CrossRef]
14. Takaki, S.; Fujioka, M.; Aihara, S.; Nagataki, Y.; Yamashita, T.; Sano, N.; Adachi, Y.; Nomura, M.; Yaguchi, H. Effect of Copper on Tensile Properties and Grain-Refinement of Steel and its Relation to Precipitation Behavior. *Mater. Trans.* **2004**, *7*, 2239–2244. [CrossRef]
15. Han, G.; Xie, Z.J.; Li, Z.Y.; Lei, B.; Misra, R.D.K. Evolution of crystal structure of Cu precipitates in a low carbon steel. *Mater. Design* **2017**, *135*, 92–101. [CrossRef]
16. Wei, R.; Enomoto, M.; Hadian, R.; Zurob, H.S.; Purdy, G.R. Growth of austenite from as-quenched martensite during intercritical annealing in an Fe–0.1C–3Mn–1.5Si alloy. *Acta Mater.* **2013**, *61*, 697–707. [CrossRef]

17. International Organization for Standardization. *ISO 15653*; International Organization for Standardization: Geneva, Switzerland, 2010.
18. Gourgues, A.F.; Flower, H.M.; Lindley, T.C. Electron Backscattering Diffraction Study of Acicular Ferrite, Banite, and Martensite Steel Microstructures. *Mater. Sci. Technol.* **2000**, *16*, 26–40. [CrossRef]
19. Hanamura, T.; Yin, F.; Nagai, K. Ductile-Brittle Transition Temperature of Ultrafine Ferrite/Cementite Microstructure in a Low Carbon Steel Controlled by Effective Grain Size. *Iron Steel Inst. Jpn. Int.* **2004**, *44*, 610–617. [CrossRef]

applied sciences

Article

Numerical Study of the Effect of Inclusions on the Residual Stress Distribution in High-Strength Martensitic Steels During Cooling

Chao Gu [1,3], Junhe Lian [2,3,*], Yanping Bao [1,*], Wei Xiao [1] and Sebastian Münstermann [3]

1 State Key Laboratory of Advanced Metallurgy, University of Science and Technology Beijing, Beijing 100083, China; b20150485@xs.ustb.edu.cn (C.G.); b20170512@xs.ustb.edu.cn (W.X.)
2 Department of Mechanical Engineering, Aalto University, Puumiehenkuja 3, 02150 Espoo, Finland
3 Steel Institute, RWTH Aachen University, Intzestrasse 1, 52074 Aachen, Germany; sebastian.muenstermann@iehk.rwth-aachen.de
* Correspondence: junhe.lian@aalto.fi (J.L.); baoyp@ustb.edu.cn (Y.B.); Tel.: +358-50-477-0765 (J.L.); +86-138-1076-8855 (Y.B.)

Received: 31 December 2018; Accepted: 26 January 2019; Published: 29 January 2019

Abstract: In high-strength martensitic steels, the inclusions significantly affect the material performance especially in terms of fatigue properties. In this study, a numerical procedure to investigate the effect of the inclusions types and shapes on the residual stresses during the cooling process of the martensitic steels is applied systematically based on the scanning electronic microscopy (SEM) and energy dispersive spectrometer (EDS) results of different types of inclusions. The results show that the maximum residual stress around the interface between Mg-Al-O inclusion and the matrix is the largest, followed by TiN, Al-Ca-O-S, and MnS when the inclusions are assumed as perfect spheres for simplicity. However, these results are proved to be 28.0 to 48.0% inaccurate compared to the results considering actual shapes of inclusions. Furthermore, the convex shape of inclusion will lead to stress concentration in the matrix while the concave shape of inclusion will lead to stress concentration in the inclusion. The residual stress increases with the increase of inclusion edge angle. The increase rate is the largest for TiN inclusions on the concave angle, which leads to extreme stress concentration inside TiN inclusion.

Keywords: martensitic high strength steel; inclusion; residual stress; Abaqus

1. Introduction

Driven by the increasing demand on the higher mechanical performance of engineering structures, high-strength steels are steadily developed and widely applied in recent decades in multiple areas, e.g., automotive, high-speed trains, and aerospace. The high strength is mainly obtained by the in-depth and complicated design of the microstructure of steels, which also leads to new damage mechanisms for multiphase steels [1–5]. In terms of fatigue properties, a fatigue limit is normally evident for the conventional steels developed some decades ago [6–8]. Murakami et al. [9] reported that the fatigue limit increases with the increase of steel hardness and decreases with the increase of the defect area. Therefore, increasing the strength of materials is leading to a higher hardness and eventually a higher fatigue strength limit assuming the same defect size. However, for the recently developed high-strength steels, it is found that even when stress amplitude is lower than fatigue limit, the fatigue crack still will occur when fatigue life reaches a certain level [9–12]. For higher stress and lower fatigue life, fatigue crack usually occurs from the component surface; for lower stress and longer fatigue life, fatigue crack usually occurs from inner microstructure [13].

The deformation heterogeneity in microstructure level, which is responsible for the damage and failure in general for steels, can be generally caused by two different mechanisms: the matrix inhomogeneity and the matrix-inclusion inhomogeneity [14,15]. Many researchers have focused on the effect of matrix inhomogeneity on mechanical properties, their damage mechanism with experiments and modelling. Lian et al. investigated the damage mechanism of a DP600 steel under different stress states experimentally [16] and via micromechanical simulation [17] and found that damage is initiated at the boundary of ferrite and martensite due to their strong deformation incompatibility. It was further revealed that the damage initiation is by nature stress state dependent due to the strain localization dependency of stress states. Krupp et al. [18] studied the fatigue crack initiation process of duplex stainless steel and quantified the interactions of cracks with the first microstructural barrier. Alharbi et al. [19] investigated the stress state of martensite and ferrite before cracking of dual phase DP1000 steel and estimated for crack initiation in the martensite. He et al. [20] reported that for a ferritic–pearlitic steel, the initiation of the cleavage fracture is located in the pearlitic phase where the ferrite and the cementite show a lamella structure with different properties.

In addition to steel matrix inhomogeneity, inclusions also play a significant role during the failure process. Fairchild et al. [21] reported the influence of inclusions on cleavage initiation and the relation between inclusion crack and steel matrix crack. Yang et al. [22] studied 42CrMo steel with different amounts of inclusions and found that the fatigue property was better with fewer inclusions. Sakai et al. [23] found that internal inclusions could lead to the formation of the penny-shape fine granular layer around inclusions during long-term cyclic loadings, which would cause small cracks inside the steel matrix. While investigating inclusions in steel, it is also found that there are multiple types of inclusions in steels [24,25] generated in different processes during the steel production [26,27]. For example, the Al_2O_3 inclusion is produced during melting of steel while the TiN inclusions are produced during the casting of steels. Besides, the shape, size, and composition of inclusions, affected by the production method, also differ, leading to different effects on the mechanical properties of steels [28,29].

The production processes additionally induce another level of inhomogeneity, residual stresses, based on the microstructure-level inhomogeneity. For the steel matrix, it is found that the mechanical deformation, such as rolling [30] and indentation or surface treatment [31] could introduce the grain-level residual stresses and they are grain orientation dependent. For the interaction between the steel matrix and inclusions, the thermal treatment, e.g., cooling, could induce large local residual stress, which would affect the mechanical properties of the steel significantly [32]. These residual stresses are generally generated due to the different thermal expansion properties of inclusions and steel matrix during temperature variation. Brooksbank and Andrews [33,34] developed a model to calculate the residual stress around spherical and cylindrical inclusions and reported the residual stresses around the boundaries between different inclusions and steel matrix were different. This model was widely applied by researchers to calculate residual stresses around inclusions in different steels [35,36]. Murakami and Uchida [37] claimed that the residual stresses induced by different thermal expansion properties around Al_2O_3 inclusions will lead to debonding of inclusions from steel matrix based on calculations, which seriously affects the mechanical properties of steel. In addition to common inclusions, carbides, which are close to the surface, will also cause irregular stress distribution [38]. In terms of finite element-based modelling, there are researchers simulating the residual stresses around inclusions and analyze their effects on mechanical properties. Pineau and Forest [39] performed an elastic–plastic finite element calculation to determine the residual stresses in the vicinity of the inclusions. The results show that the residual equivalent plastic flow occurs all around the inclusion during cyclic loading, which plays an essential role in crack initiation. Gu et al. [40,41] proposed a microstructure-based model considering residual stress induced by temperature variation. This model offers a more arcuate prediction on crack initiation site during fatigue process compared to the model neglecting residual stress.

Although all these above-mentioned studies offer good illustrations of the numerical residual stress distributions around inclusions and their strong effects on the mechanical behaviour, the gap is also evident: Actual inclusion shape is not considered. For simplicity, these models and calculations only considered the inclusion as a simple geometrical shape, such as spherical and cylindrical. For actual inclusions in steel, the shape varies dramatically and contains multiple details, which will greatly affect the residual stress distribution around inclusions. In terms of experiments, quantifying residual stress around inclusions still faces challenges due to the redistribution of residual stress while machining specimens. Therefore, it is aimed in this study to apply a numerical procedure to investigate the effect of the inclusions types and shapes on the residual stresses systematically during the cooling process of the martensitic steels. It is emphasized that by using a numerical approach, the shape effect and the effect of the inclusions type in terms of the mechanical properties can be distinguished and individually studied. The ideal spherical shape is first assumed for all the inclusions, and the numerical results are compared with the analytical model. Furthermore, models were established with the actual shapes of inclusions in steel, and the residual stress distributions were analyzed numerically to clarify the relation between shape variation and residual stress distribution of different inclusions. A further quantitative comparison study of the residual stress result between the inclusions with ideal shape and the actual shape was conducted. This study provides the basis for further researches on numerical prediction of crack behaviour and cracks mechanism modelling for high-strength steels.

2. Materials and Methods

The material in the present study is high carbon martensitic bearing steel with 1.12% C and 1.41% Cr. The main chemical composition is shown in Table 1. This material was hot rolled into a steel bar with a diameter of 35 mm, vacuum oil quenched after holding for 20 min at 835 °C and tempered for 120 min at 180 °C.

Table 1. Main chemical composition (unit: weight %).

C	Cr	Si	Mn	P	S	Cu	Al
1.12	1.41	0.37	0.42	0.017	0.004	0.078	0.02

The mechanical properties were tested with the specimen represented in Figure 1. The specimen was perpendicular to the cross section of the steel bar. According to recent studies [42–45], the manufacturing methods of the specimen can affect the results of the tensile test. In this study, the cutting method of a wire electro discharge machining was adopted. With this method, the geometric-dimensional requirements can be easily met without any significant change in the mechanical properties. The extension rate of the tensile test was 1.0 mm/min. A metallographic specimen was also taken from the steel bar and polished by SiC paper and diamond suspensions in the cross-section of steel bar. Nonmetallic inclusions were observed with scanning electronic microscopy (SEM) and the compositions of inclusions were analyzed with energy dispersive spectrometer (EDS).

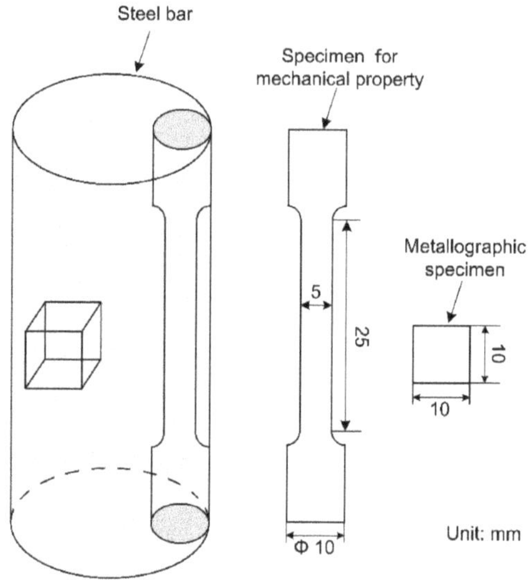

Figure 1. Specimen geometries and their taken positions.

3. Modelling and Simulation

The geometrical profiles of typical inclusions in the steel observed with SEM were captured as coordinates with MATLAB (version 2016a, MathWorks, Natick, MA, USA, 1984). Due to the differences in contrast between inclusion and steel matrix in the SEM picture, the MATLAB codes can assist to recognize the details of inclusion shape in a desirable way. Thereafter, the coordinates of inclusion shape were processed by a Python script using the Abaqus (Dassault Systèmes Simulia Crop., Providence, RI, USA, 1994) extension of the Python programming language to create the inclusion component of the Abaqus model. The version of Abaqus is Abaqus 2017. The steel matrix was set as a square area (30×30 μm^2) and assumed to be a homogeneous material with a single phase and elastic only. The inclusions were positioned in the middle of the matrix area so that the residual stress around the inclusion was not affected by the matrix interface.

The simulated process was the cooling procedure during oil quenching. The temperature variation was from 835 °C to the room temperature, which was set at 20 °C. The temperature was defined with pre-defined field in Abaqus/Standard. The thermal expansion coefficients and the mechanical property parameters of different inclusions and steel matrix were obtained from references and experiments in the present study.

During the cooling process, the thermal stress of every node during the elastic deformation in the inclusion-matrix system was calculated with Equation (1).

$$\varepsilon = \alpha_i(T - T_0) \tag{1}$$

where i is the material definition number, which represents steel matrix and different inclusions; α_i is the thermal expansion coefficient; T_0 is the initial temperature, and T is the current temperature.

To avoid over constraining the thermal expansion of the system and guarantee the accuracy of the simulation, it is important to define proper interface conditions. In the present study, the left side of the square steel area was fixed in the x-direction and still can move in the y-direction. The element type is set as "continuum-plane stress-3 nodes (CPS3)". The simulation type is static.

4. Results and Discussion

4.1. Properties of Inclusion and Steel Matrix

The stress vs. strain curve of the martensitic steel is shown in Figure 2 based on the tensile tests described in Section 2. According to the figure, the mechanical properties of the steel matrix are shown in Table 2.

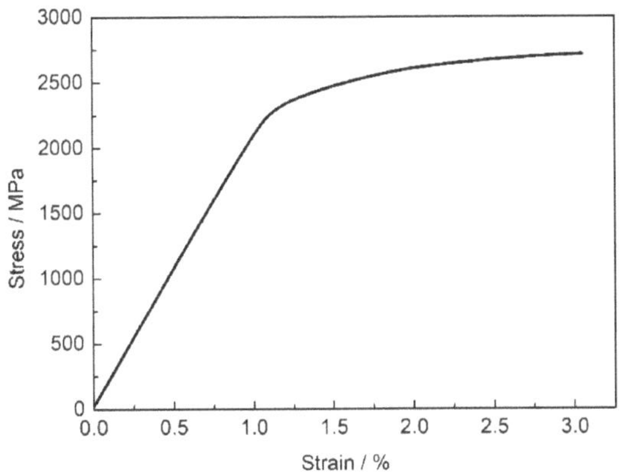

Figure 2. Stress vs. strain curve of the martensitic steel.

Table 2. Mechanical properties of the martensitic steel.

Material	Yield Stress (MPa)	Tensile Stress (MPa)	Young's Modulus, *E* (GPa)
Martensitic steel	2369	2633	206

There are four main types of inclusions in the martensitic steel: spinel (Mg-Al-O), calcium aluminum (Al-Ca-O-S), titanium nitride (TiN), and manganese sulfide (MnS). The typical morphologies are shown in Figure 3. The Mg-Al-O inclusions are usually polygon. The Al-Ca-O-S inclusions are usually spherical. The TiN inclusions are usually irregular in shape. The MnS inclusions usually precipitate in the grain interface as spherical, cylindrical or irregular. All these different shape features will contributed to the variation of residual stress distribution around inclusions, which will be discussed in Section 4.3. The thermal expansion coefficients and the mechanical properties of different inclusions, as well as the steel matrix, are presented in Table 3, which are the required material parameters of the simulation in the following sections.

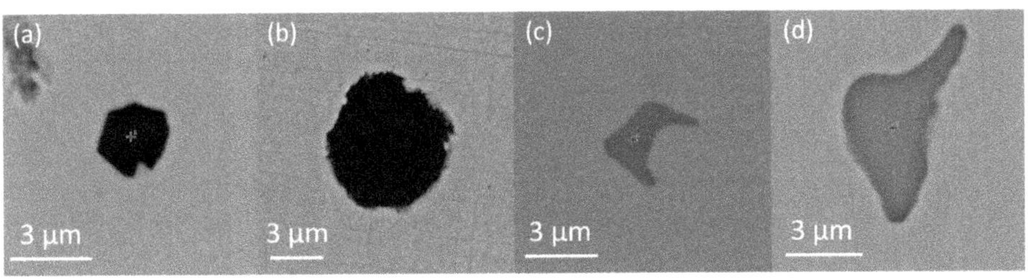

Figure 3. Scanning electronic microscopy (SEM) micrographs of typical inclusions: (**a**) Mg-Al-O; (**b**) Al-Ca-O-S; (**c**) TiN; (**d**) MnS.

Table 3. Mechanical properties of the steel matrix and the inclusions in the martensitic steel [33,34,46].

Material	Coefficient of Linear Expansion, α ($10^{-6}\cdot{}^{\circ}C$)	Young's Modulus, E (GPa)	Poisson's Ratio, v
Mg-Al-O	8.4	271	0.26
Al-Ca-O-S	5.0	113	0.23
MnS	14.8	103	0.30
TiN	9.4	320	0.19
Steel matrix	23.0	206	0.30

4.2. Residual Stress Distribution Around Spherical Inclusions

The residual stress distribution around inclusions was already calculated by Brooksbank and Andrews [33]. In their model, inclusions were simplified as perfect spherical. The radial stress in the surface between inclusion and steel matrix can be calculated with Equations (2) and (3).

$$\sigma_R = \frac{(\alpha_M - \alpha)\Delta T}{\frac{0.5(1+v_M)+(1-2v_M)d^3}{E_M(1-d^3)} + \frac{(1-2v)}{E}} \quad (2)$$

$$d = R/R_M \quad (3)$$

where α_M and α are the coefficients of linear expansion of the steel matrix and the inclusion, respectively; v_M and v are the Poisson's ratios of the steel matrix and the inclusion, respectively; E_M and E are the Young's modulus of the steel matrix and inclusion, respectively; ΔT is the difference between the holding temperature before vacuum quenching and the room temperature; R is the radius of the inclusion, and R_M is the radius of the steel matrix around a single inclusion, which is half of the distance between two inclusions with the same type.

To clarify the influence of inclusion type on residual stress distribution and make a comparison of the calculation results in the present study with the model of Brooksbank and Andrews, perfect spherical Mg-Al-O, Al-Ca-O-S, TiN, and MnS inclusions were inserted in the square steel matrix. The geometry model and the calculation results of residual stress distribution are shown in Figure 4. It is obvious that the residual stress in the steel matrix around the inclusion decreases when the distance to inclusion surface increases. The position of the maximum stress is on the surface between the inclusion and steel matrix. When the inclusion is spherical, the stress on the surface between inclusion and steel matrix keeps constant around the inclusion. This constant maximum stress varies with inclusion type. The value of Mg-Al-O inclusion is the largest, followed by TiN and Al-Ca-O-S inclusions while the value of MnS inclusion is the smallest. These simulation results are also compared with the calculation results with Equations (2) and (3) in Figure 5. Except for TiN inclusions, the order of the values of the maximum residual stress on the interface between inclusions and steel matrix calculated with the two methods keeps consistent, while the results of TiN inclusions differ more greatly. It is noted that all the residual stresses around perfect spherical Mg-Al-O, Al-Ca-O-S, TiN, and MnS inclusions are compressive due to smaller α of inclusions than steel matrix.

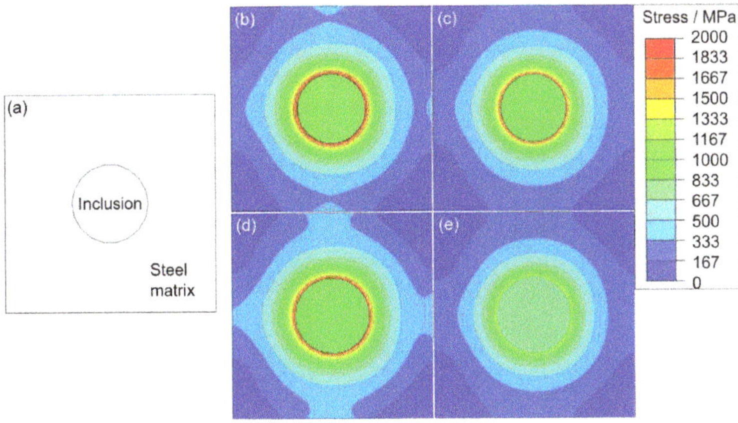

Figure 4. (**a**) Geometric illustration of the model for the residual stress simulation during cooling; residual stress distributions around the spherical inclusion of (**b**) Mg-Al-O; (**c**) Al-Ca-O-S; (**d**) TiN; (**e**) MnS.

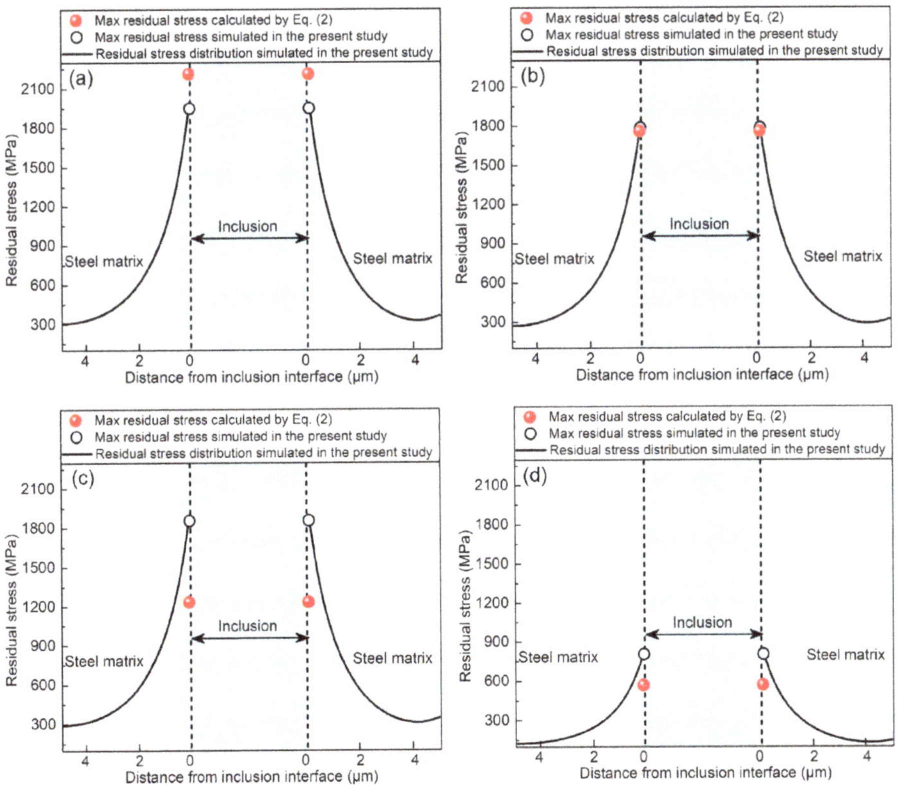

Figure 5. Comparison of simulated residual stresses and the maximum residual stress on the interface between steel matrix and inclusions calculated with Equation (2): (**a**) Mg-Al-O; (**b**) Al-Ca-O-S; (**c**) TiN; (**d**) MnS.

4.3. Residual Stress Distribution Around Actual Inclusions

When inclusions are assumed as spherical, the shape details are actually neglected. Nevertheless, as shown in Figure 3, the inclusions of different types usually possess different geometry shapes, which will lead to important stress concentration around interfaces between the inclusions and the steel matrix. In this Section, four types of inclusions in Figure 3 were inserted in the square steel area, and the residual stress distributions were simulated. The simulation results are shown in Figure 6. Generally, the residual stress also decreases when the distance to inclusion surface increases, which is consistent with the case of the perfect spherical inclusion, while the maximum residual stresses for these four types of inclusions all increased compared to the maximum stress of spherical inclusions. Besides, the residual stress on the surface between the inclusion and the steel matrix changes with the shape of inclusions.

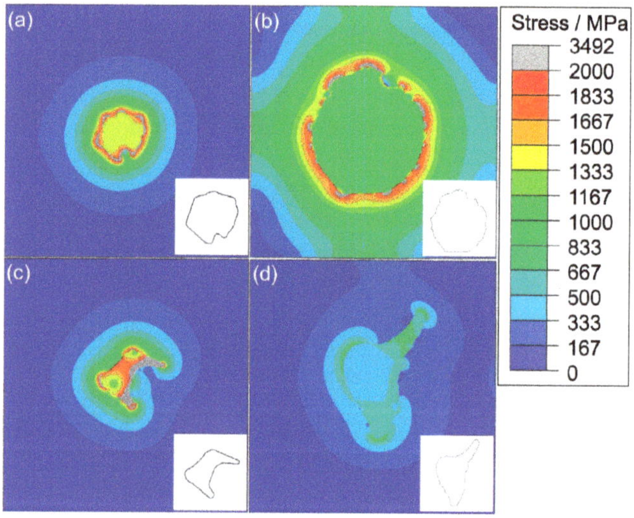

Figure 6. Residual stress distributions around the inclusions with their actual shapes: (**a**) Mg-Al-O; (**b**) Al-Ca-O-S; (**c**)TiN; (**d**) MnS.

To quantitatively analyze the effect of the inclusion shape on residual stress changes, two parameters were introduced: the edge angle $\theta_{sv,i}$ and the local residual stress $\sigma_{sv,i}$. These two parameters represent the smooth of the inclusion edge and the corresponding local residual stress for each $\theta_{sv,i}$. The calculation method of $\theta_{sv,i}$ is shown in Equation (4) and the schematic is shown in Figure 7.

$$\theta_{sv,i} = \theta_{i+1} - \theta_i \quad (4)$$

where i is the identification number of digital point on the inclusion edge. All the digital point were sequenced continually in an anticlockwise direction. θ_i is the angle between the x-axis and the straight line through the $(i\ 1)^{th}$ and the ith digital point on the inclusion edge, which can be calculated with Equation (5).

$$\theta_i = \arctan\left[(y_i - y_{i\text{-}1})/(x_i - x_{i\text{-}1})\right]\ (0 \le \theta_i \le 180°) \quad (5)$$

where x_i and y_i are the coordinates of the digital i^{th} point on the inclusion edge. When θ_i is a positive value, the local part of inclusion is convex; when θ_i is a negative value, the local part of inclusion is concave. The smaller the absolute value of $\theta_{sv,i}$ is, the smoother the inclusion edge is. When $\theta_{sv,i}$ is zero, the local shape of the inclusion is a straight line. For a right angle, $\theta_{sv,i}$ is 90°. It is to be noted that, the distance between each neighboring point will affect the result of $\theta_{sv,i}$. Therefore, the distance between each neighboring point is kept consistent in this study which offers a value of 3.6° for $\theta_{sv,i}$ in a perfect circle.

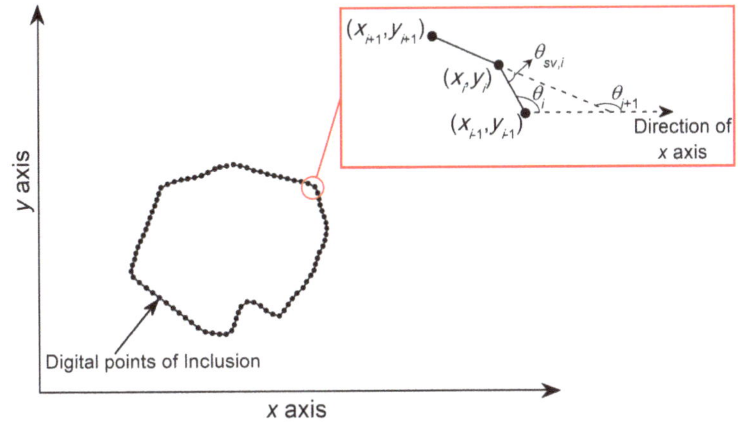

Figure 7. Schematic of the edge angle $\theta_{sv,i}$ for inclusions.

These two parameters of each digital point on inclusion edges of Mg-Al-O, Al-Ca-O-S, TiN and MnS shown in Figure 6 were extracted and calculated. The relation between these two parameters is shown in Figure 8. To avoid the interaction of neighboring points, the data with $\theta_{sv,i}$ lower than 5° were neglected. As shown in the figure, the local residual stress $\sigma_{sv,i}$ increases with the absolute value of edge angle $\theta_{sv,i}$ for all four types of inclusions, which indicates that no matter the local shape of the inclusion is convex or concave, the sharp edge will lead to stress concentration. When the local shape is convex, the stress concentration is in the steel matrix side; when the local shape is concave, the stress concentration is in the inclusion side (see Figure 6).

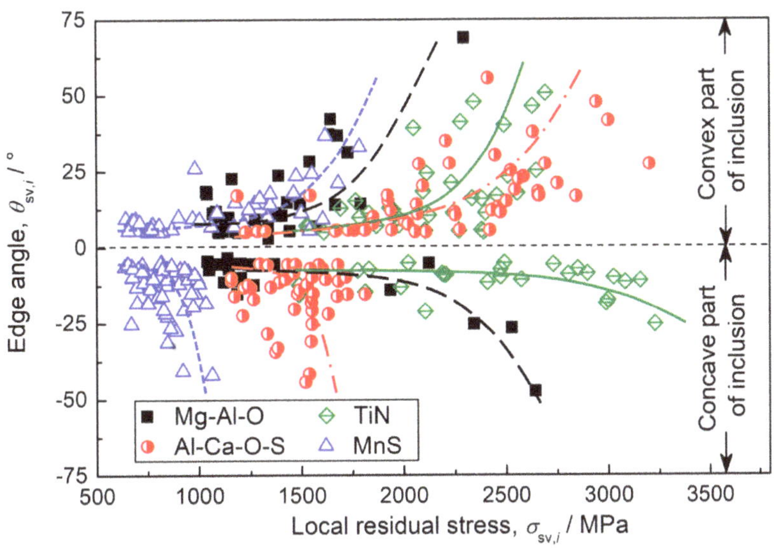

Figure 8. Relation between edge angle $\theta_{sv,i}$ and local residual stress $\sigma_{sv,i}$ for different inclusions.

In addition, $\theta_{sv,i}$, inclusion type also plays a significant role in the effect on local residual stress variation with $\theta_{sv,i}$. For the convex part of inclusion, the local residual stress of MnS inclusion increases the fastest with $\theta_{sv,i}$ among these four types of inclusions. When there is a convex angle of 25° on the inclusion surface, the local residual stress will increase about 1.64 times for MnS, 1 time for Al-Ca-O-S, 0.75 times for Mg-Al-O, and 0.53 times for TiN. For concave part of inclusion, the local residual stresses of Mg-Al-O and TiN inclusions increase the fastest, followed by MnS and Al-Ca-O-S. When there is the concave angle of 25° on the inclusion surface, the local residual stress will increase about 0.48 times for Al-Ca-O-S, 0.51 times for MnS, 1.32 times for Mg-Al-O, and 1.25 times for TiN. Although the local residual stress increases the fastest for convex part of MnS inclusion, the concentrated residual stress on the steel matrix side still does not exceed the other three types of inclusion under the same value of $\theta_{sv,i}$. However, for the concave part of inclusion, large $\theta_{sv,i}$ on the surface of TiN inclusion will lead to extremely large stress concentration on the inclusion side. This is also the reason that TiN inclusion itself often cracks in steel. These cracks inside inclusions can extend into steel matrix during the service of steel, causing failure of component [4]. Therefore, to obtain better performance it is important to avoid sharp edges of TiN inclusion.

To clarify the effect of the shape on the maximum residual stress around different types of inclusions, the maximum residual stresses of perfect spherical inclusions and inclusions with actual shape are shown in Figure 9. The maximum residual stresses of perfect spherical inclusions of all types are obviously smaller than the cases considering inclusion shapes. When we assume the inclusion shape as a perfect sphere for simplicity, there will be a non-negligible error of around 28.0 to 48.0%.

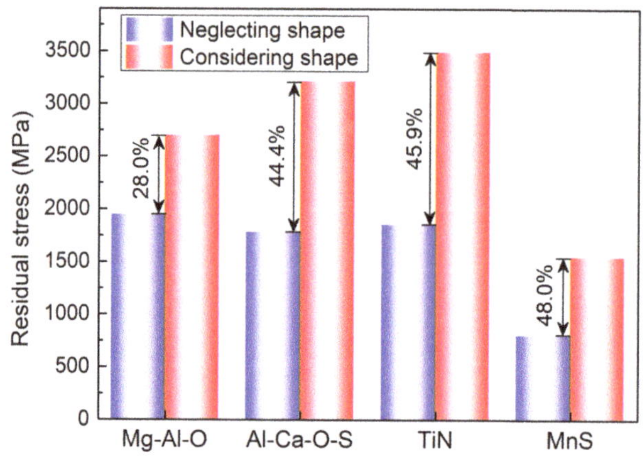

Figure 9. Comparison between the maximum residual stress around different types of inclusions neglecting shapes and considering shapes.

5. Conclusions

The cooling process during heat treatment will lead to residual stress around the interface between inclusion and steel matrix due to different thermal expansion coefficients and mechanical properties. In this study, the effect of the inclusion types and shapes on the residual stresses during the cooling process of the martensitic steels was investigated with a numerical procedure. The following results were obtained:

(1) This residual stress varies with inclusion type and shape. For perfect spherical inclusion, the residual stress around the interface between Al-Mg-O inclusion and steel matrix is the largest, followed by TiN and Al-Ca-O-S inclusion while the residual stress around MnS inclusion is the smallest.

(2) When the inclusion shape is assumed as a perfect sphere for simplicity, there will be a non-negligible error of the maximum residual stress of around 28.0 to 48.0% compared with the residual stresses considering the actual shapes.

(3) The convex local shape of inclusion will lead to stress concentration in the steel matrix side, and the concave local shape of inclusion will lead to stress concentration in the inclusion side.

(4) The local residual stress of MnS inclusion increases the fastest with the increase of inclusion edge angle $\theta_{sv,i}$ for the convex part of inclusion. For the concave part, the local residual stress of TiN inclusions increases the fastest, which leads to extreme stress concentration inside TiN inclusion and contributes greatly to material failure.

Author Contributions: Investigation, C.G. and J.L.; methodology, C.G.; data curation, C.G. and W.X.; writing-original draft preparation, C.G.; writing-review and editing, L.J. and S.M.; supervision, L.J., S.M. and Y.B.; funding acquisition, Y.B.

Funding: This research was funded by the State Key Laboratory for Advanced Metallurgy Foundation (No. 41602014).

Conflicts of Interest: The authors declare no conflict of interest.

References

1. Uthaisangsuk, V.; Prahl, U.; Bleck, W. Modelling of damage and failure in multiphase high strength DP and TRIP steels. *Eng. Fract. Mech.* **2011**, *78*, 469–486. [CrossRef]
2. Münstermann, S.; Lian, J.; Bleck, W. Design of damage tolerance in high-strength steels. *Int. J. Mater. Res.* **2012**, *103*, 755–764. [CrossRef]
3. Li, S.; He, J.; Gu, B.; Zeng, D.; Xia, Z.C.; Zhao, Y.; Lin, Z. Anisotropic fracture of advanced high strength steel sheets: Experiment and theory. *Int. J. Plast.* **2018**, *103*, 95–118. [CrossRef]
4. Pornputsiri, N.; Kanlayasiri, K. Experimental Study on Failure Mechanism of Advanced High Strength Steels in Air Bending Process. *IOP Conf. Ser. Mater. Sci. Eng.* **2018**, *361*, 012015. [CrossRef]
5. Fras, T.; Roth, C.C.; Mohr, D. Fracture of high-strength armor steel under impact loading. *Int. J. Impact Eng.* **2018**, *111*, 147–164. [CrossRef]
6. Murakami, Y.; Usuki, H. Quantitative evaluation of effects of non-metallic inclusions on fatigue strength of high strength steels. II: Fatigue limit evaluation based on statistics for extreme values of inclusion size. *Int. J. Fatigue* **1989**, *11*, 299–307. [CrossRef]
7. Collins, J.A. *Failure of Materials in Mechanical Design: Analysis, Prediction, Prevention*; John Wiley & Sons: Hoboken, NJ, USA, 1993.
8. Pascual, F.G.; Meeker, W.Q. Estimating Fatigue Curves with the Random Fatigue-Limit Model. *Technometrics* **1999**, *41*, 277–289. [CrossRef]
9. Murakami, Y.; Nomoto, T.; Ueda, T. Factors influencing the mechanism of superlong fatigue failure in steels. *Fatigue Fract. Eng. Mater. Struct.* **1999**, *22*, 581–590. [CrossRef]
10. Bathias, C. There is no infinite fatigue life in metallic materials. *Fatigue Fract. Eng. M* **1999**, *22*, 559–565. [CrossRef]
11. Marines, I. An understanding of very high cycle fatigue of metals. *Int. J. Fatigue* **2003**, *25*, 1101–1107. [CrossRef]
12. Pyttel, B.; Schwerdt, D.; Berger, C. Very high cycle fatigue—Is there a fatigue limit? *Int. J. Fatigue* **2011**, *33*, 49–58. [CrossRef]
13. Li, S.X.; Weng, Y.Q.; Hui, W.J.; Yang, Z.G. *Very High Cycle Fatigue Properties of High Strength Steels-Effects of Nonmetallic Inclusions*, 1st ed.; Metallurgical Industry Press: Beijing, China, 2010.
14. Gu, C.; Bao, Y.P.; Gan, P.; Lian, J.H.; Münstermann, S. An Experimental Study on the Impact of Deoxidation Methods on the Fatigue Properties of Bearing Steels. *Steel Res. Int.* **2018**, *89*, 1800129. [CrossRef]
15. Zhao, P.; Gao, G.; Misra, R.D.K.; Bai, B. Effect of microstructure on the very high cycle fatigue behavior of a bainite/martensite multiphase steel. *Mat. Sci. Eng. A Struct.* **2015**, *630*, 1–7. [CrossRef]
16. Lian, J.; Sharaf, M.; Archie, F.; Münstermann, S. A hybrid approach for modelling of plasticity and failure behaviour of advanced high-strength steel sheets. *Int. J. Damage Mech.* **2013**, *22*, 188–218. [CrossRef]
17. Lian, J.; Yang, H.; Vajragupta, N.; Münstermann, S.; Bleck, W. A method to quantitatively upscale the damage initiation of dual-phase steels under various stress states from microscale to macroscale. *Comp. Mater. Sci.* **2014**, *94*, 245–257. [CrossRef]

18. Krupp, U.; Giertler, A.; Söker, M.; Fu, H.; Dönges, B.; Christ, H.J.; Istomin, K.; Hüsecken, A.; Pietsch, U.; Fritzen, C.P.; et al. Significance and Mechanism of the Crack Initiation Process during Very High Cycle Fatigue of Duplex Stainless Steel. *Procedia Eng.* **2014**, *74*, 143–146. [CrossRef]
19. Alharbi, K.; Ghadbeigi, H.; Efthymiadis, P.; Zanganeh, M.; Celotto, S.; Dashwood, R.; Pinna, C. Damage in dual phase steel DP1000 investigated using digital image correlation and microstructure simulation. *Model. Simul. Mater Sci.* **2015**, *23*, 085005. [CrossRef]
20. He, J.; Lian, J.; Golisch, G.; He, A.; Di, Y.; Münstermann, S. Investigation on micromechanism and stress state effects on cleavage fracture of ferritic-pearlitic steel at −196 °C. *Mater. Sci. Eng. A* **2017**, *686*, 134–141. [CrossRef]
21. Fairchild, D.P.; Howden, D.G.; Clark, W.A.T. The mechanism of brittle fracture in a microalloyed steel-Part I Inclusion-induced cleavage. *Metall. Mater. Trans A* **2000**, *31*, 641–652. [CrossRef]
22. Yang, Z.G.; Li, S.X.; Zhang, J.M.; Zhang, J.F.; Li, G.Y.; Li, Z.B.; Hui, W.J.; Weng, Y.Q. The fatigue behaviors of zero-inclusion and commercial 42CrMo steels in the super-long fatigue life regime. *Acta Mater.* **2004**, *52*, 5235–5241. [CrossRef]
23. Sakai, T.; Oguma, N.; Morikawa, A. Microscopic and nanoscopic observations of metallurgical structures around inclusions at interior crack initiation site for a bearing steel in very high-cycle fatigue. *Fatigue Fract. Eng. M* **2015**, *38*, 1305–1314. [CrossRef]
24. Cai, X.F.; Bao, Y.P.; Lin, L.; Gu, C. Effect of Al content on the evolution of non-metallic inclusions in Si–Mn deoxidized steel. *Steel Res. Int.* **2016**, *87*, 1168–1178. [CrossRef]
25. Deng, Z.; Zhu, M. Evolution mechanism of non-metallic inclusions in Al-killed alloyed steel during secondary refining process. *ISIJ Int.* **2013**, *53*, 450–458. [CrossRef]
26. Gu, C.; Bao, Y.P.; Lin, L. Cleanliness distribution of high-carbon chromium bearing steel billets and growth behavior of inclusions during solidification. *Rev. Metal.* **2017**, *53*, 089. [CrossRef]
27. Cai, X.F.; Bao, Y.P.; Wang, M.; Lin, L.; Dai, N.C.; Gu, C. Investigation of precipitation and growth behavior of Ti inclusions in tire cord steel. *Metall. Res. Technol.* **2015**, *112*, 407. [CrossRef]
28. Spriestersbach, D.; Grad, P.; Kerscher, E. Influence of different non-metallic inclusion types on the crack initiation in high-strength steels in the VHCF regime. *Int. J. Fatigue* **2014**, *64*, 114–120. [CrossRef]
29. Karr, U.; Schuller, R.; Fitzka, M.; Schönbauer, B.; Tran, D.; Pennings, B.; Mayer, H. Influence of inclusion type on the very high cycle fatigue properties of 18Ni maraging steel. *J. Mater. Sci.* **2017**, *52*, 5954–5967. [CrossRef]
30. Xie, Q.; Gorti, S.; Sidor, J.J.; An, Y.G.; Wang, Y.D.; Lian, J.; Lan, H.; An, K. Grain Orientation Dependence of the Residual Lattice Strain in a Cold Rolled Interstitial-Free Steel. *Steel Res. Int.* **2018**, *89*, 1700408. [CrossRef]
31. Xie, Q.; Li, R.; Wang, Y.D.; Su, R.; Lian, J.; Ren, Y.; Zheng, W.; Zhou, X.; Wang, Y. The in-depth residual strain heterogeneities due to an indentation and a laser shock peening for Ti-6Al-4V titanium alloy. *Mater. Sci. Eng. A* **2018**, *714*, 140–145. [CrossRef]
32. Li, S.X. Effects of inclusions on very high cycle fatigue properties of high strength steels. *Int. Mater. Rev.* **2012**, *57*, 92–114. [CrossRef]
33. Brooksbank, D.; Andrews, K.W. Tessellated stresses associated with some inclusions in steel. *J. Iron Steel Inst.* **1969**, *207*, 474–483.
34. Brooksbank, D.; Andrews, K. Thermal expansion of some inclusions found in steels and relation to tessellated stresses. *J. Iron Steel Inst.* **1968**, *206*, 595–599.
35. Ma, Y.; Cui, Y. Study of the effect of sulfur contents on fracture toughness of railway wheel steels for high speed train. *Acta Metall. Sin.* **2011**, *47*, 978–983. [CrossRef]
36. Gu, C.; Bao, Y.-P.; Gan, P.; Wang, M.; He, J.-S. Effect of main inclusions on crack initiation in bearing steel in the very high cycle fatigue regime. *Int. J. Min. Metall. Mater.* **2018**, *25*, 623–629. [CrossRef]
37. Murakami, Y.; Uchida, M. Effects of thermal residual stress and external stress on stress intensity factors for a penny-shaped crack emanating from an ellipsoidal nonmetallic inclusion. In Proceedings of the JSME Meeting, Tokyo, Japan, 1992; Volume B, pp. 239–241.
38. Murakami, Y. *Metal Fatigue: Effects of Small Defects and Nonmetallic Inclusions*; Elsevier: London, UK, 2002.
39. Pineau, A.; Forest, S. Effects of inclusions on the very high cycle fatigue behaviour of steels. *Fatigue Fract. Eng. M.* **2017**, *40*, 1694–1707. [CrossRef]
40. Gu, C.; Lian, J.; Bao, Y.; Münstermann, S. Microstructure-based fatigue modelling with residual stresses: Prediction of the microcrack initiation around inclusions. *Mater. Sci. Eng. A* **2019**, under review.

41. Gu, C.; Lian, J.; Bao, Y.; Xie, Q.; Münstermann, S. Microstructure based fatigue modelling with residual stresses: Prediction of the fatigue life for various inclusion sizes. *Int. J. Fatigue* **2019**, under review.
42. Krahmer, D.M.; Polvorosa, R.; López de Lacalle, L.N.; Alonso-Pinillos, U.; Abate, G.; Riu, F. Alternatives for Specimen Manufacturing in Tensile Testing of Steel Plates. *Exp. Tech.* **2016**, *40*, 1555–1565. [CrossRef]
43. Silva, C.M.A.; Rosa, P.A.R.; Martins, P.A.F. Innovative Testing Machines and Methodologies for the Mechanical Characterization of Materials. *Exp. Tech.* **2016**, *40*, 569–581. [CrossRef]
44. Dixit, U.S.; Joshi, S.N.; Davim, J.P. Incorporation of material behavior in modeling of metal forming and machining processes: A review. *Mater. Des.* **2011**, *32*, 3655–3670. [CrossRef]
45. De Lacalle, L.N.L.; Lamikiz, A.; Salgado, M.A.; Herranz, S.; Rivero, A. Process planning for reliable high-speed machining of moulds. *Int. J. Prod. Res.* **2002**, *40*, 2789–2809. [CrossRef]
46. Brooksbank, D. Thermal expansion of calcium aluminate inclusions and relation to tessellated stresses. *J. Iron Steel Inst.* **1970**, *208*, 495–499.

applied sciences

MDPI

Article

Effects of Temperature and Time of Isothermal Holding on Retained Austenite Stability in Medium-Mn Steels

Adam Grajcar [1,*], Paweł Skrzypczyk [2] and Aleksandra Kozłowska [1]

1 Institute of Engineering Materials and Biomaterials, Silesian University of Technology, 18A Konarskiego Street, 44-100 Gliwice, Poland; aleksandra.kozlowska@polsl.pl
2 Z.U.H. LT Skrzypczyk, 10 Obywatelska Street, 43-100 Tychy, Poland; skrzypczykp@o2.pl
* Correspondence: adam.grajcar@polsl.pl; Tel.: +48-32-2372933

Received: 25 September 2018; Accepted: 1 November 2018; Published: 4 November 2018

Abstract: Effects of isothermal holding time and temperature on the stability of retained austenite in medium manganese bainitic steels with and without Nb microaddition were investigated. The amount of retained austenite for various variants of thermomechanical processing was determined by X-ray diffraction. Relationships between processing conditions and microstructure were revealed using light microscopy and scanning electron microscopy techniques. The isothermal holding temperatures changed from 500 to 300 °C and the time was from 60 to 1800 s. The optimal time and temperature of isothermal holding for all the investigated steels were 400 °C and 300 s, respectively. The relationships between the Mn content, amount of retained austenite, and carbon enrichment of the retained austenite (RA) were observed. The noticeable effect of Nb microaddition on the amount of retained austenite was not observed. In general, the carbon content in RA was slightly lower for the steels containing Nb. The optimum gamma phase amount was up to 18% for the 3% Mn steels, whereas it was c.a. 13% for the steels with 5% Mn. It was found that the morphology of blocky/interlath retained austenite depends substantially on the isothermal holding temperature.

Keywords: medium Mn steel; automotive sheet steel; bainitic steel; physical simulation; retained austenite; isothermal holding

1. Introduction

Medium-Mn bainitic steels with retained austenite belong to a group of steels dedicated to the automotive industry due to a great combination of mechanical and technological properties. These steels can be produced as cold-rolled—when the required microstructure is obtained during heat treatment after cold rolling—or they can be obtained as thermomechanically rolled [1]. A higher amount of retained austenite can be achieved for hot-rolled steels; however, the optimal ductility of steel sheets is not always obtained for the steels containing the highest fraction of γ phase. The critical factor is the optimal kinetics of strain-induced martensitic transformation related to the mechanical stability of retained austenite [2].

Medium-Mn sheet steels offer the best combination of strength and ductility at reasonable cost. They are much cheaper compared to fully austenitic high-Mn steels. The intermediate Mn content allows to stabilize a fraction of retained austenite between the high-Mn steels (2nd generation Advanced High-Strength Steels) and the low-Mn automotive multiphase steels (1st generation Advanced High-Strength Steels). Moreover, the mixture of ultra-fine ferrite and austenite or carbide-free bainite-austenite enables us to reduce the hardness difference between microstructural constituents compared to a mixture of ferrite, bainite, and retained austenite [2,3]. It results in better edge formability, stretch flangeability, and mechanical properties of medium-Mn sheet steels.

In order to provide the highest number of mechanical properties of steel, it is crucial to select proper conditions of bainitic transformation, which ensure obtaining the optimal amount and stability of the retained austenite. The conditions (time and temperature) of isothermal holding in a bainitic transformation region have a significant effect on the stability of this structural constituent. If the duration of isothermal holding is too short, the retained austenite is characterized by low stability due to the small carbon content. This leads to an increase in the temperature, at which point martensitic transformation begins (M_s temperature). Conversely, if the time of isothermal holding is too long, carbon is trapped in carbides, which also leads to a negative increase in M_s [3–5].

Incorrect time and temperature of isothermal holding during bainitic transformation result in obtaining some fraction of martensite, usually as martensite–austenite (MA) constituents. Sugimoto et al. [6] reported that the temperature of isothermal holding during bainitic transformation for C–Mn–Si steels strongly affects the amount of retained austenite and carbon concentration in this phase. They found that the maximum amount of RA was obtained at ~425 °C, while the maximum C content in RA was detected at ~375 °C for the steels containing 0.1–0.2 wt. % C. Girault et al. [7] noted that the carbon content in retained austenite depends on the chemical composition of steel, that is, silicon and aluminum contents.

Our previous research [8] regarding the effect of bainitic transformation temperature on the thermodynamic stability of retained austenite was carried out using a medium-C bainitic steel containing 1.5% Mn. Currently, there is a large amount of interest in bainitic steels with manganese content from 3 to 12%. The increased Mn amount enables to obtain a high fraction of retained austenite (~10–30%) [9,10]. In order to improve the mechanical properties of steels with a TRIP (TRansformation Induced Plasticity) effect, microadditions of Nb and Ti can be also added [11]. The influence of bainitic transformation conditions on the microstructure consisting of ferrite, bainite, and retained austenite have been analyzed by many authors [12–14]. Garcia-Mateo et al. [15,16] analyzed the factors affecting the stability of retained austenite in nanostructured bainitic steels. There are only few reports addressing the temperature and time aspects of bainitic transformation in medium-Mn steels with retained austenite. Therefore, the aim of the present study is to determine the effects of isothermal holding time and temperature on the thermodynamic stability of retained austenite in medium-Mn bainitic steels. Additionally, the effect of Nb microaddition on the microstructure of the investigated steels was analyzed.

2. Materials and Methods

The investigations were carried out on four medium-Mn steels characterized by manganese content c.a. 3% and 5%. The increased Mn content was added to stabilize the retained austenite. A silicon addition was partially replaced by aluminum to improve the susceptibility of sheet steel to hot dip galvanizing and to prevent carbide precipitation. Al also accelerates bainite nucleation, which is advantageous due to the possibility of the short duration of isothermal holding at the bainitic area [17]. Table 1 shows the chemical composition of the investigated steels. They are characterized by high metallurgical purity and low contents of S and P. Nb microaddition was intended to increase the strength through grain refinement and precipitation strengthening [18]. Mo was added for solid solution strengthening. A relatively low carbon content (max. 0.17%) should not deteriorate the weldability of the steels [19,20] and ensures the optimal enrichment of the austenite by this element during isothermal bainite transformation. Lun et al. [21] showed that the 0.15C–10Mn–1.5Al steel is weldable despite the medium Mn content.

Table 1. Chemical composition of investigated steels in wt. %.

Steel Type	C%	Mn%	Al%	Si%	Mo%	Nb%	S%	P%	N%
3Mn	0.17	3.3	1.7	0.22	0.23	-	0.014	0.010	0.0043
3MnNb	0.17	3.1	1.6	0.22	0.22	0.04	0.005	0.008	0.0046
5Mn	0.16	4.7	1.6	0.20	0.20	-	0.004	0.008	0.0039
5MnNb	0.17	5.0	1.5	0.21	0.20	0.03	0.005	0.008	0.0054

The investigated steel was melted in a Balzers VSG-50 vacuum induction furnace under Ar atmosphere. After melting and casting, the ingots were hot forged between 1200 to 900 °C. The specimens 15 × 20 × 35 mm for the thermomechanical processing using the Gleeble 3800 simulator were prepared. Detailed information on deformation conditions performed in 7 steps (T_1–T_7) in a temperature range 1200–850 °C (Figure 1) was listed in work [8]. The 5Mn and 5MnNb steels were continuously cooled to the bainitic transformation temperature T_B and isothermally treated in temperatures: 350, 400, 450, 475, and 500 °C. The duration of isothermal holding was: 300 s, and for 450 °C it was also 60, 600, and 1800 s, respectively. The aim was to determine the effect of isothermal holding time on the thermodynamic stability of retained austenite. 3Mn and 3MnNb steels after deformation were cooled according to the parameters given in Table 2. The duration and temperatures of isothermal holding at the bainitic region were the same as for 5Mn and 5MnNb steels (Figure 1).

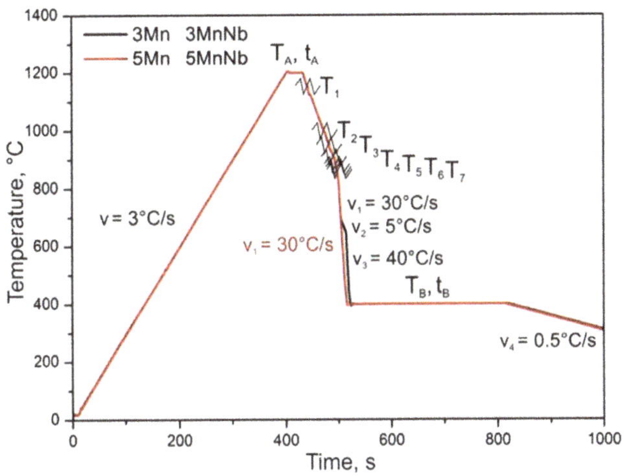

Figure 1. Thermomechanical processing conditions of investigated steels (for specimens isothermally held at 400 °C).

Table 2. Parameters of cooling after deformation for 3Mn and 3MnNb steels.

Operation Number	Temperature Range, °C	Cooling Rate, °C/s	Duration of Isothermal Holding, s
1	850→700	30	-
2	700→650	5	-
3	650→T_B	40	-
4	350	-	300
	400		300
	450		60, 300, 600, 1800
	475		300
	500		300
5	T_B→RT	0.5	-

The X-ray investigations were done using the X′ Pert PRO diffractometer with cobalt radiation and a graphite monochromator on a diffracted beam. The phase identification was carried out according to the data from International Centre for Diffraction Data ICDD. X-ray data obtained from the measurements served in the quantitative determination of the volume fraction of retained austenite. For this purpose, the Averbach–Cohen method recommended by The Society of Automotive Engineers for X-Ray diffraction measurements was used [22,23]. This method enables to determine a retained austenite amount using integrated intensities (area under a peak above a background) of X-ray diffraction peaks [24]. Using this method enables us to analyze the individual peaks, which allows to take into account the texture effects, at the same time minimizing background noise. This method is often applied to determine the retained austenite in TRIP-assisted steels [7,12–14]. Another method applied for TRIP steels is the Rietveld method [25]. In this method, a theoretical line profile is calculated from a structure model that is refined using a least-squares approach until it matches the discrete data from neutron or X-ray diffraction patterns [26]. The Rietveld analysis is widely used in powder phase quantitative analysis [27]. Another technique used for retained austenite detection is neutron diffraction. This method is a useful method due to the high penetration ability of a neutron beam [28–30]. The positions of maxima of the diffractions lines of austenite were used to determine the lattice constant of retained austenite. This parameter is necessary to calculate the concentration of carbon in the retained austenite. The dependence often applied for TRIP-type steels was used [31]:

$$a_\gamma = 3.578 + 0.033C_\gamma, \quad (1)$$

where: a_γ—lattice parameter of the austenite (Å), C_γ—carbon content in the austenite (wt. %).

To characterize the microstructure of the thermomechanically processed specimens, optical and SEM observation were performed. The analysis was carried out using a Leica MEF 4A optical microscope. The microstructural details were revealed with a scanning electron microscope Zeiss SUPRA 25 operating at 20 kV. The specimens were prepared in the plane, consistent with the direction of plastic flow. A first step of sample preparation included mechanical grinding with SiC paper up to 1500 grid. Then, they were polished with a diamond paste and etched using 5% nital to reveal the microstructure. For the purpose of the best identification of retained austenite, etching in 10% water solution of sodium metabisulfite was also applied. The hardness of steel samples was measured using the Vickers method (HV 10) to follow microstructure changes.

3. Results and Discussion

3.1. X-ray Results

X-ray diffraction analysis was carried out to identify the phases and to determine a retained austenite amount and its carbon content. It was necessary to assess the stability of retained austenite. Figure 2 shows selected X-ray diffraction patterns of the examined steels. The phase identification revealed the presence of diffraction lines from α phase and retained austenite. The detailed information on the fraction of retained austenite and carbon content in this phase are listed in Table 3. The obtained results are also presented in Figures 3–5 as a function of temperature and time.

Table 3. Fraction of retained austenite and carbon content in this phase for different variants of thermomechanical processing.

Steel Type	Thermomechanical Treatment Conditions	Fraction of Retained Austenite, %	Carbon Content In Retained Austenite, wt. %	Steel Type	Thermomechanical Treatment Conditions	Fraction of Retained Austenite, %	Carbon Content in Retained Austenite, wt. %
3Mn	850-700-10s-650-450-60s	10.4	1.26	3MnNb	850-700-10s-650-450-60s	11.5	1.18
	850-700-10s-650-450-300s	13.3	1.23		850-700-10s-650-450-300s	15.9	1.21
	850-700-10s-650-450-600s	13.2	1.09		850-700-10s-650-450-600s	12.4	1.08
	850-700-10s-650-450-1800s	4.7	1.03		850-700-10s-650-450-1800s	7.8	0.91
	850-700-10s-650-350-300s	9.4	1.34		850-700-10s-650-350-300s	9.3	1.39
	850-700-10s-650-400-300s	16.8	1.35		850-700-10s-650-400-300s	17.7	1.26
	850-700-10s-650-500-300s	11.1	1.14		850-700-10s-650-500-300s	9.8	1.07
5Mn	850-450-60s	7.8	1.08	5MnNb	850-450-60s	7.1	0.99
	850-450-300s	10.9	1.19		850-450-300s	11.0	1.17
	850-450-600s	8.1	1.14		850-450-600s	8.0	1.12
	850-450-1800s	7.0	1.09		850-450-1800s	6.8	1.07
	850-400-300s	9.8	1.27		850-400-300s	13.4	1.32
	850-350-300s	9.5	1.28		850-350-300s	12.2	1.24
	850-500-300s	8.2	1.06		850-500-300s	9.5	0.99

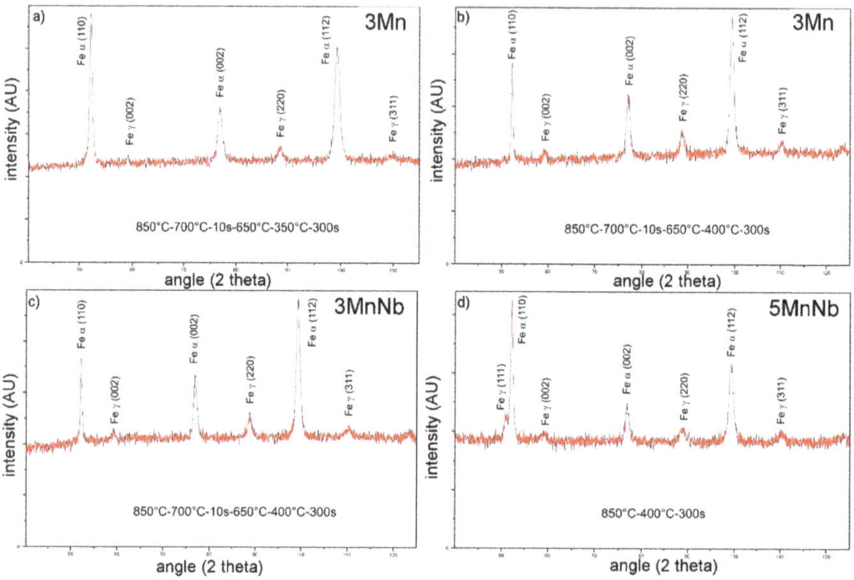

Figure 2. (**a**) X-ray diffraction pattern of 3Mn type steel isothermally held at 350 °C for 300 s; (**b**) X-ray diffraction pattern of 3Mn type steel isothermally held at 400 °C for 300 s; (**c**) X-ray diffraction pattern of 3MnNb type steel isothermally held at 400 °C for 300 s; (**d**) X-ray diffraction pattern of 5MnNb type steel isothermally held at 400 °C for 300 s.

In the case of 3Mn-type steel, the highest fraction of retained austenite was obtained during isothermal holding at 400 °C (16.8%). 3MnNb steel possessed a slightly higher fraction of γ phase—17.7%. Similar results were reported by Sugimoto et al. [32] in 0.2C–1.5Mn–1.5Si type steels with and without Nb microaddition. Lowering the bainitic transformation temperature to 350 °C caused a decrease in the amount of retained austenite to c.a. 10% for both steels containing 3% Mn (Table 3). This was due to the decrease in the diffusion rate while the temperature dropped. The amount of retained austenite slightly decreased when the temperature of bainitic transformation increased to 450 °C. Further increasing the temperature to 500 °C caused a significant decrease in the amount of retained austenite (Figure 3a).

A similar tendency was observed in 0.1C–1.5Mn–1.5Si and 0.2C–1.5Mn–1.5Si steels with lower Mn content [6]. The maximum amount of retained austenite was obtained at 425 °C. However, it was lower (~7%) than the values obtained in the present study. A small number of points did not allow to interpolate the data. However, from the physical point of view, it seems that the points should form a parabolic tendency with a peak value at a particular temperature. Such a behavior was observed by Wang et al. [12] in 0.2C–1.5Mn–0.7Si–1Al–0.5Cu steel. In their steel, peak temperature was 440 °C. This was related to the initiation of the carbide precipitation and the resulting transformation to martensite of some austenite fraction with the smallest carbon content. In our case, the drop in the amount of retained austenite is earlier, that is, above 400 °C. This is due to the chemistry of the steels. The increased manganese content seems to play a major role in this behavior.

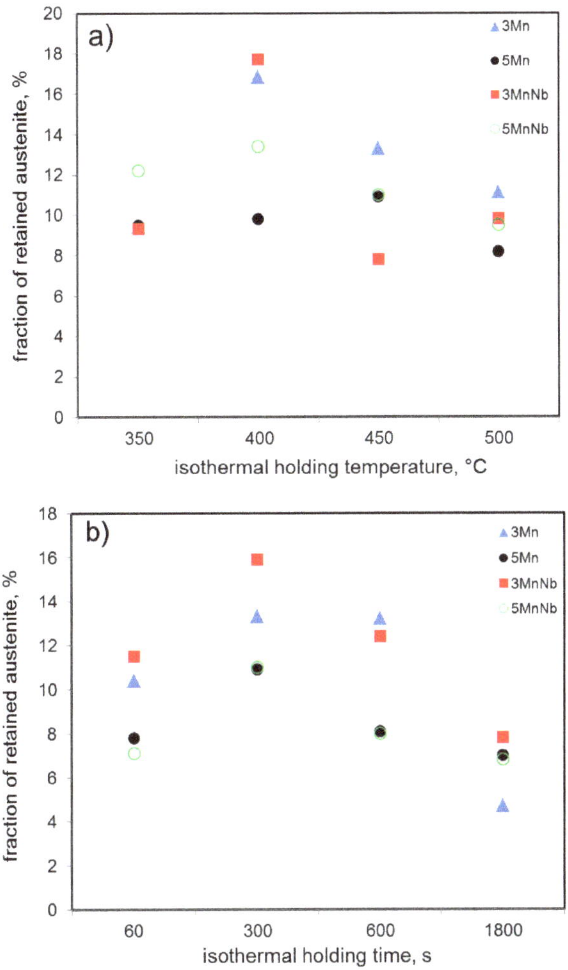

Figure 3. (**a**) Influence of temperature on the amount of retained austenite for steels isothermally held for 300 s; (**b**) Influence of isothermal holding time on the amount of retained austenite for steels treated at 450 °C.

Steels containing 5% Mn are less vulnerable to temperature changes. The amount of retained austenite in these steels is almost constant at the bainitic transformation temperature range from 350 to 450 °C (Figure 3a). The amount of retained austenite (RA) in this case was from 9 to 12% (Table 3) and it was higher for the steel with the Nb microaddition. The effect of isothermal holding time at 450 °C on the amount of retained austenite is shown in Figure 3b. 5Mn and 5MnNb steels showed fewer changes in the amount of γ phase due to variable duration of bainitic transformation. The optimal time of isothermal holding for all the steels seems to be 300 s. The highest fraction of retained austenite was detected for 3MnNb steel (Table 3)—17.7%. It was found that the γ phase was not enriched in enough carbon during isothermal holding at 450 °C for 60 s (Table 3). Therefore, the thermodynamic stability of the austenite in the steel containing Nb addition seems to be slightly lower. This is especially true for the steels with the higher Mn content. The stability of retained austenite is higher in the case of 3MnNb steel in comparison to 5MnNb steel, due to the higher carbon content. In general, the Nb-containing steels show slightly smaller carbon contents (compare the values in Table 3). However, there are also some disturbances in this tendency. For example, the 3MnNb steel sample treated at 350 °C and 5MnNb steel sample treated at 400 °C show a little bit higher carbon content in the RA compared to the reference steels without Nb. This indicates the complex interactions of the chemistry in TRIP steels containing microadditions [12,17] and it will be analyzed in future investigations. Increasing the time of bainitic transformation up to 1800 s caused a drop in the amount of RA (Figure 3b).

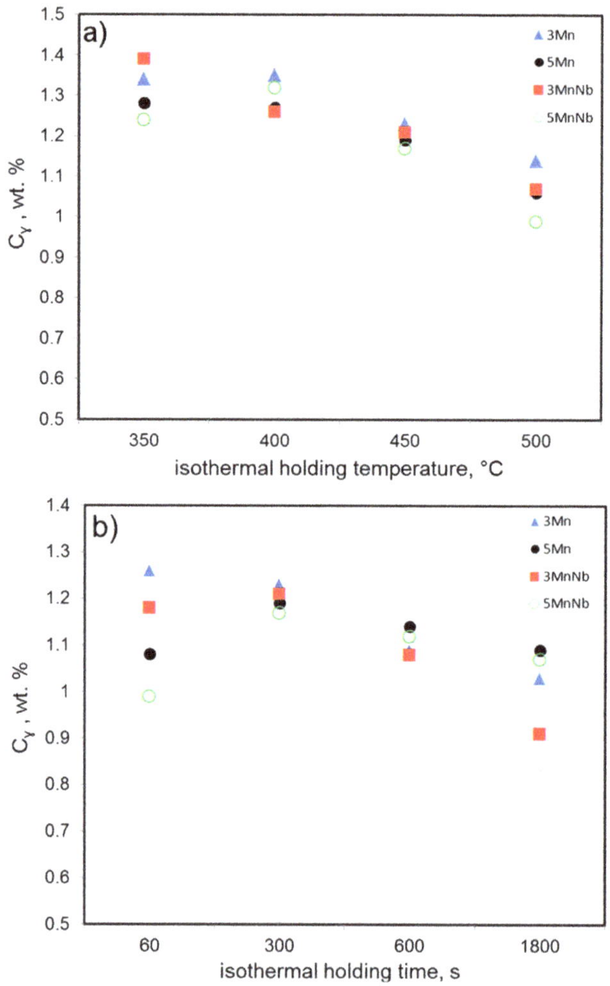

Figure 4. (**a**) Influence of isothermal holding temperature on the carbon content in retained austenite for the steels treated for 300 s; (**b**) Influence of the isothermal holding time at 450 °C on the carbon content in retained austenite.

In some cases, X-ray diffraction cannot provide the correct retained austenite content values because strain-induced martensitic transformation takes place easily near the specimen surface. In such cases, the metastable retained austenite fraction should be determined using neutron diffraction. This can be found in works [28–30].

The amount of retained austenite is closely related to the carbon content in this phase (Figure 4a). The obtained results showed that increasing the temperature of isothermal bainitic transformation leads to lowering the carbon content in retained austenite (Figure 4a). This tendency is similar for all the tested steels. However, the highest carbon content in retained austenite was detected for the 3Mn and 3MnNb steels. The lowest C content was obtained for the 5MnNb steel. For the steel containing 5% Mn, isothermal holding for 60 s is too short to enrich the austenitic phase in carbon (Figure 4b). The optimal duration of isothermal bainitic transformation for these steels is 300 s, similar to the steels, which contain 3% Mn.

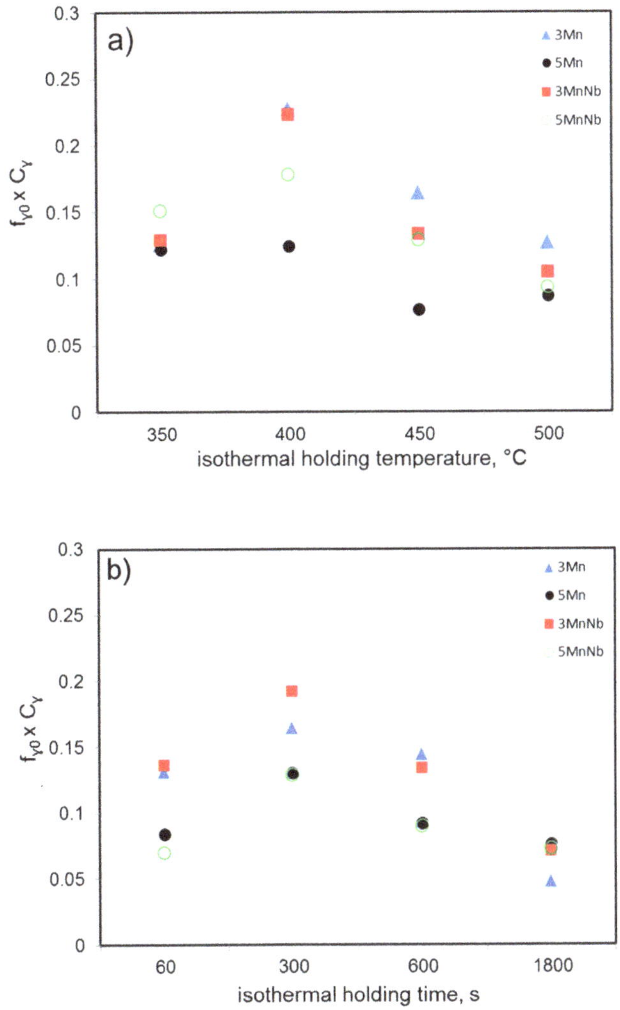

Figure 5. (**a**) Influence of isothermal holding temperature on the product of the retained austenite amount and its carbon content for the steels treated for 300 s; (**b**) Influence of isothermal holding time at 450 °C on the product of the retained austenite amount and its carbon content.

In fact, there is a lack of literature data concerning the effects of isothermal holding time on the fraction of retained austenite in medium-Mn steels and its carbon content. The data available in the literature concern mostly the 1st generation AHSS, with a Mn content below ~2%. Girault et al. [7] reported that for 0.1C–1.5Mn–1.5 Al steel isothermally held at 375 °C, the maximum carbon content in retained austenite was obtained after 1000 s. According to Wang et al. [12], the maximum carbon content in the 0.2C–1.5Mn–0.7Si–1Al–0.5Cu steel was measured at a level of 0.91% at 440 °C.

Fu et al. [33] compared the microstructure of Fe–0.2C–1.49Mn–0.98Al–0.5Si steel after bainitic transformation at 400 °C and 450 °C. They found that the stability of retained austenite is lower for the higher bainitic transformation temperature (450 °C) due to a lower carbon enrichment of the γ phase. A similar effect was observed in our study—increasing the temperature of the isothermal bainitic transformation above 400 °C leads to lowering the carbon content in the retained austenite. Due to the lower stability of RA, martensitic transformation occurred for relatively lower strain levels, which results in lower elongation [34].

Based on the data listed in Figures 3 and 4, it can be concluded that the maximum carbon content in RA was not detected at 400 °C, when the amount of the γ phase was the highest. The maximum carbon content in RA was detected in 350 °C; however, a fraction of this phase was not the highest. Hence, it seems that the optimal parameter which characterizes the thermodynamic stability of retained austenite is not the volume fraction or carbon content in the austenite, but the $f_{\gamma 0} \times C_{\gamma}$ product. It is interesting that the values of the $f_{\gamma 0} \times C_{\gamma}$ product were higher for the steels containing 3% Mn and the optimal time and temperature of isothermal holding were 400 °C (Figure 5a) and 300 s (Figure 5b), respectively. The values of the $f_{\gamma 0} \times C_{\gamma}$ product are similar for the steels with or without Nb microaddition.

3.2. Hardness Changes

The mechanical properties of the steels were assessed by hardness tests, which are sensitive to any possible phase transformation of the austenite upon cooling. On the one hand, this allows to indicate when the austenite is thermodynamically stable and remains at room temperature. On the other hand, one can easily assess when partial martensitic transformation upon cooling takes place and martensitic–austenitic (M–A) constituents are formed. Hence, the microstructure changes are reflected in sensitive hardness measurements.

The hardness of the steel samples was measured using the Vickers method. The obtained results are presented in Figure 6 as a function of time and temperature. The data presented in Figure 6a show that the lowest hardness values were obtained for specimens isothermally held within 300 s, both for 3Mn and 5Mn type steels. However, it is worth to note that higher hardness values were detected for the steels containing 5% Mn, especially for the steels containing the Nb microaddition. This can suggest the higher degree of the transformation into martensite. Figure 6b shows the effect of the temperature of isothermal holding during bainitic transformation on the hardness changes. In the case of 3Mn, 3MnNb, and 5Mn type steels, the lowest hardness was obtained at 450 °C, whereas for the 5MnNb steel at 400 °C. Higher hardness values were detected for the steels with 5% Mn.

Generally, the steels containing the Nb microaddition were characterized by higher hardness than the specimens without this element. However, the effect of Nb is not clear, probably due to its presence in the solid solution [11], as the manganese content increases the solubility of Nb in the austenite raises. It seems that the prior austenite grain size is similar for all the steels investigated. Hence, niobium is dissolved and its effect is at least partially masked (not so clear).

The increase in hardness is directly related to the higher amount of martensite or M–A constituents. Generally, the steels containing 3% Mn showed lower hardness than the 5% Mn steels. This is due to the higher fraction of retained austenite observed in the 3Mn and 3MnNb steels. Increasing the temperature to 500 °C caused a significant increase in martensite fraction, which was reflected in the higher hardness values in Figure 6b. The same effect is observed when the time of the isothermal holding is too short, that is, 60 s (Figure 6a).

In our previous study, Grajcar et al. [5] analyzed the mechanical stability of the retained austenite in the investigated steels using larger samples based on the work hardening behavior and elongation changes. The yield stress of 3Mn steels was near 700 MPa and above 200 MPa higher for the steels containing 5% Mn. These properties are slightly higher for Nb-containing steels. The total elongation was about 14–15% for the steels containing 3% Mn, whereas the 5Mn type steels showed lower elongation values. The tensile test data are in good agreement with the hardness changes presented in the current paper. For further mechanical stability data, the reader is referred to the text in Reference [5].

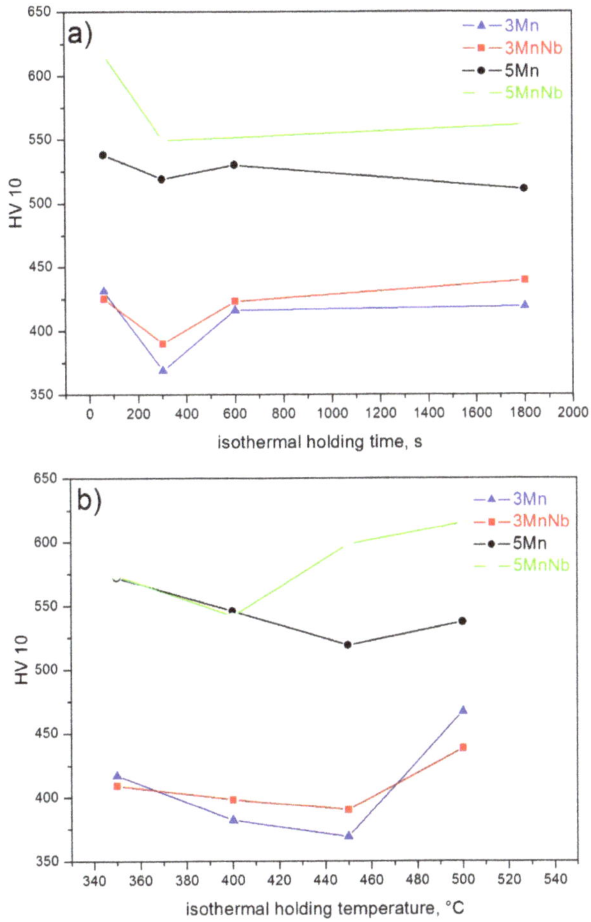

Figure 6. (**a**) Influence of isothermal holding time at 450 °C on the hardness; (**b**) Influence of isothermal holding temperature for 300 s on the hardness.

3.3. Microstructure Changes

Microstructures of samples for different temperatures of isothermal holding during bainitic transformation are shown in Figures 7–9. The obtained microstructures are fine-grained due to finishing the plastic deformation below the recrystallization temperature of austenite (850 °C) [5]. The microstructures were also affected by a high deformation level ($\varepsilon = 1.7$). The microstructures of 3Mn type steels were composed of fine-grained bainitic-martensitic plates. The highest fraction of retained austenite was observed in 3Mn and 3MnNb steels after isothermal holding at 400 °C and 450 °C (Figure 7c–f). This was earlier confirmed by the X-ray diffraction method. The most optimal duration of isothermal holding in the bainitic region seems to be 300 s. The clear influence of Nb microaddition on the microstructure and the amount of retained austenite was not detected. It seems that the Nb microaddition remained dissolved in solid solution due to the relatively high manganese content [5]. Thus, its effect can be smaller than that expected for NbC precipitates [17]. Based on the obtained diffraction patterns, it was found that the amount of retained austenite in steels with the Nb microaddition was lower by c.a. 2% (Table 3).

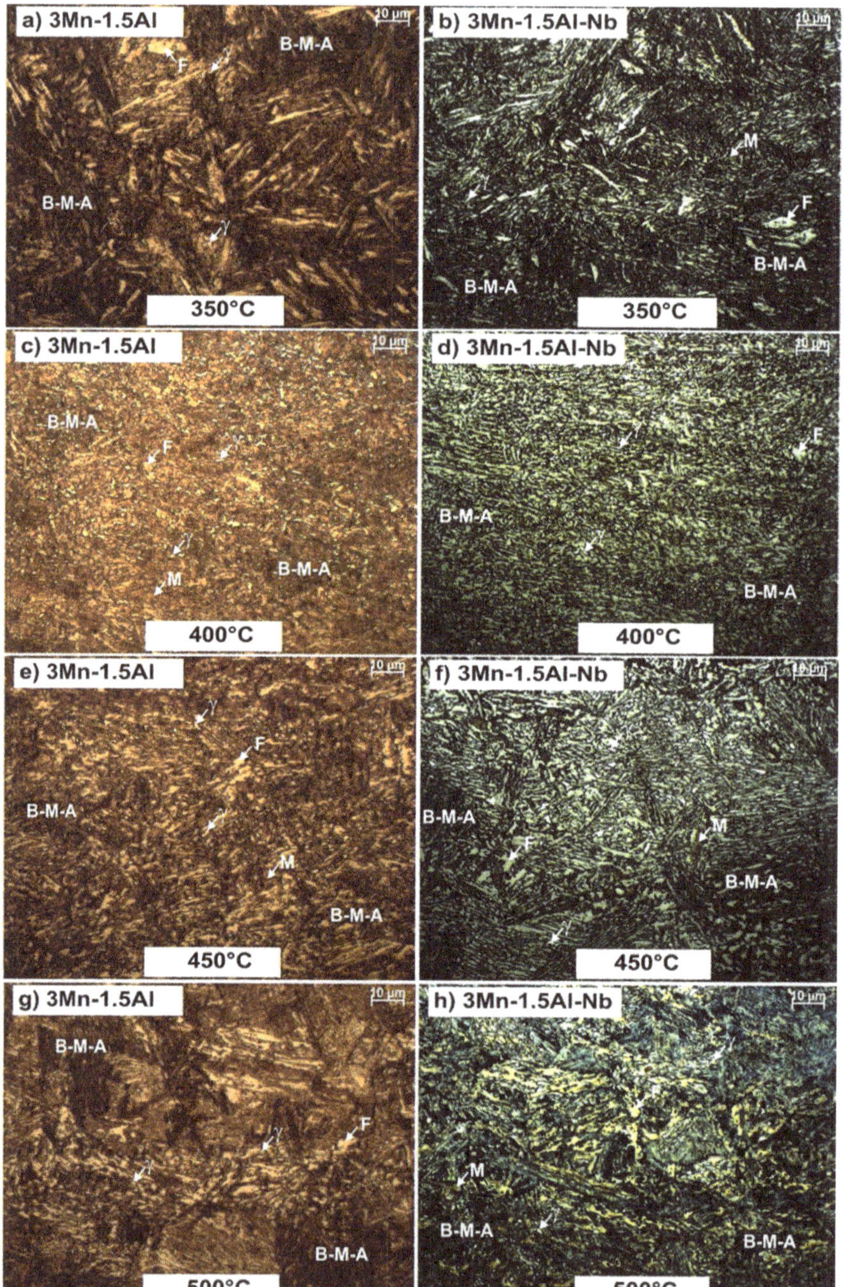

Figure 7. Multiphase microstructures of 3Mn and 3MnNb steels isothermally held at: (**a**,**b**) 350 °C, (**c**,**d**) 400 °C, (**e**,**f**) 450 °C, (**g**,**h**) 500 °C, for 300 s (LM).

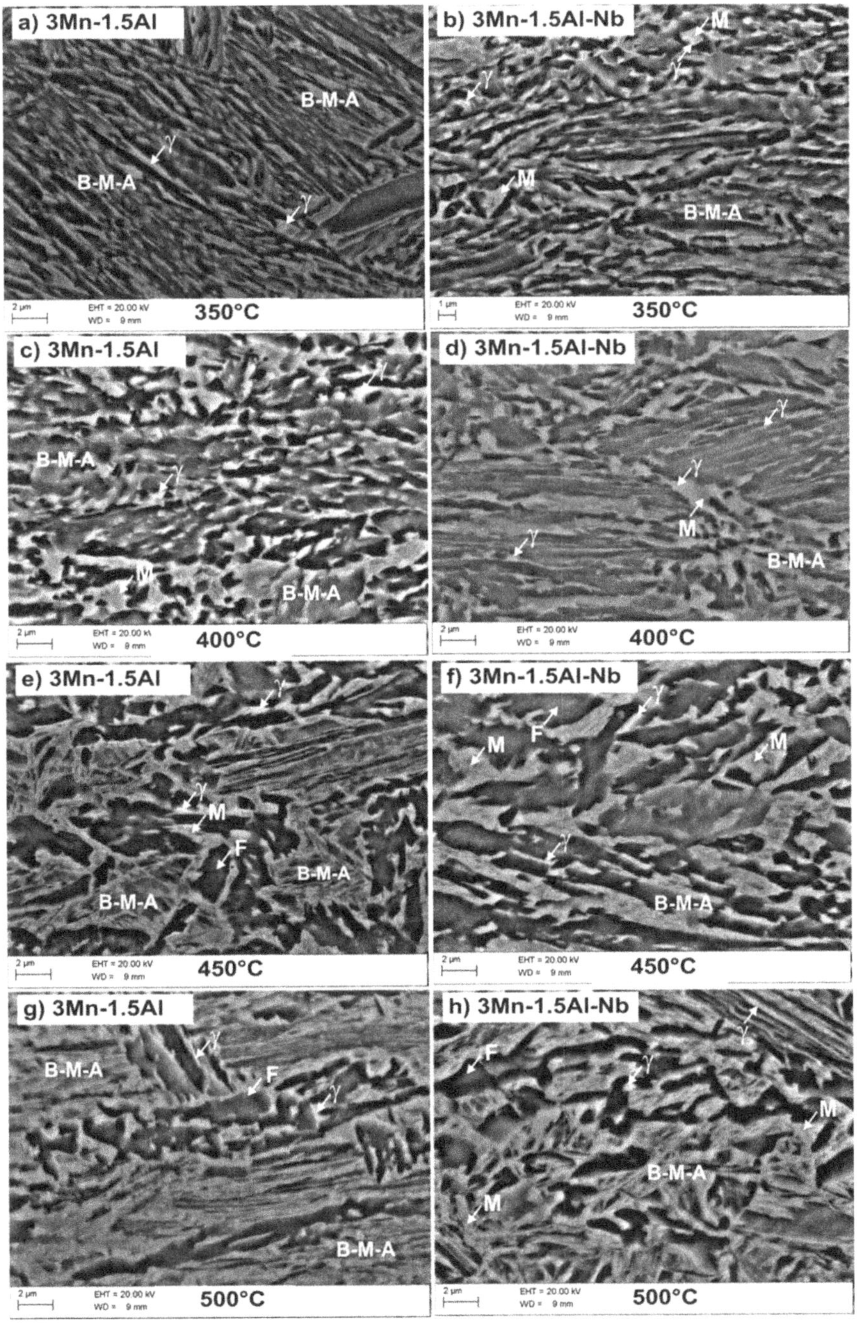

Figure 8. Multiphase microstructure of 3Mn and 3MnNb steels isothermally held at: (**a**,**b**) 350 °C, (**c**,**d**) 400 °C, (**e**,**f**) 450 °C, (**g**,**h**) 500 °C, for 300 s (SEM).

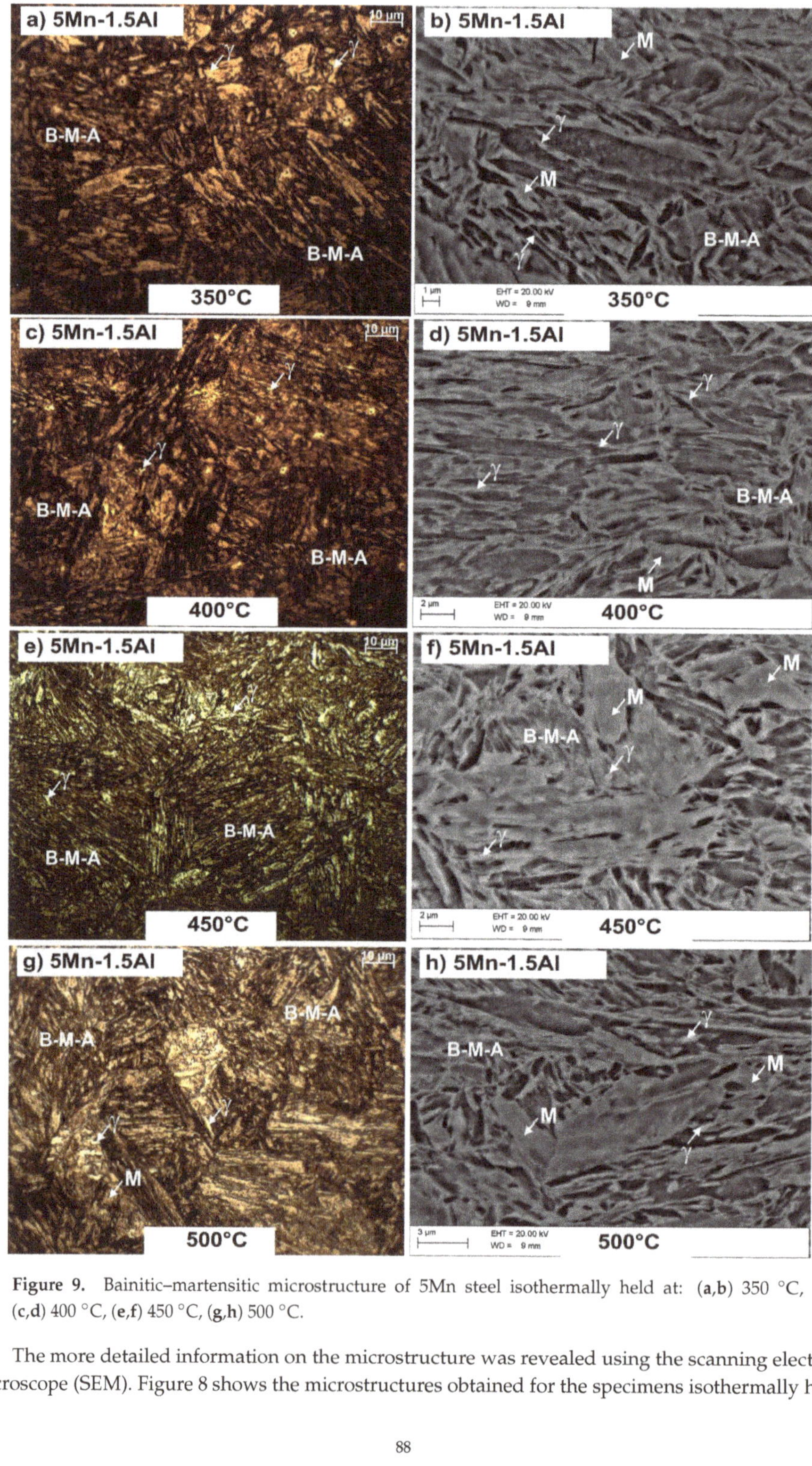

Figure 9. Bainitic–martensitic microstructure of 5Mn steel isothermally held at: (**a**,**b**) 350 °C, (**c**,**d**) 400 °C, (**e**,**f**) 450 °C, (**g**,**h**) 500 °C.

The more detailed information on the microstructure was revealed using the scanning electron microscope (SEM). Figure 8 shows the microstructures obtained for the specimens isothermally held between 350 and 500 °C. At 350 °C, the bainitic–martensitic laths were clearly observed and the retained austenite formed thin layers between the bainitic ferrite laths (Figure 8a,b). Increasing the bainitic transformation temperature up to 400 °C caused changes in the morphology of retained austenite. The fine-grained blocky retained austenite was observed at the boundary of bainitic–martensitic blocks (Figure 8c,d). Larger grains were subjected to partial martensitic transformation during cooling. As a consequence, they formed martensitic–austenitic regions (M–A). The confirmation of the phases using the EBSD (Electron Back-Scattered Diffraction) method can be found in our previous paper [5].

This indicates that the carbon content in the γ phase varies and depends on its size and morphology. In addition to the blocky-type grains of retained austenite, that phase also occurs in the form of continuous or intermittent layers characterized by various thicknesses, located inside the bainitic areas. This kind of microstructure is composed of degenerate upper bainite, where austenite occurs instead cementite along bainite laths [35]. While the bainitic transformation temperature was increased to 450 °C, the decomposition of the austenite into martensite began (Figure 8e,f). Under these conditions, only thin layers of retained austenite were thermodynamically stable. Bainitic–martensitic–austenitic (B–M–A) and martensitic–austenitic (M–A) areas occurred. Increasing the temperature to 500 °C resulted in the almost complete disappearance of the γ phase (Figure 8g,h).

For the steels containing a higher Mn content (~5%), mostly for the specimens isothermally held at 350 and 400 °C, the retained austenite occurred as layers characterized by various thicknesses (Figure 9a–d). Inside the bainitic laths, the martensitic–austenitic (M–A) particles and cementite were located. This is a typical morphology of degenerate lower bainite [35]. While the temperature of isothermal holding increased to 450–500 °C, retained austenite layers became thicker and the amount of blocky-type grains located mostly between martensitic–bainitic packets increased (Figure 9e–h). The blocky-type grains are characterized by lower thermodynamic stability, due to a smaller carbon content compared to the interlath retained austenite [6,13]. That is why they transformed partially or completely into the martensite during final cooling to the room temperature. This was confirmed by the previous indications of a carbon content in the austenitic phase (Table 3) and corresponding hardness changes (Figure 6). It means that the stability of the austenite is determined by the size and morphology of these areas. Due to the increased hardenability, 5Mn and 5MnNb steels are more susceptible to martensitic transformation. Therefore, the amount of stable retained austenite is lower than that in the steels containing 3% Mn.

4. Conclusions

The effects of temperature and time of isothermal holding in the bainitic range on the retained austenite stability were investigated in four medium-Mn steels with and without Nb microaddition. The following conclusions are drawn:

- The optimal time and temperature of isothermal holding for all the steels were 400 °C and 300 s, respectively. Under these conditions, 16–18% of γ phase was obtained for the 3Mn type steels, whereas 10–13% of retained austenite was detected in 5Mn type steels. The amount of retained austenite was similar for the steels with and without Nb microaddition;
- as the concentration of Mn increases, the amount of retained austenite decreases due to the smaller C enrichment of this phase. For 3Mn type steels, the carbon content in retained austenite was 0.9–1.4%, and it was up to 1.3% for 5Mn type steels. The carbon content in retained austenite was slightly lower in the steels with Nb microaddition;
- at 350 °C, the highest C enrichment of the retained austenite was obtained. Increasing the bainitic transformation temperature to 450 °C and extending the time over 300 s initiated the austenite decomposition and reduced the C content in this phase;
- in steels containing 3% Mn, isothermally held at 400 °C, mostly blocky-type retained austenite occurred, whereas at 350 °C the RA was located as layers between bainitic ferrite laths. As the temperature rose, the thickness of the γ phase layers increased and the blocky grains partially underwent martensitic transformation, forming M–A islands.

Author Contributions: A.G. and P.S. conceived and designed the experiments; A.G., P.S., and A.K. performed the experiments and wrote the paper; A.G. reviewed and supervised the work; P.S. and A.K. visualized the results.

Funding: This research was funded by the rector grant in the area of scientific research and development works. Silesian University of Technology, grant number 10/010/RGJ17/0143.

Conflicts of Interest: The authors declare no conflict of interest.

References

1. Kyeong-Won, K.; Kyung-Il, K.; Chang-Hoon, L.; Jun-Yun, K.; Tae-Ho, L.; Kyung-Mox, C.; Kyu Hwan, O. Control of retained austenite morphology through double bainitic transformation. *Mater. Sci. Eng. A* **2016**, *673*, 557–561.
2. Grajcar, A.; Radwanski, K. Microstructural comparison of the thermomechanically treated and cold deformed Nb-microalloyed TRIP steel. *Mater. Tehnol.* **2014**, *48*, 679–683.
3. Radwanski, K. Application of FEG-SEM and EBSD methods for the analysis of the restoration processes occurring during continuous annealing of dual-phase steel strips. *Steel Res. Int.* **2015**, *86*, 1379–1390. [CrossRef]
4. Marcisz, J.; Garbarz, B.; Burian, W.; Adamczyk, M.; Wisniewski, A. New generation maraging steel and high-carbon bainitic steel for armours. In Proceedings of the 26th International Symposium on Ballistics, Miami, FL, USA, 12–16 September 2011; pp. 1595–1606.
5. Grajcar, A.; Skrzypczyk, P.; Wozniak, D. Thermomechanically rolled medium-Mn steels containing retained austenite. *Arch. Metall. Mater.* **2014**, *59*, 1691–1697. [CrossRef]
6. Sugimoto, K.; Yu, B.; Mukai, Y.; Ikeda, S. Microstructure and formability of aluminium bearing TRIP-aided steels with annealed martensite matrix. *ISIJ Int.* **2005**, *45*, 1194–1200. [CrossRef]
7. Girault, E.; Mertens, A.; Jacques, P.; Hubaert, Y.; Verlinden, B.; Van Humbeeck, J. Comparison of the effect of silicon and aluminium on the tensile behavior of multiphase TRIP-assisted steels. *Scr. Mater.* **2001**, *44*, 885–892. [CrossRef]
8. Grajcar, A.; Krzton, H. Effect of isothermal holding temperature on retained austenite fraction in medium-carbon Nb/Ti-microalloyed TRIP steel. *J. Achiev. Mater. Manuf. Eng.* **2011**, *49*, 391–399.
9. Sun, B.; Fazeli, F.; Scott, C.; Guo, B.; Aranas, C., Jr.; Chu, X.; Jahazi, M.; Yue, S. Microstructural characteristics and tensile behavior of medium manganese steels with different manganese additions. *Mater. Sci. Eng. A* **2018**, *729*, 496–507. [CrossRef]
10. Aydin, H.; Jung, I.H.; Essadiqi, E.; Yue, S. Twinning and Tripping in 10% Mn steels. *Mater. Sci. Eng. A* **2014**, *591*, 90–96. [CrossRef]
11. Grajcar, A. Thermodynamic analysis of precipitation processes in Nb-Ti-microalloyed Si-Al TRIP steel. *J. Therm. Anal. Calorim.* **2014**, *118*, 1011–1020. [CrossRef]
12. Wang, C.; Ding, H.; Tang, Z.Y.; Zhang, J. Effect of isothermal bainitic processing on microstructures and mechanical properties of novel Mo and Nb microalloyed TRIP steel. *Ironmak. Steelmak.* **2015**, *42*, 9–16. [CrossRef]
13. Varshney, A.; Sangal, S.; Kundu, S.; Mondal, K. Super strong and highly ductile low alloy multiphase steels consisting of bainite, ferrite and retained austenite. *Mater. Des.* **2016**, *95*, 75–88. [CrossRef]
14. Mohamadizadeh, A.; Zarei-Hanzaki, A.; Mehtones, S.; Porter, D.; Moallemi, M. Effect of intercritical thermomechanical processing on austenite retention and mechanical properties in a multiphase TRIP-assisted steel. *Metall. Mater. Trans. A* **2016**, *47A*, 436–449. [CrossRef]
15. Garcia-Mateo, C.; Jimenez, J.A.; Lopez-Ezquerra, B.; Rementeria, R.; Morales-Rivas, L.; Kuntz, M.; Caballero, F.G. Analyzing the scale of the bainitic ferrite plates by XRD, SEM and TEM. *Mater. Charact.* **2016**, *122*, 83–89. [CrossRef]
16. Garcia-Mateo, C.; Caballero, F.G.; Bhadeshia, H.K.D.H. Acceleration of low-temperature bainite. *ISIJ Int.* **2003**, *43*, 1821–1825. [CrossRef]
17. Krizan, D.; De Cooman, B.C. Analysis of the strain-induced martensitic transformation of retained austenite in cold rolled micro-alloyed TRIP steel. *Steel Res. Int.* **2008**, *79*, 513–522. [CrossRef]
18. Opiela, M. Thermomechanical treatment of Ti-Nb-V-B micro-alloyed steel forgings. *Mater. Tehnol.* **2014**, *48*, 587–591.

19. Kurc-Lisiecka, A.; Piwnik, J.; Lisiecki, A. Laser welding of new grade of advanced high strength steel strenx 1100 MC. *Arch. Metall. Mater.* **2017**, *62*, 1651–1657. [CrossRef]
20. Gorka, J. Welding thermal cycle-triggered precipitation processes in steel S700MC subjected to the thermo-mechanical control processing. *Arch. Metall. Mater.* **2017**, *62*, 321–326. [CrossRef]
21. Lun, N.; Saha, D.C.; Macwan, A.; Pan, H.; Wang, L.; Goodwin, F.; Zhou, Y. Microstructure and mechanical properties of fibre laser welded medium manganese TRIP steel. *Mater. Des.* **2017**, *131*, 450–459. [CrossRef]
22. Gnaupel-Herolda, T.; Creuziger, A. Diffraction study of the retained austenite content in TRIP steels. *Mater. Sci. Eng. A* **2011**, *528*, 3594–3600. [CrossRef]
23. *ASTM E975—3: Standard Practice for X-ray Determination of Retained Austenite in Steel with Near Random Crystallographic Orientation*; ASTM International: West Conshohocken, PA, USA, 2013.
24. Jatczak, C.F.; Larson, J.A.; Shin, S.W. *Retained Austenite and Its Measurements by X-ray Diffraction*; SAE Special Publication 453; SAE Technical Paper: Warrendale, PA, USA, 1980.
25. Xu, P.G.; Tomota, Y.; Arakaki, Y.; Harjo, S.; Sueyoshi, H. Evaluation of austenite volume fraction in TRIP steel sheets using neutron diffraction. *Mater. Charact.* **2017**, *127*, 104–110. [CrossRef]
26. Feng, Z.; Hou, Q.; Zheng, Y.; Ren, W.; Ge, J.Y.; Li, T.; Cheng, C.; Lu, W.; Cao, S.; Zhang, J.; et al. Method of artificial intelligence algorithm to improve the automation level of Rietveld refinement. *Comput. Mater. Sci.* **2019**, *156*, 310–314. [CrossRef]
27. Ramkumar, T.; Selvakumar, M.; Vasanthsankar, R.; Sathishkumar, A.S.; Narayanasamy, P.; Girija, G. Rietveld refinement of powder X-ray diffraction, microstructural and mechanical studies of magnesium matrix composites processed by high energy ball milling. *J. Magnes. Alloys* **2018**, in press. [CrossRef]
28. Koo, M.; Xu, P.; Tomota, Y.; Suzuki, H. Bainitic transformation behavior studied by simultaneous neutron diffraction and dilatometric measurement. *Scr. Mater.* **2009**, *61*, 797–800. [CrossRef]
29. Voothaluru, R.; Bedekar, V.; Xie, Q.; Stoica, D.A.; Hyde, R.S.; An, K. In-situ neutron diffraction and crystal plasticity finite element modeling to study the kinematic stability of retained austenite in bearing steels. *Mater. Sci. Eng. A* **2018**, *711*, 579–587. [CrossRef]
30. Zrnik, J.; Muransky, O.; Lukas, P.; Novy, Z.; Sittner, P.; Hornak, P. Retained austenite stability investigation in TRIP steel using neutron diffraction. *Mater. Sci. Eng. A* **2006**, *437*, 114–119. [CrossRef]
31. Dyson, D.J. Effect of alloying additions on the lattice parameter of austenite. *J. Iron Steel Inst.* **1970**, *5*, 469–474.
32. Sugimoto, K.; Muramatsu, T.; Hashimoto, S.; Makai, Y. Formability of Nb bearing ultra-high-strength TRIP-aided sheet steels. *J. Mater. Process. Technol.* **2006**, *177*, 390–395. [CrossRef]
33. Fu, B.; Yang, W.Y.; Li, L.F.; Sun, Z.Q. Effect of bainitic transformation temperature on the mechanical behavior of cold-rolled TRIP steels studied with in-situ high-energy X-ray diffraction. *Mater. Sci. Eng. A* **2014**, *603*, 134–140. [CrossRef]
34. Kammouni, A.; Saikaly, W.; Dumont, M.; Marteau, C.; Bano, X.; Chari, A. Effect of the bainitic transformation temperature on retained austenite fraction and stability in Ti microalloyed TRIP steels. *Mater. Sci. Eng. A* **2009**, *518*, 89–96. [CrossRef]
35. Zajac, S.; Schwinn, V.; Tacke, K.H. Characterization and quantification of complex bainitic microstructures in high and ultra-high strength linepipe steels. *Mater. Sci. Forum* **2005**, *500–501*, 387–394. [CrossRef]

applied sciences

MDPI

Article

Failure Analysis on a Collapsed Flat Cover of an Adjustable Ballast Tank Used in Deep-Sea Submersibles

Fang Wang [1], Mian Wu [2], Genqi Tian [3], Zhe Jiang [1], Shun Zhang [1], Jian Zhang [2] and Weicheng Cui [1,4,*]

1 Shanghai Engineering Research Center of Hadal Science and Technology, Shanghai Ocean University, Shanghai 201306, China; wangfang@shou.edu.cn (F.W.); zjiang@shou.edu.cn (Z.J.); Shun_zhang@rainbowfish11000.com.cn (S.Z.)

2 Jiangsu Provincial Key Laboratory of Advanced Manufacture and Process for Marine Mechanical Equipment, Jiangsu University of Science and Technology, Zhenjiang 212003, China; wumianmail@126.com (M.W.); zhjian127@163.com (J.Z.)

3 School of Materials Science and Engineering, Shanghai Jiao Tong University, Shanghai 200240, China; sjtutgq@sina.com

4 School of Engineering, Westlake University, Hangzhou 310024, China

* Correspondence: wccui@shou.edu.cn

Received: 14 October 2019; Accepted: 28 November 2019; Published: 3 December 2019

check for updates

Abstract: A flat cover of an adjustable ballast tank made of high-strength maraging steel used in deep-sea submersibles collapsed during the loading process of external pressure in the high-pressure chamber. The pressure was high, which was the trigger of the collapse, but still considerably below the design limit of the adjustable ballast tank. The failure may have been caused by material properties that may be defective, the possible stress concentration resulting from design/processing, or inappropriate installation method. The present paper focuses on the visual inspections of the material inhomogeneity, ultimate cause of the collapse of the flat cover in pressure testing, and finite element analysis. Special attention is paid to the toughness characteristics of the material. The present study demonstrates the importance of material selection for engineering components based on the comprehensive properties of the materials.

Keywords: flat cover; adjustable ballast tank; fracture; high-strength maraging steel; toughness

1. Introduction

Deep-sea manned/unmanned submersibles are the necessary high-tech equipment for ocean exploration. It is used to carry crews or equipment to various deep-sea complex environments for efficient exploration, scientific investigation and resource exploitation [1,2]. Deep-sea submersibles contain multiple complex systems. Buoyancy regulation system is one of the important subsystems of the submersible. The adjustable ballast tank is one of the key components, which can ensure that the submersible has a good ability of depth setting and weight fine-tuning [3–5]. Through the high-pressure seawater pump, seawater is pumped in and out of the ballast tank, and the buoyancy balance of the submersible in seawater is adjusted within a certain range.

Reducing the weight for better performances such as range, speed, and payload is a significant consideration for the designers of the submersibles, as the weight-to-displacement ratio is used to evaluate the structural efficiency of underwater vehicles [1,2]. The weight of the submersible is distributed among the main components. At present, a spherical adjustable ballast tank with its promising application to a sea depth of 11,000 m and a volume of 300 L is designed, where domestic ultra-high strength maraging steel 18Ni(350) was used for first time for this purpose, as shown in Figure 1. The adjustable ballast tank is composed of upper and lower hemispheres, which are connected by a Huff clamp. An O-ring seals the hemispheres at the end. Fine-tuning makes the weight of the submersible maintain positive buoyancy or produce enough negative buoyancy for safely sitting at bottom. A flat cover is beneficial for installation of different cabin piercing parts. Sealing is carried out through a radial O-ring and is evenly locked with the sphere by screws. The selection of maraging steel for the current ballast tank is exactly based on the principle of strength enhancement for lower weight. This is also the first time that this material has been used in deep-sea pressure vessels. Some of the candidate materials for underwater pressure hulls, such as titanium and high strength steel, and their main properties can be found in literature [6–9]. The candidate material for pressure hulls using 18Ni grade maraging steels has been preliminarily investigated in terms of their application history, performance, and manufacturing capability by the authors [9]. The damage tolerance related performances of 18Ni grade maraging steels, including yield ratio (σ_y/σ_b) and fracture toughness, have been evaluated. There are basically four wrought commercial maraging steels of the 18 percent nickel family i.e., 18Ni (200), 18Ni (250), 18Ni (300), and 18Ni (350) with yield strength ranging from 1400 MPa to 2400 MPa [10]. In the development of 'MIR' submersibles in the 1980s [11], new techniques to produce high strength, high nickel-content steel 18Ni(250) for pressure hulls are applied for its two pressure spheres. Up to now, other grades of 18Ni series of maraging steels have no application experience in deep-sea pressure hulls.

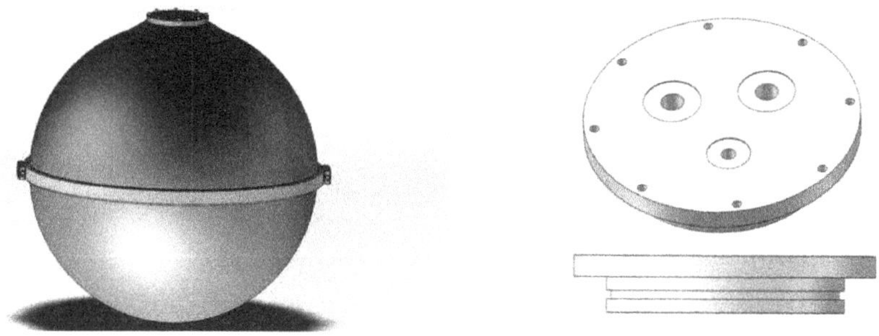

(a) (b)

Figure 1. Sketch of the adjustable ballast tank and its sealing cover: (**a**) front view of the adjustable ballast tank; (**b**) description of the sealing cover.

The increase in strength is very powerful in reducing the weight of pressure hulls, but the only concern is that increased strength generally leads to a decrease in toughness [12,13] while it is generally not considered in the design of deep-sea compressive components. Plane strain fracture toughness is an important factor to represent the crack resistance property of the material. Wherein, 18Ni(250) with yield strength of 1700 MPa has the plane-strain fracture toughness level of 85–110 MPa·m$^{1/2}$, but the value of 18Ni(350) with yield strength of 2400 MPa rapidly reduces to 30–50 MPa·m$^{1/2}$ [9,12,13]. Improving the toughness and plasticity of maraging steel can be done in various ways, such as reducing harmful elements or gas content in steel by a double vacuum smelting process, controlling inclusions morphology, adjusting microstructure by special processing, and heat treatment processes [14]. Results have shown that when the harmful elements in 18Ni(350) maraging steel are reduced to 10^{-5} magnitude at the same time, the number and volume of inclusions are greatly reduced, which is an important reason for the remarkable improvement of fracture toughness of ultra-pure 18Ni(350) maraging steel. The influence of the toughness on the resistance to cracking for deep-sea components needs to be examined.

The hydraulic pressure test on the adjustable ballast tank with its sealing flat cover was carried out to examine its performance. The ultimate goal of pressure test was to check whether it can endure the external pressure of 126.5 MPa (i.e., 115 MPa for 11,000 m deep-sea environments times a safety factor of 1.1 for pressure testing. Note, the actual pressure value for the 11,000 m deep-sea environment is 113.8 MPa and this value has been used in the design calculation, but here for the test, a more conservative value of 115 MPa is used). There are sensor mounting holes and high pressure pipe mounting holes on the sealing cover. However, the flat cover collapsed during the loading process of external pressure in the high-pressure chamber. The pressure was high, which was the trigger of the collapse, but still considerably below the design limit of the hull. The failure can be caused by unexpected defective material properties, the possible stress concentration resulting from design/processing, or inappropriate installation method. In this paper, the mechanical properties of the actual materials used in the collapsed flat cover are re-examined by sampling and testing of the broken parts. Non-metallic inclusions analysis, micro-structure analysis, and fracture surface analysis are conducted to acquire the possible fracture cause. Furthermore, finite element analysis based on fracture mechanics is conducted to understand the ultimate cause of destruction. The analysis results demonstrate the importance of material selection for engineering components based on the comprehensive properties of the materials.

2. Description of the Material and Structure

2.1. Chemical Composition and Material Properties

The preparation process of the material consists of vacuum induction furnace smelting and vacuum consumable furnace remelting. Solution heat treatment was conducted at 820 °C and kept for 1 h; aging treatment was conducted at 510 °C and kept for 4 h.

Chemical composition and mechanical properties of 18Ni (350) material are tested after sequential steps of melting, melt treatment practice, forging process, heat treatments, machining operations, and inspection, which is the same way as the cabin shell material was treated. Their chemical compositions and mechanical properties are given in Tables 1 and 2 and the stress–strain curve is depicted in Figure 2. This material has high tensile strength, the yield strength of which can reach about 2258 MPa and there is no obvious yield platform. However, its elongation is only 5.4%. The current design code for the spherical hull used in deep-sea submersibles and the design guideline for its flat cover have not specified the requirements for elongation of the material. The designers of the current spherical adjustable tank select 18Ni (350) instead of traditional material considering its high strength to reduce the weight of the tank, which will benefit the capability of the whole submersible.

Table 1. Chemical composition of 18Ni(350), (wt.%).

Ni	Co	Mo	Ti	Al	C	Si	Mn	P	S
18.80	11.72	4.41	1.32	0.125	0.006	0.02	0.01	0.004	0.002

Table 2. Mechanical properties of 18Ni(350).

σ_y, (MPa)	σ_b, (MPa)	A, (%)	Z, (%)	E, (MPa)	υ	KV at 20 °C, (J)
2258	2324	5.4	31.0	188	0.3	13.5

where σ_y: Yield strength; σ_b: Ultimate tensile strength; E: Young's modulus; υ: Poisson's ratio; A (%): Total elongation; Z (%): Area reduction.

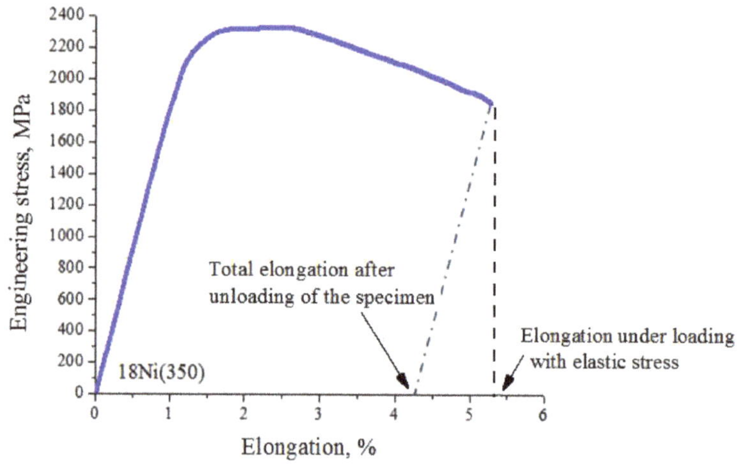

Figure 2. Stress–strain curve of 18Ni(350) (solution heat treatment: 820 °C for 1 h; aging treatment: 510 °C for 4 h).

2.2. Design and Geometry of the Flat Cover

The opening of the ballast tank is sealed by flange connection through the flat cover. The calculation of the thickness of the flat cover is carried out according to the requirements of the circular flat cover specified in the guidelines of steel pressure vessels [15].

$$\delta_p = D_c \sqrt{\frac{k\,P_c}{[\sigma]^t \phi}}, \tag{1}$$

where D_c = 176 mm is the calculated diameter of the flat cover; k = 0.25 is the structural characteristic coefficient; P_c is the calculated pressure, which can be 1.4 times the designed working pressure, i.e., 1.4 × 113 = 158.2 MPa; $[\sigma]^t = \sigma_y/1.5$. In order to make the strength safety factor larger, the yield stress applied by the designer presently is reduced to a lower level, 1900 MPa, then $[\sigma]^t$ = 1266.7 MPa; ϕ = 1.0 is the welding coefficient. This value of δ_p = 31 mm is obtained in calculation but was 32 mm in practice, including the thickness of 12 mm in the part of Φ190 mm and the thickness of 20 mm in the part of Φ150 mm, as shown in Figure 3. Necessary verification has been conducted by designers to check the strength and stiffness of the structure and it is considered to meet the requirements.

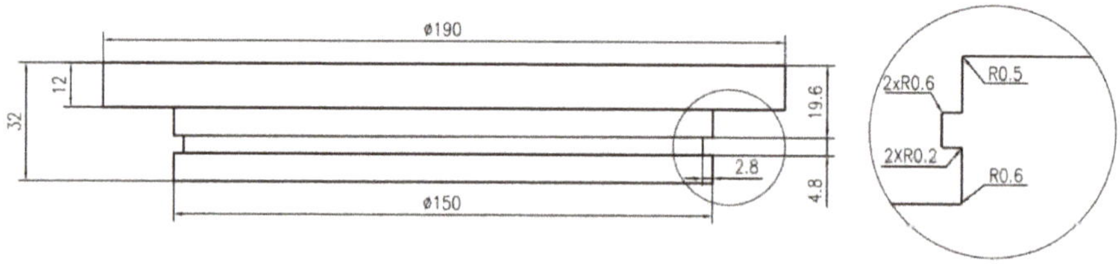

Figure 3. Dimension of the flat cover (unit: mm).

2.3. Failure of the Flat Cover During Testing

The spherical adjustable tank with its sealing cover was put into a high pressure chamber to examine its strength by pressure test before application. The pressure test was conducted in the high pressure chamber of the HAST laboratory at Shanghai Ocean University. There was no real-time video monitoring equipment in the pressure chamber and the test was carried out in several steps of cyclic pressure in order to stop and examine it periodically. The scheduled loading scheme is shown in Figure 4. A shorter load cycle was done before formal test. The maximum pressure set for this cycle was 75 MPa. The ultimate goal of the pressure test was to check if the tank could endure the external pressure of 126.5 MPa (i.e., 115 MPa for the 11,000 m deep-sea environment times a safety factor of 1.1 for pressure testing). After test preparation, the tank with strain gauges pasted outside the shell in three typical positions was sent to the pressure chamber, then the first cycle for watertight and insulation tests combining with strain and stress measurements was conducted. The loading phase was carried out gradually with the step of pressure increase of 5 MPa and maintained for 3 min in the loading and unloading stage. It was presumed that when the first cycle was finished, the tank should be taken out from the pressure chamber for a further inspection. If there was no problem, the tank was sent to the chamber for a final test.

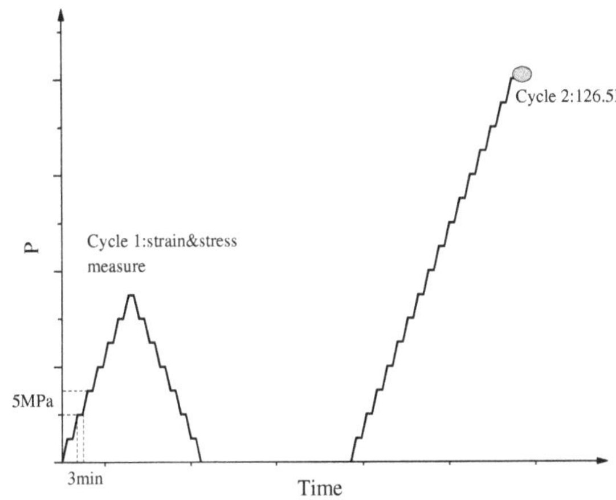

Figure 4. Loading scheme of pressure test.

However, when the tank was taken from the chamber after the first load cycle, it was found that the flat cover is cracked as shown in Figure 5. Local fracture occurred at the top part of the flat cover. The upper surface and fracture surface of the flat cover are respectively shown in Figure 5a,b. Based on the macroscopic observation, it was found that the flat cover bore large stress from outside to inside, and stress concentration occurred at the right-angle transition root. The crack morphology presented intergranular cracking, which was a brittle fracture.

Fracture occurred before the maximum pressure reached 75 MPa; however, it is difficult to assess at what pressure level the failure starts. The strain–load curves in Figure 6 show some abnormal phenomenon. Position 1 is at the opening reinforcement part of the spherical hull; position 2 is above the equatorial flange of the spherical hull; position 3 is at the bottom of the spherical hull. At a pressure of 40 MPa, there is a relatively obvious change of strain rate in the curves for three recording positions. This indicates that under this loading condition, the whole stress state of the spherical shell changes. At this point, the obvious stress concentration may appear on the flat cover. Because the external load is distributed both on the outer surface of the spherical shell and the flat cover, the change of stress state of the flat cover will cause stress concentration on the shell. From the fracture surface position, the stress concentration may come from the existence of crack-like defects, such as inadequate chamfer and scratches when the sharp part was processed. Therefore, in the next section, defects and fracture analysis will be carried out by means of microscopic observation and finite element calculation.

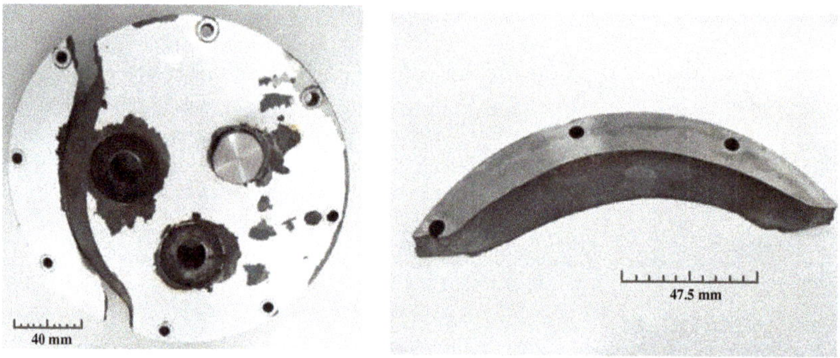

(**a**) (**b**)

Figure 5. Photography of the cracked flat cover: (**a**) upper surface of the cover; (**b**) fracture surface in the outer edge of the cover.

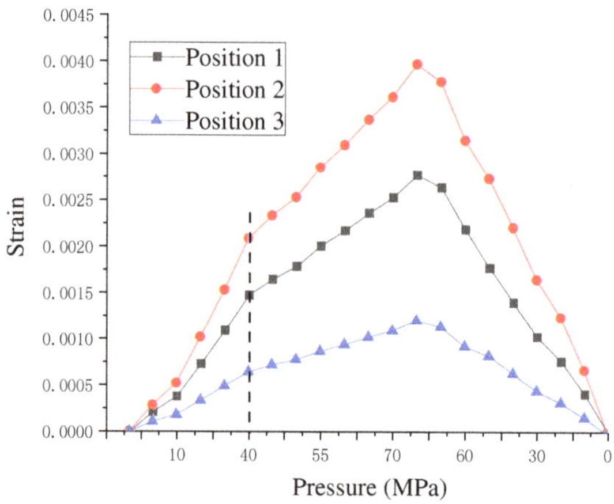

Figure 6. Recorded strains in the three positions outside the shell with variations of pressure.

3. Fracture Examination of the Flat Cover

Before the tank was manufactured, the material from the same furnace was tested to verify its conformance to the specified material standard and also the requirements of the designer. However, the same furnace material also had sampling uncertainty. Then in this section the tensile properties of the actual materials used in the flat cover were re-examined by sampling from the broken parts. The non-metal inclusions analysis, micro-structure analysis, and fracture surface analysis were conducted to find out the possible cracking cause.

3.1. Re-Examination of Tensile Properties

The static tensile test of a standard specimen with a diameter of 5 mm and gauge length of 30 mm [16] was carried out on SANS HST5106 hydraulic universal testing machine. Table 3 shows the results of the tests. It was found that the material has a higher tensile strength, yield strength, elongation, and reduction of area compared to the material presented in Table 2.

3.2. Re-Examination of Impact Properties

Standard Charpy V-notched impact specimens with length of 55 mm and square cross-sections of 10 mm × 10 mm [17] were used to determine impact toughness at room temperature on the PIT-752H test machine. Table 4 presents the room temperature impact test results of the flat cover material. The fluctuation between the test values is small, and the average impact energy is 14 J. However, the material has low impact toughness. The minimum value of impact toughness of 18Ni(350) is 12 J while the impact toughness of 18Ni(250) is about 24 J, which may be one of the reasons for the cracking of the flat cover during the test.

Table 3. Re-examination results of tensile properties of flat cover material at room temperature.

Specimen	σ_y (MPa)	σ_b (MPa)	A (%)	Z (%)	E (MPa)	υ
L1	2341	2421	6.4	47	186	0.300
L2	2355	2423	6.6	42	188	0.310
Average	2348	2423	6.5	45	187	0.305

Table 4. Re-examination results of impact toughness of flat cover material at room temperature.

Test Point	1	2	3	Average
KV, J	12	16	14	14

3.3. Re-Examination of Hardness

Rockwell and Brinell Hardness (HRC) hardness test was carried out with Rockwell hardness tester. The load was 150 KGF with holding time of 10 s. The average hardness of the flat cover material was about 54 HRC.

3.4. Non-Metallic Inclusions Observation

Non-metallic inclusions in flat cover samples were inspected [18]. The results of non-metallic inclusions rating and the inclusion morphology in the most important part of the flat cover showed that there were fewer kinds of non-metallic inclusions. The inclusions specified in the standards include sulfide inclusions, alumina inclusions, silicate inclusions, and spherical oxides. The original state of the material meets the design requirements. Figure 7 shows the inclusion morphology of the sample.

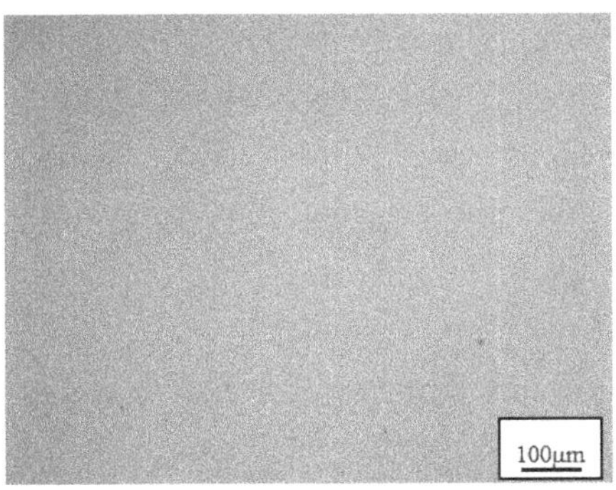

Figure 7. Inclusion morphology of the sample.

3.5. Microstructure Observation

Metallographic specimens were taken from the sealing cover. After mechanical grinding and polishing along the longitudinal section, the specimens were etched by an alcohol solution of $FeCl_3$ + hydrochloric acid at room temperature. The structure was observed under a ZEISS Axiovert 200MAT optical microscope (OM). The grading method of the grain size of samples was carried out according to grain size rating determination standard [19]. Figure 8 shows the structure of the flat cover material which is martensite.

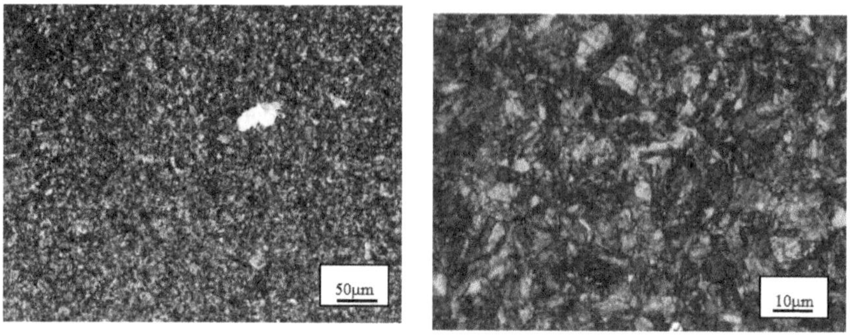

(a) (b)

Figure 8. Metallographic structure of the cover sample: (**a**) low magnification; (**b**) high magnification.

Microstructure and morphology of the flat cover material were observed and analyzed by TESCAN VEGA3 XMU scanning electron microscopy (SEM). Figure 9 shows the SEM low and high magnification microstructures of the specimens. The martensite structure of the material was uniform, and a small amount of precipitated phase particles could be seen at the grain boundary.

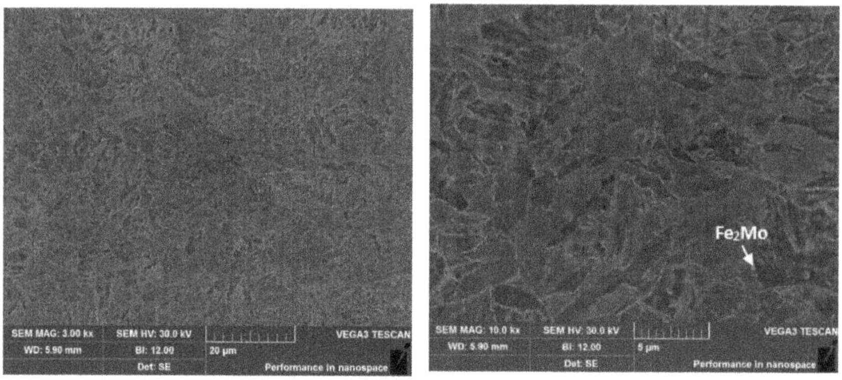

(a) (b)

Figure 9. SEM microstructure of the cover sample: (**a**) low magnification; (**b**) high magnification.

Further observation was conducted under JEOL JSM-2100 transmission electron microscopy (TEM). Figure 10 shows the TEM morphology of the specimen. It can be seen that the width of martensite lath is about 0.2–0.8 μm, and a large number of precipitations were uniformly distributed in martensite matrix with high density. Electron diffraction analysis results show that a large number of rod-like particles are $Ni_3(Al, Ti)$-type γ' precipitates, as shown in Figure 10b,c. In addition, large spherical granules in Figure 10c are Fe_2Mo precipitates. Some studies [20] have shown that the precipitations of Fe_2Mo occur with the increase of aging time and aging temperature.

3.6. Fracture Surface Observation

The fracture surface morphology was observed and analyzed by scanning electron microscopy after ultrasonic cleaning of the sample. Figure 11 shows the fracture morphology in the vicinity of the fracture source. It can be seen at the low magnification (Figure 11a,b) that the crack source is located at the root of the transition between the upper cover and the edge liner. It can be seen at high magnification (Figure 11c,d) that the crack source has an intergranular morphology and shows brittle fracture. During the external pressure test, the flat cover was subjected to a large stress from the outside to the inside, and a large stress concentration was generated at the right-angle transition root in which a crack source was easily generated and resulted in fracture.

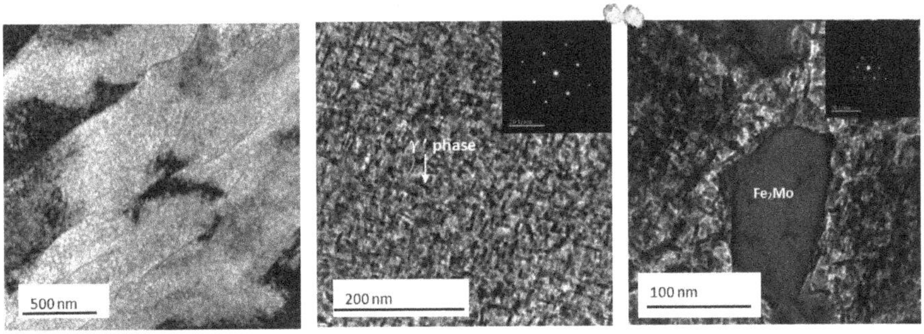

(a) (b) (c)

Figure 10. TEM microstructure of sealed top cover sample: (**a**) low magnification; (**b**) high magnification; (**c**) precipitated phase Fe_2Mo morphology.

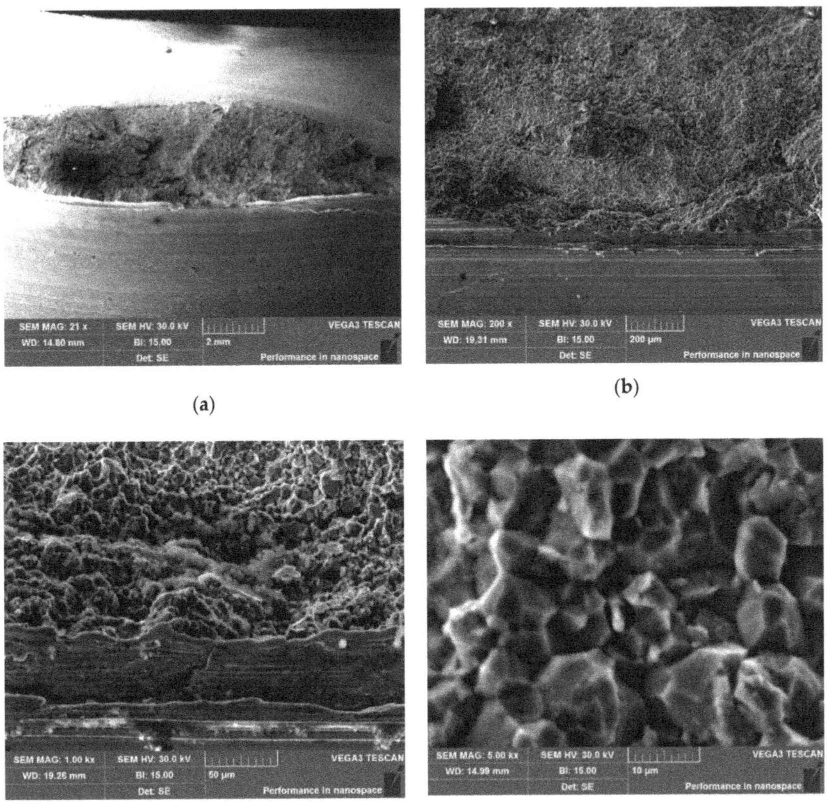

(c) (d)

Figure 11. SEM morphology of crack source at the fracture surface of the flat cover: (**a**) low magnification morphology; (**b**) low magnification morphology at the crack source; (**c**) high magnification morphology at the crack source; (**d**) partially magnified morphology at the crack source.

Figure 12 shows the SEM topography of the tearing area of the flat cover. The tearing surface has a dimple morphology, which shows ductile fracture characteristics.

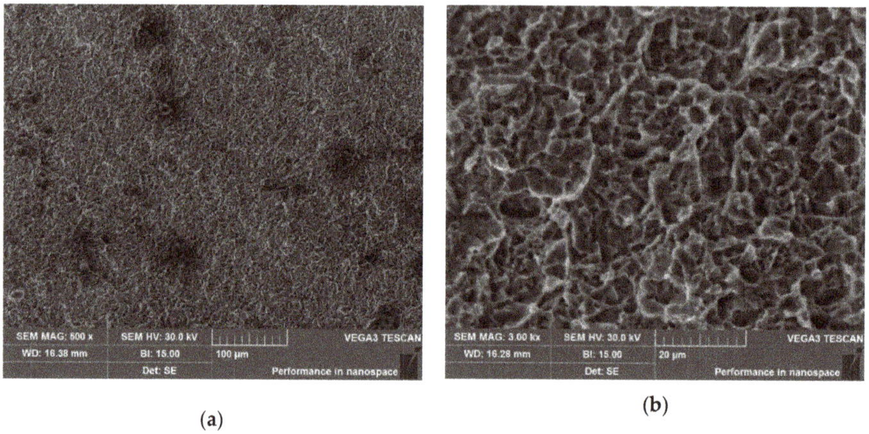

Figure 12. Topography of cracking area of the flat cover: (**a**) low magnification morphology; (**b**) high magnification morphology.

4. Finite Element Analysis

Based on the observations in Section 3, it can be concluded that the final fracture of the flat cover is due to the large stress concentration generated at the right-angle transition root of the thicker central part and the thinner edge. A crack initiated from this point is shown in Figure 13. It was initiated from improper manufacture and has the length of 3 mm and the depth of 1.5 mm. Final failure easily occurs because the material has low impact and toughness properties. A presumed semi-elliptical surface crack is considered in this section to analyze the failure mode of the flat cover during loading process. The fracture criterion is that the stress intensity factor at the crack tip reaches the plane strain fracture toughness of the material.

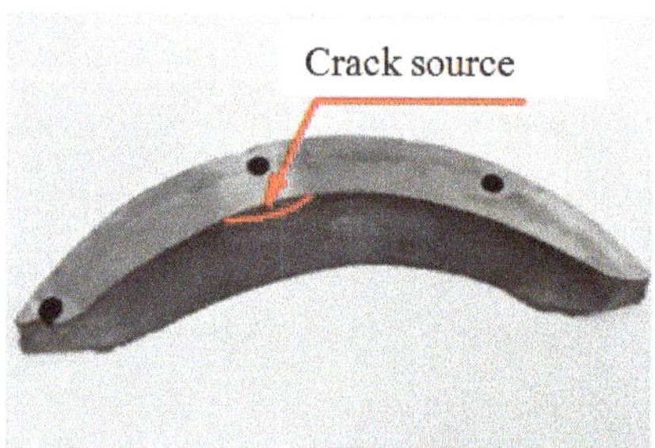

Figure 13. Crack position at the right-angle transition root.

4.1. Fracture Toughness of the Material

Fracture toughness tests are not conducted in the present study; however, it can be deduced from the widely accepted assumptions that the fracture toughness is taken to be a critical value of the maximum stress intensity factor for unstable crack growth [21]. According to this definition, the unstable crack growth will occur when K_{max} reaches K_{IC}. In order to reduce test costs, this method was used to determine the fracture toughness of the material conveniently on the basis of the crack growth rate test for data over a wide range of growth rates [9] at four load ratios, i.e., 0.1, 0.3, 0.5, and 0.7.

The crack growth rate curves (*da/dN*~Δ*K* curves) at four load ratios of 0.1, 0.3, 0.5, and 0.7 were experimentally obtained in [9]. At each load ratio, two groups of tests were carried out. When the crack growth curve began to transit from stable growth stage to unstable growth stage, the requirement of $K_{max} = K_{IC}$ should be satisfied. Then a value of K_{IC} can be calculated based on each crack growth rate curve. The average of these values can be taken as the final value of K_{IC} as shown in Figure 14. The data of $\Delta K(1-R)$ are within a relatively stable range, and the fracture toughness of 18Ni (350) can approach 37~40 MPa·m$^{1/2}$, which agrees well with the data range in literature [12,13] and can be applied for fracture analysis. The dotted line shows the average level of the calculated data of K_{IC}, which was about 38 MPa·m$^{1/2}$. It must be emphasized that the toughness of this material is greatly reduced compared to 18Ni (250), which is traditionally used in the pressure hulls of deep-sea manned submersible [9].

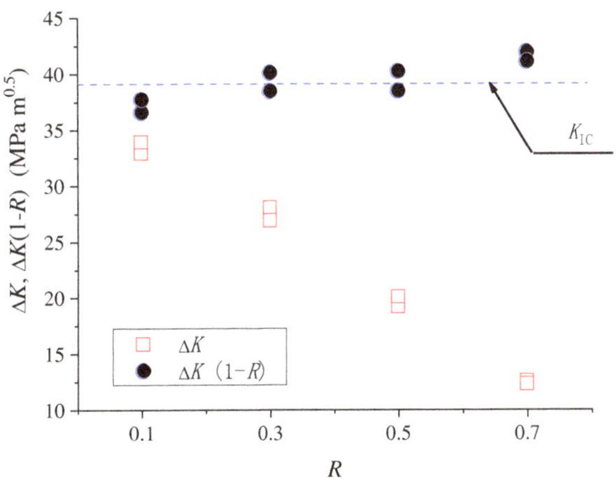

Figure 14. ΔK and $\Delta K(1-R)$ values of 18Ni(350) determined by unstable crack growth properties.

4.2. Finite Element Analysis

SolidWorks software was used for three-dimensional modeling, and ABAQUS software is used for finite element analysis after modeling. C3D8R element is used for meshing. Local mesh encryption was made in the vicinity of key chamfers of the flat cover and contacts the part between the spherical model and the flat cover, with mesh size of 1 mm, compared to the mesh size of 20 mm in other parts. The small chamfer was not included in the calculation model, because when the crack was inserted into the model, the stress concentration at the crack tip would be the dominant factor, and the stress concentration at the small chamfer was no longer important. The whole structure model was used in finite element calculation to obtain the critical stress area and the surface crack was inserted into the local model after stress analysis. Figure 15 gives the whole model for stress analysis. The ballast tank used in the deep-sea submersible was subjected to uniformly distributed hydrostatic pressure, which was applied both in the spherical shell and the flat cover. In order to eliminate the overall rigid body displacement without hindering the relative deformation, the boundary conditions were set symmetrically with six displacement components constrained by the three-point support method recommended in [22]. The contact surface between the flat cover and the sphere was set as the friction contact with the friction coefficient of 0.1.

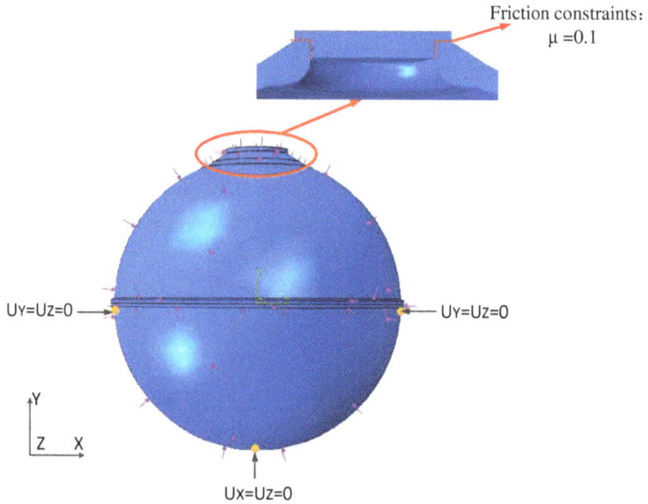

Figure 15. Finite element model for stress analysis.

When a crack is simulated, the crack will be inserted into a sub-model. The sub-model was taken out separately for the study. The calculation can be conducted by using the connection between the global model and the sub-model, thus greatly reducing unnecessary computation and time costs. According to the results of previous failure observations, the defective part of the cover is treated as a sub-model and the crack with the length of 3 mm and the depth of 1.5 mm is inserted into the model. After static analysis, the stress intensity factor of crack front can be obtained. The position relationship between the sub-model and the global model is shown in Figure 16.

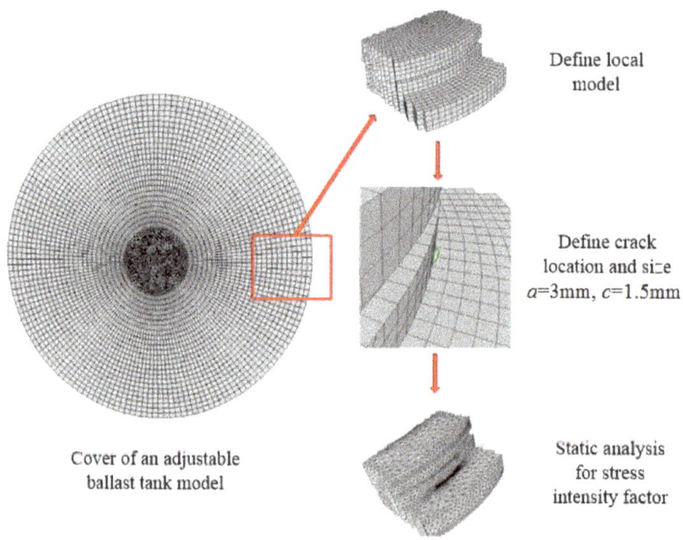

Figure 16. The position relationship between the sub-model and the global model.

When the flat cover is subjected to different external pressures from 5 to 85 MPa, the stress intensity factors of the crack can be obtained. Figure 17 depicts the stress intensity factors along the crack front with increasing the external pressure. Figure 18 shows the changing of the stress intensity factors at the surface points and the deepest point with the external pressure. The stress intensity factor in the length direction is larger than that in the depth direction. The maximum value of the stress intensity factor occurs at the surface points. When the external load increases to about 75–85 MPa, the stress intensity factor at the crack tip reaches 35–42 MPa·m$^{1/2}$, i.e., approximating to the value of fracture toughness, 38 MPa·m$^{1/2}$, shown in Figure 14, which is the ultimate cause of the damage.

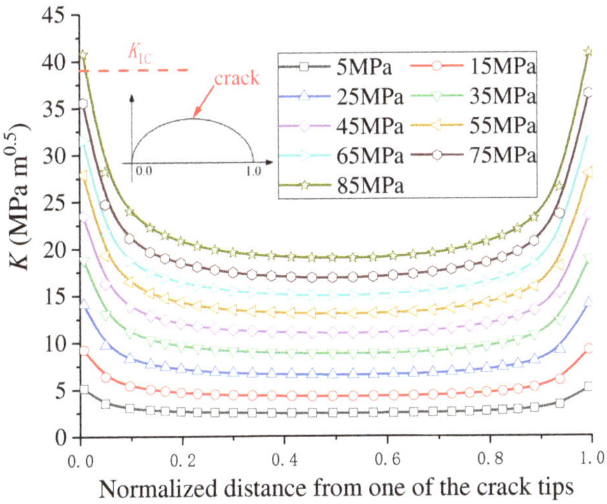

Figure 17. Stress intensity factor along crack front with increasing external pressure.

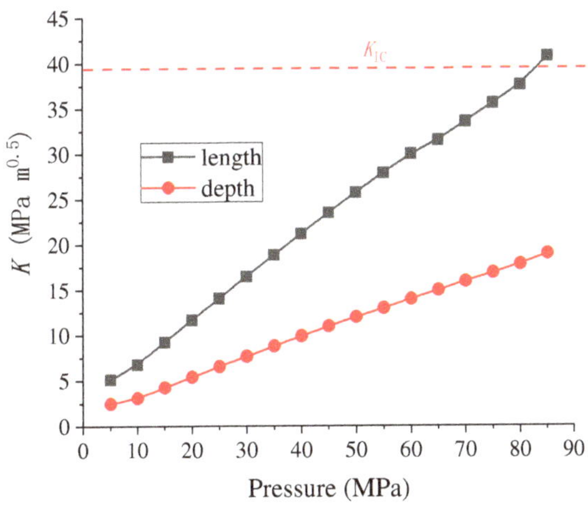

Figure 18. Stress intensity factor at the surface points and the deepest point.

5. Summary and Conclusions

In this paper, a destroyed flat cover made of 18Ni (350) used in an adjustable ballast tank during high pressure testing was analyzed. To find out the cause of the damage, the mechanical properties of the material used in the flat cover were re-examined by sampling and testing of the broken parts. Non-metallic inclusions analysis, microstructure analysis, and fracture surface analysis were conducted to find out the possible fracture cause. Finite element analysis based on fracture mechanics to understand the ultimate cause of destruction was conducted. The following conclusions and recommendations can be made:

(1) Maraging steels are expected to still have high toughness when improving their ultra-high strength level. 18Ni (250) grade is currently an acceptable candidate material for deep-sea pressure hulls, which has achieved satisfactory application experience. To obtain better weight-to-displacement ratio, the designers applied 18Ni (350) for an adjustable ballast tanks used in a deep-sea submersible. The strength of 18Ni (350) is higher than that of 18Ni (250), however its toughness is lower. In the current design guideline for deep-sea pressure hulls, the requirements for material toughness have not been emphasized. The trial of applying 18Ni (350) was made in the present study but unexpected failure of the flat cover occurred when load was relatively small.

(2) Non-metal inclusions analysis, microstructure analysis, and fracture surface analysis showed that the original state of the material meets the design requirements with fewer kinds of non-metallic inclusions. The material has low impact toughness, which is one of the reasons for the cracking of the flat cover during the test.

(3) From the design point of view, it is suggested to optimize the transition from right angle to chamfer at variable cross-sections, and to increase the thickness of the outer edge liner of the cover appropriately, which will reduce the possibility of damage. During processing, the surface finish of structural parts should be ensured, especially the root surface of the variable cross-section, so as to avoid surface processing defects.

(4) The reliability of deep-sea pressure hulls, especially the pressure hulls of manned submersibles, guarantees the safety of personnel and equipment. There must be attention paid to the comprehensive performance of materials. From the point of view of material selection, the application of 18Ni (350) maraging steel to the pressured structure in deep-sea environments is a risky choice. On the other hand, the 18Ni series of maraging steels have broad application prospects in deep-sea environments, so improving the toughness and plasticity together with increasing the strength level will be an important research direction [23].

Author Contributions: The individual contributions are specified as follows: conceptualization, F.W. and W.C.; methodology, F.W. and J.Z.; software, M.W.; validation, M.W., S.Z., G.T., and Z.J.; formal analysis, F.W.; investigation, S.Z.; resources, Z.J.; data curation, M.W. and G.T.; writing—original draft preparation, F.W.; writing—review and editing, W.C.; visualization, F.W.; supervision, W.C.; project administration, F.W. and W.C.; funding acquisition, F.W. and W.C.

Funding: This research was funded by the General Program of National Natural Science Foundation of China, grant number 51679133, the State Key Program of National Natural Science of China, grant number 51439004, and the "Construction of a Leading Innovation Team" project by the Hangzhou Municipal government.

Conflicts of Interest: The authors declare no conflict of interest.

References

1. Nanba, N.; Morihana, H.; Nakamura, E.; Watanabe, N. Development of Deep Sub-mergence Research Vehicle "Shinkai 6500". *Tech. Rev.* **1990**, *27*, 157–168.
2. Walden, B.B.; Brown, R.S. A replacement for the Alvin submersible. *Mar. Technol. Soc. J.* **2004**, *38*, 85–91. [CrossRef]
3. Liu, Y.S.; Wu, D.F.; Li, D.L.; Zhao, X.F. Seawater hydraulic buoyancy adjusting system for large-depth submersible. *Chin. Hydraul. Pneum.* **2014**, *10*, 1–10.
4. Qiu, Z.L. Design and research on a variable ballast system for deep-sea manned submersibles. *J. Mar. Sci. Appl.* **2008**, *7*, 255–260. [CrossRef]
5. Tangirala, S.; Dzielski, J. A variable buoyancy control system for a large AUV. *IEEE J. Ocean. Eng.* **2007**, *32*, 762–771. [CrossRef]
6. Jame, J.G.; Lisa, L.L. Submersible pressure hull design parametrics. *SNAME Trans.* **1991**, *99*, 119–146.
7. Ross, C.T.F. A conceptual design of an underwater vehicle. *Ocean Eng.* **2006**, *33*, 2087–2104. [CrossRef]
8. Qu, W.X. Structure Design and Analysis of Pressure Hull in Manned Submersible. Ph.D. Thesis, Harbin Engineering University, Harbin, China, 2013.
9. Wang, F.; Hu, Y.; Cui, W.C. Preliminary evaluation of maraging steels on its application to full ocean depth manned cabin. *J. Ship Mech.* **2016**, *20*, 1557–1572.

10. *18 Percent Nickel Maraging Steels-Engineering Properties, Distributed by the Nickel Development Institute, Courtesy to Inco Europe Limited*; NiDI Publication: Groningen, The Netherlands, 1976; No.4419.
11. Sagalevitch, A.M. From the Bathyscaph Trieste to the Submersibles Mir. *Mar. Technol. Soc. J.* **2009**, *43*, 79–86. [CrossRef]
12. Fan, S.J.; Dai, T. Properties and applications of 18Ni grade maraging steels. *Foreign Met. Heat Treat.* **1995**, *3*, 41–47.
13. Gangloff, R.P. Hydrogen assisted cracking of high strength alloys. In *Comprehensive Structural Integrity Vol. 6*; Milne, I., Ritchie, R.O., Karihaloo, B., Eds.; Elsevier Science: New York, NY, USA, 2003.
14. He, Y.; Liu, K.; Yang, K. Effect of solution temperature on fracture toughness and microstructure of ultra-purified 18ni(350) maraging steel. *Acta Met. Sin.* **2003**, *39*, 381–386.
15. *ISO 16528-1-2007. Boilers and Pressure Vessels—Part 1: Performance Requirements*; ISO: Geneva, Switzerland, 2007.
16. *ISO 6892-1-2016. Metallic Materials-Tensile Testing—Part 1: Method of Test at Room Temperature*; ISO: Geneva, Switzerland, 2016.
17. *ISO 14556-2000. Steel-Charpy V-notch Pendulum Impact Test-Instrumented Test Method*; ISO: Geneva, Switzerland, 2000.
18. *ISO4967-1998. Steel-determination of Content of Nonmetallic Inclusions-micrographic Method Using Standards Diagrams*; ISO: Geneva, Switzerland, 1998.
19. *GB/T 6394-2017. Determination of Estimating the Average Grain Size of Metal*; China National Standards: Beijing, China, 2017.
20. Mu, B. Effect of Trace Elements on the Mechanical Properties and Microstructure of 18 wt.%Nickel Maraging Steel. Master's Thesis, Jiangsu University of Science and Technology, Zhenjiang, China, 2011.
21. Farahmand, B.; Nikbin, K. Predicting fracture and fatigue crack growth properties using tensile properties. *Eng. Fract. Mech.* **2008**, *75*, 2144–2155. [CrossRef]
22. China Classification Society (CCS). *Rules for the Classification and Construction of Diving Systems and Submersibles*; China Classification Society: Beijing, China, 2013.
23. Marcisz, J.; Adamczyk, M.; Garbarz, B. Optimisation of Mechanical Properties of 18%Ni350 Grade Maraging Steel Using Novel Heat Treatment. *Arch. Metall. Mater.* **2017**, *62*, 73–84. [CrossRef]

applied sciences

MDPI

Article

Fatigue Crack Growth in Maraging Steel Obtained by Selective Laser Melting

Fernando Antunes [1,*], Luís Santos [2], Carlos Capela [2], José Ferreira [1], José Costa [1], Joel Jesus [1] and Pedro Prates [1]

1 CEMMPRE, Department of Mechanical Engineering, University of Coimbra, 3030-788 Coimbra, Portugal; martins.ferreira@dem.uc.pt (J.F.); jose.domingos@dem.uc.pt (J.C.); joel.jesus@dem.uc.pt (J.J.); pedro.prates@dem.uc.pt (P.P.)

2 Department of Mechanical Engineering, Instituto Politécnico de Leiria, ESTG, 2411-901 Leiria, Portugal; luis_lms@sapo.pt (L.S.); ccapela@ipleiria.pt (C.C.)

* Correspondence: fernando.ventura@dem.uc.pt; Tel.: +351-239-790-722

Received: 10 September 2019; Accepted: 15 October 2019; Published: 18 October 2019

check for updates

Abstract: Selective Laser Melting (SLM) is an additive manufacturing technology, ideal for the production of complex-shaped components. Design against fatigue is fundamental in the presence of cyclic loads, particularly for these materials which typically have significant porosity, high surface roughness and residual stresses. The main objective here is to study fatigue crack growth (FCG) in the 18Ni300 steel obtained by SLM. Typical da/dN-ΔK curves were obtained in C(T) specimens, indicating that cyclic plastic deformation may be the controlling mechanism. A complementary analysis, based on plastic CTOD range, showed a relatively low level of crack tip plastic deformation, and consequently a reduced level of plasticity induced crack closure. The curve da/dN versus plastic CTOD range is clearly above the curves for other materials.

Keywords: fatigue crack growth; Selective Laser Melting (SLM); Crack tip Opening Displacement (CTOD); crack tip plastic deformation; ΔK

1. Introduction

Selective Laser Melting (SLM) is a technology for additive manufacturing consisting of the fusion of a fine metal power layer by layer. It is an iterative procedure consisting of (1) deposition of a thin layer of powder; (2) selective melting with a laser; and (3) descent of the manufacturing platform. The laser moves in the build area with controlled speed and scan pattern. The unmelted powder is collected and therefore there is no waste of material. The geometry is defined in a CAD model which is read by the SLM equipment. This procedure is very interesting to generate components with complex geometry and eventually composed of different materials. Therefore, SLM is becoming prominent in the automotive, aerospace, medical and injection molds industries. Various materials, such as titanium alloys [1], nickel-based alloys, iron, aluminum, bronze, copper, stainless steels [2] and high-speed steels of metal matrix, can be considered in this process.

Several studies, primarily focused on the influence of the selection of metal powder and sintering parameters on the microstructure of the sintered components, state that SLM components can provide static mechanical properties competitive with those of conventional bulk materials. Nevertheless, in service the components are typically dynamically loaded; therefore, the design must include the analyses of fatigue performance. Studies focused on the fatigue behavior of sintered materials are relatively scarce. Most previous studies developed tested smooth specimens under stress control [1,3,4] or strain control [2]. Notched specimens have also been tested under stress control in order to obtain S-N curves [5]. The presence of porosity, high surface roughness and thermal residual stresses, typical of these materials, reduce the fatigue initiation life, increasing the importance of the crack propagation regime. Pores result from powder contaminations, evaporation or local voids after powder-layer deposition. However, few studies on fatigue crack growth can be seen in the literature. Greitemeier et al. [6] obtained da/dN-ΔK curves in compact tension (CT) specimens made of TiAl6V4 (width, W = 40 mm and thickness, t = 10 mm). Wang et al. [7] presented da/dN-ΔK curves obtained in four-point bending specimens and using replica technique to measure the crack length.

The main objective of this work is to study the fatigue crack propagation in AISI 18Ni300 maraging steel obtained by SLM. The mechanisms behind the phenomenon were analyzed using two different approaches. First, the fatigue crack growth rate, da/dN, was obtained using compact tension specimens, in agreement with ASTM E647 standard, and da/dN-ΔK curves are plotted. Second, the crack tip plastic deformation was analyzed, and da/dN is presented versus plastic Crack Tip Opening Displacement (CTOD), predicted numerically for the C(T) specimen. The material plastic behavior was studied using results obtained from low-cycle fatigue tests on smooth specimens under constant amplitude strain range. The stress-strain hysteresis loops were used for fitting the hardening models.

2. Fatigue Crack Growth Analysis Based on ΔK

2.1. Material

Table 1 presents the chemical composition of the AISI 18Ni300 maraging steel. This steel contains nickel as the primary strengthening source rather than carbon, promoting superior strength and toughness. In addition to its high strength, the18Ni300 steel can be easily machined or formed; afterwards it can undergo an aging heat treatment step, which forms intermetallic precipitates involving cobalt, molybdenum and titanium that aid in increasing the tensile strength. Samples were manufactured using Lasercusing® technology, with layers growing in the same direction as the load application in the mechanical tests. The equipment for selective laser melting is of the mark "Concept Laser" and model "M3 Linear". This apparatus comprises a laser type Nd:YAG with a maximum power of 100 W in continuous wave mode and a wavelength of 1064 nm. The scan speed was 200 mm/s. The material layers were found to have about 40 μm of thickness. The analysis by scanning electron microscope (SEM) showed the presence of a significant amount of small porosities and the formation of martensitic needles [8]. Additionally, the increase of laser speed was found to increase the level of porosity.

Table 1. Chemical composition of the 18Ni300 steel (weight percentage).

Element	Percentage
C	0.01
Ni	18.2
Co	9.0
Mo	5.0
Ti	0.6
Al	0.05
Cr	0.3
P	0.01
Si	0.1
Mn	0.04
Fe	Balance

2.2. *da/dN-ΔK Curves*

Fatigue crack growth tests were carried out according to the recommendations outlined in the ASTM E647 standard. Figure 1 shows the geometry of the C(T) specimens, obtained with thicknesses of 3 or 6 mm. The surface finishing of the C(T) specimens was achieved by high-speed mechanical polishing. All fatigue crack growth (FCG) experiments were carried out using a DARTEC servo-hydraulic testing machine, equipped with a 100 kN load cell. The tests were performed at room temperature in ambient air, under load control mode, at a frequency of 25 Hz, using a PC-based data acquisition system. The load cases studied are presented in Table 2. Three stress ratios were considered: R = 0.05, 0.3 and 0.6. The crack length was measured with a travelling microscope, with magnification of 45× and an accuracy of 10 μm. The data was collected and recorded for crack lengths, a, greater than 7 mm, as indicated in Table 2. The fatigue crack growth rates (FCGR) were obtained from the five-point incremental polynomial method.

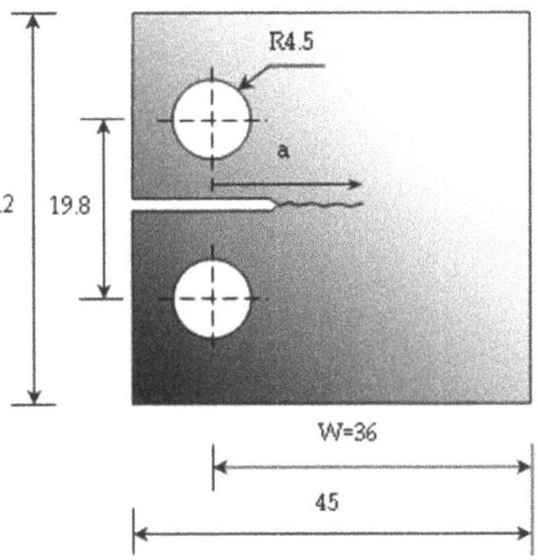

Figure 1. Geometry of C(T) specimen.

Table 2. Load parameters.

R	Thickness, t [mm]	F_{min} [N]	F_{max} [N]	A [mm]	ΔK [MPa.m$^{0.5}$]
0.05	3	74.4	1488	7.1–27.3	8.1–58.1
0.05	6	114	2284.2	7.2–26.6	7.6–47.8
0.3	6	870	2900	7.1–24.0	7.1–27.6
0.6	6	2850	4750	47.5	79.17

Figure 2a shows da/dN-ΔK plots for the 18Ni300 steel, in log-log scales. The increase of ΔK increases the FCG rate significantly, as is well known. Paris law regime is evident, and for relatively low values of da/dN, there is a significant reduction of da/dN, typical of regime I of FCGR. There is also a trend for the increase of da/dN with stress ratio, as could be expected. Anyway, the influence of stress ratio is relatively small, which indicates a reduced level of crack closure. The results for thicknesses of 3 and 6 mm are nearly coincident, which also indicates a small level of crack closure. Figure 2b compares FCGR for the 18Ni300 steel with results for other materials. The 7050-T6 aluminum alloy (AA) has the highest crack growth rate, followed by the AA2050-T8. The 304L stainless steel has the lowest values. Finally, the 18Ni300 is between the SS304L and the AA2050-T8. The slopes of da/dN-ΔK curves in the Paris law regime are similar. Please note that only the 18Ni300 steel was obtained by SLM.

2.3. Discussion

In this Paris law regime, cyclic plastic deformation is the mechanism usually used to explain FCG. The model of striation formation by crack tip plastic blunting of Laird [9,10] is widely accepted to generally describe the propagation mechanism of fatigue cracks in regime II of da/dN-ΔK curves. According to this model, plastic deformation at the crack tip is highly concentrated along the 45° direction, producing blunting and creation of a new fracture surface. Compression stresses at the crack tip reverse slipping, the fracture surfaces approach, but the new surface cannot be removed by re-connection of the atomic bonds, which is in accordance with the entropy law of thermodynamics.

The results for the SLM steel are very typical in terms of linearity and slope of the Paris law regime; therefore, it can be expected that FCG is controlled by cyclic plastic deformation at the crack tip. The analysis of fracture surface by Scanning Electronic Microscopy (SEM) showed that fatigue crack growth occurs mainly at the interfaces between layers of the SLM material. In some cases where the adhesion between the deposited layers is higher, the failure can occur through the grain itself, in transgranular mode. On the contrary, in many other cases, the failure occurs at the layer boundaries producing steps corresponding to the thickness of one or two grains. Please note that the layer plane is normal to the loading direction, which is expected to have a negative effect. In fact, Edwards and Ramulu [11] showed that a layer plane parallel to the loading direction gives higher fatigue life. The weakness of the interfaces is responsible for a relatively tortuous crack path, and therefore for a relatively high fracture surface roughness.

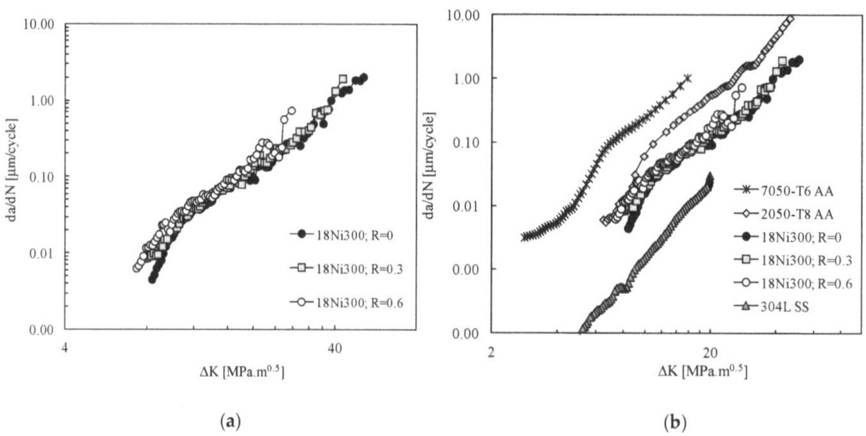

Figure 2. (**a**) da/dN-ΔK curves (18Ni300 steel; t = 6 mm). (**b**) Effect of material on da/dN-ΔK plots (R = 0.1; R = 0.05 for the 18Ni300).

It is, therefore, questionable whether da/dN is linked to a brittle mechanism or to crack tip plastic deformation. Llanes et al. [12] studied FCG in WC-Co cemented carbides, which are brittle materials. They observed that da/dN strongly depends on the variation of K_{max}. A law was proposed for FCGR which was a function of K_{max} and ΔK ($da/dN = C(\Delta K)^m(K_{max})^n$). The values of n were in the range 5–18, while the values of m were in the range 3–5. Intergranular crack growth is also observed in nickel base superalloys [13]. However, in this case the brittle crack growth is associated with oxidation of grain boundaries, which work as highways for the diffusion of oxygen. Tong et al. [14] resorted to the progressive accumulation of tensile strains occurring near the crack tip to predict fatigue crack growth rate in RR1000 nickel base superalloy. They used vacuum in order to remove oxidation, so that viscoplastic deformation controls fatigue crack growth. Figure 3 presents da/dN versus K_{max} for the SLM steel. The comparison with Figure 2a indicates that ΔK is the driving force for FCG, instead of K_{max}. Therefore, cyclic plastic deformation is expected to be the controlling mechanism instead of a brittle mechanism as observed in cemented carbides and nickel base superalloys. For a better understanding of FCG mechanisms, an analysis based on crack tip plastic deformation was developed, as is described in the next section.

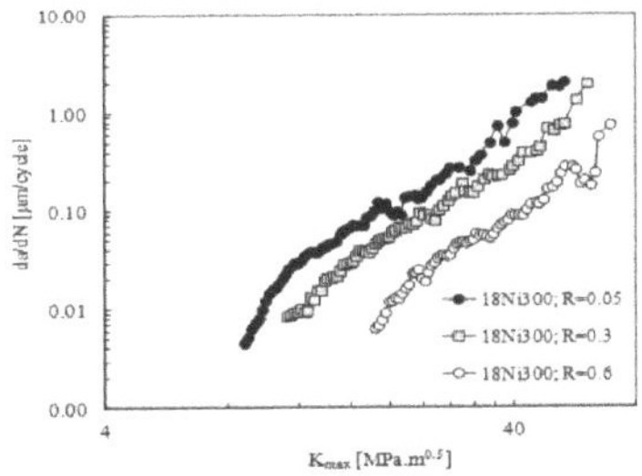

Figure 3. da/dN-K_{max} curves (18Ni300 steel; t = 6 mm).

3. FCG Analysis Based on Plastic CTOD Range

A complementary approach was followed to study FCG, making use of the plastic CTOD range, δ_p, instead of ΔK. This approach follows two assumptions: (i) fatigue crack growth is closely connected with plastic deformation at the crack tip; and (ii) the plastic deformation at the crack tip can be quantified by the plastic CTOD. δ_p was numerically determined using the finite element method. The quality of numerical predictions is dictated by the adequate modeling of the material behavior. Therefore, cyclic loading was applied to smooth specimens in order to obtain stress-strain response, which was subsequently used to fit the material constants. The capability of analyzing crack closure and fatigue crack growth using CTOD was fully demonstrated in an earlier work by the authors [15].

3.1. Low Cycle Fatigue Test

Experimental tests were performed at room temperature and in laboratory air environment, on a DARTEC servo-hydraulic testing machine equipped with a 100 kN load cell. The tests were conducted under axial total strain-controlled mode, with sinusoidal waves, using a constant strain rate (da/dt) equal to 0.008 s-1, and total strain ratios (R_ε) of zero and total strain amplitude ($\Delta\varepsilon/2$) equal to 0.8%. Specimens were produced according to the specifications outlined in ASTM E606 [16], with a gage section measuring 15 mm in length and 8 mm in diameter (Figure 4a).

The final surface finishing was obtained by high-speed mechanical polishing using a sequence of silicon carbide paper grades, P600-grit, P1200-grit, and P2500-grit, followed by 3 µm diamond paste. A 12.5-mm strain-gage extensometer was attached directly to the specimen gage section, using rubber bands, to assess the stress-strain relationship during the test. A total of 41 loading cycles were obtained to failure. 200 samples were collected for each loading cycle, using a PC-based acquisition system. Figure 4b shows the stress-strain curve obtained. The material exhibits a mild cyclic softening behaviour from the beginning of loading until about 80% of total life.

3.2. Identification of Material Parameters

The high precision in the FE simulation results of the plastic CTOD depends on the accurate modeling of the material behavior. In this work, an elastic-plastic model was used: the isotropic elastic behavior is modeled by the generalized Hooke's law; the plastic behavior is described by the von Mises yield criterion coupled with a mixed isotropic-kinematic hardening law under an associated flow rule. The von Mises yield surface is described as follows:

$$(\Sigma_{22}-\Sigma_{33})^2+(\Sigma_{33}-\Sigma_{11})^2+(\Sigma_{11}-\Sigma_{22})^2+3\Sigma_{23}^2+3\Sigma_{13}^2+3\Sigma_{12}^2 = 2Y^2 \quad (1)$$

where Σ represents the effective stress tensor ($\Sigma = \boldsymbol{\sigma}' - \mathbf{X}'$, where $\boldsymbol{\sigma}'$ and $\mathbf{X}'$ are the deviatoric components of the Cauchy stress tensor and back-stress tensor, respectively); Y is the yield stress, and its evolution during plastic deformation is modeled by the Voce isotropic hardening law [17]:

$$Y = Y_0 + (Y_{Sat} - Y_0)[1 - \exp(-C_Y\bar{\varepsilon}^p)] \quad (2)$$

where Y_0, Y_{Sat} and C_Y are material parameters and $\overline{\varepsilon}^p$ is the equivalent plastic strain. The non-linear kinematic hardening is modeled by the Armstrong-Frederick law [18], as follows:

$$\dot{\mathbf{X}} = C_X\left[X_{Sat}\frac{\mathbf{\Sigma}}{\overline{\sigma}} - \mathbf{X}'\right]\dot{\overline{\varepsilon}}^p \tag{3}$$

where C_X and X_{Sat} are material parameters and $\dot{\overline{\varepsilon}}^p$ is the equivalent plastic strain rate. An optimization procedure was carried out to obtain the set of material parameters that best model the cyclic plastic behavior of 18Ni300, by minimization of the following least-squares function:

$$F(\mathbf{A}) = \sum_{i=1}^{N}\left(\frac{\sigma^{Fit}(\mathbf{A}) - \sigma^{Exp}}{\sigma^{Exp}}\right)_i^2 \tag{4}$$

where $\sigma^{Fit}(\mathbf{A})$ and σ^{Exp} are, respectively, the analytically fitted and experimentally measured values of true stress at data point i (that corresponds to an equivalent plastic strain value); N is the number of experimental data points and **A** is the set of Voce and Armstrong-Frederick parameters to be identified. The fitting procedure was carried out for 33 loading cycles (N = 6600), representing about 80% of total life, using a non-linear gradient-based optimization algorithm available in the Microsoft Excel SOLVER tool [19]. Table 3 shows the fitted material parameters that describe the elastic-plastic behavior of 18Ni300. In this table, the Voce isotropic hardening parameters follow $Y_0 = Y_{Sat}$, and therefore the hardening is purely kinematic, which leads to a cyclically stable stress-strain fitting. This is acceptable in case of 18Ni300, where mild cyclic softening occurs during the first 33 loading cycles. Accordingly, the fitted material parameters describe adequately the experimental results, as can be seen in Figure 4c. Table 3 also shows material parameters for other five metal alloys, which were characterized by the authors in previous works.

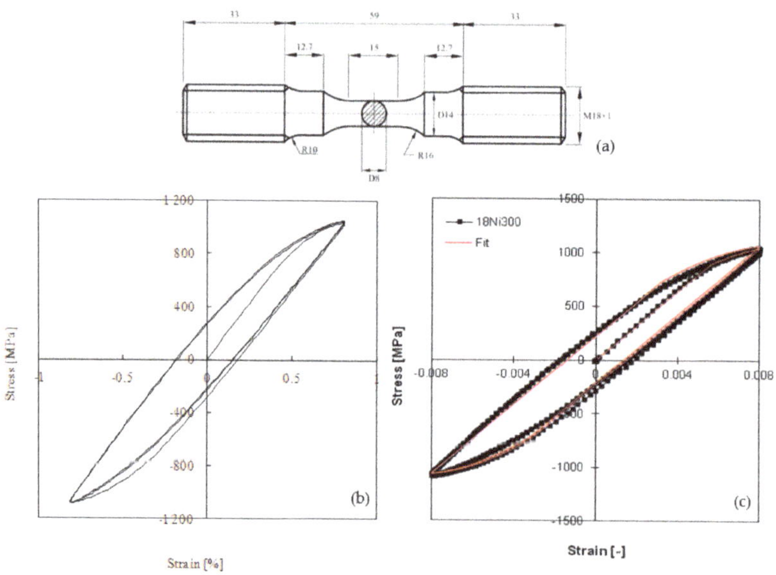

Figure 4. (**a**) Low cycle fatigue specimens. (**b**) Stress-strain plot (Δε = ± 0.8%). (**c**) Experimental versus fitted curve (18Ni300 steel).

Table 3. Elastic-plastic parameters (AA—Aluminum Alloy; SS—Stainless Steel).

Material	Hooke's Law Parameters		Isotropic Hardening (Voce)			Kinematic Hardening (Armstrong-Frederick)	
	E [GPa]	N [-]	Y_0 [MPa]	Y_{Sat} [MPa]	C_Y [-]	C_X [-]	X_{Sat} [MPa]
18Ni300	160	0.30	683.62	683.62	0	728.34	402.06
AA7050-T6 [15]	71.7	0.33	420.50	420.50	0	228.91	198.35
AA2050-T8 [20]	77.4	0.30	383.85	383.85	0	97.38	265.41
AA6082-T6 [21]	70	0.29	238.15	487.52	0.01	244.44	83.18
AA6016-T4 [22]	70	0.29	124.00	415.00	9.5	146.50	34.90
304L SS [23]	196	0.3	117	87	9	300	176

3.3. Numerical Determination of δ_p

The C(T) specimen (Figure 1) presents geometric, material and loading symmetries; therefore, 1/4 of the specimen was modeled numerically considering adequate boundary conditions, as indicated in Figure 5a,b. A pure plane stress state was simulated by assuming a small thickness equal to 0.1 mm. Several initial crack lengths a_0 were considered, equal to 7 mm, 10 mm, 13 mm, 16 mm, 19 mm, 22 mm and 24 mm, to replicate a wide range of experimental crack lengths. The simulations were done under load control, similar to the experimental tests. The maximum and minimum values of the remote load were defined considering the experimental loads listed in Table 2.

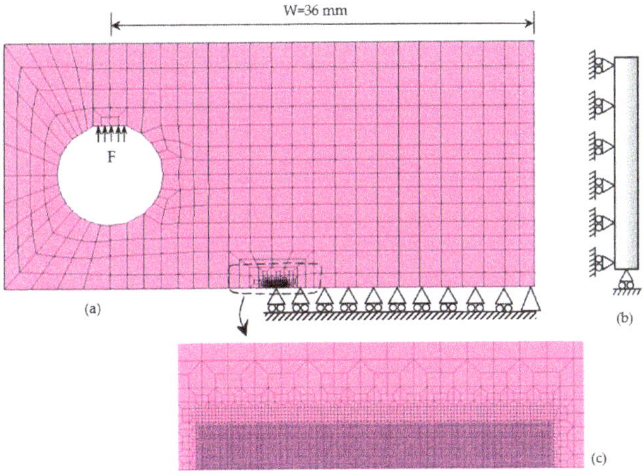

Figure 5. Model of the C(T) specimen. (**a**) Load and boundary conditions. (**b**) Boundary conditions for plane stress state. (**c**) Detail of finite element mesh.

The finite element mesh (Figure 5a,c) comprised 7142 linear isoparametric elements and 14,606 nodes, with two main regions: (i) An ultra-refined mesh near the crack tip, composed of elements with 8×8 μm side; and (ii) a coarser mesh in the remaining specimen, to reduce the computational overhead. Only one layer of elements' through-thickness was used. The crack propagation occurs at the minimum load, by successive debonding of both crack front nodes over the thickness. A total of 159 crack propagations were modeled, each propagation corresponding to the size of one finite element (= 8 μm), with a total crack advance (Δa) of 1272 μm (i.e., Δa = (160 − 1) × 8 μm). Five load cycles were applied between each crack increment.

The numerical simulations were performed using the DD3IMP (Deep-Drawing 3D IMPlicit) in-house code, originally developed to model deep-drawing processes [24]. The evolution of the deformation is modeled by an updated Lagrangian scheme, assuming a hypoelastic-plastic material model. The material plastic behavior was modeled considering the set of elastic-plastic parameters shown in Table 3. The contact between crack flanks is modeled considering a rigid plane surface aligned with the crack symmetry plane. A master–slave algorithm is used; an augmented Lagrangian approach is used for the contact problem treatment.

3.4. Numerical Results

Figure 6 presents a typical numerical result of CTOD as a function of the remote stress, obtained for an initial crack length a_0 = 24 mm, followed by 20 crack propagations (Δa = 160 μm); the CTOD was assessed at the first node located behind the crack tip, at a distance of 8 μm from the tip, as is schematically indicated, which presents the most sensitivity to crack tip phenomena. The crack is closed between A and B, for relatively low loads, i.e., the CTOD equals zero. The load increase opens the crack at point B. After point B, the crack opening evolves linearly with load increase up to point C, which is the boundary of the linear elastic regime. The range of loads between the opening load and the onset of plastic deformation was used to predict the fatigue threshold, as will be described later. A progressive increase of plastic deformation is found between points C and D, achieving its maximum value for the maximum load. The load decrease generates reversed elastic deformation with the same rate observed during loading. The subsequent load decrease produces reversed plastic deformation. Figure 6 also plots the variation of plastic CTOD. Plastic deformation initiates at point C, achieving its maximum value at maximum load. The plastic CTOD range, δ_p, which is correlated with fatigue crack propagation rate, is shown in Figure 6. It should be noted that crack closure is implicit in the value of δ_p. Increasing the crack closure phenomenon reduces the effective range of stress, also reducing the total CTOD and the plastic CTOD. In the absence of crack closure, the entire load cycle is felt by the crack tip. The plastic CTOD range also excludes the elastic deformation, which is not supposed to affect FCG.

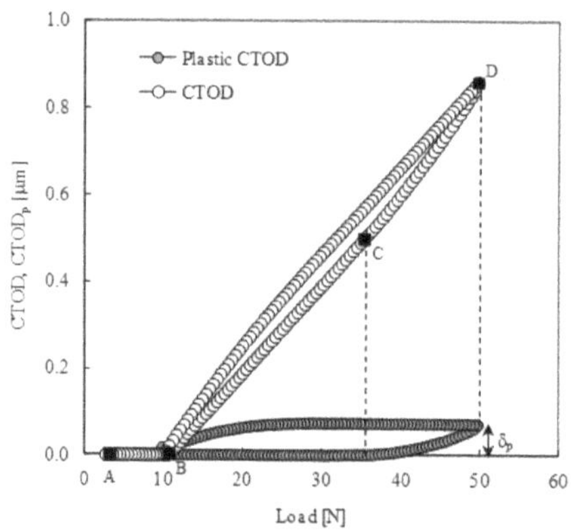

Figure 6. Evolution of Crack Tip Opening Displacement (CTOD) with load (a_0 = 24 mm; plane stress).

Figure 7 presents the evolution of plastic CTOD range, δ_p, with load range quantified by ΔK_{eff} (= $K_{max} - K_{open}$), i.e., the effective range of stress intensity factor. The remote loads applied were those used in the experimental FCG tests of the SLM material. The numerical tests were repeated for different materials keeping constant the specimen geometry, the crack lengths and loading. The increase in the load level increases δ_p, as might be expected. However, for the SLM material, the plastic deformation level is relatively low, being about one order of magnitude lower than the values obtained for the other materials. In fact, the values of plastic CTOD range, δ_p, are lower than 0.1 μm. As can be seen in Table 3, this material has a relatively high yield stress, which explains the low level of plastic deformation. This smallness makes the results very sensitive to errors of material modeling or geometry. Anyway, there is a well-defined variation of δ_p with load range.

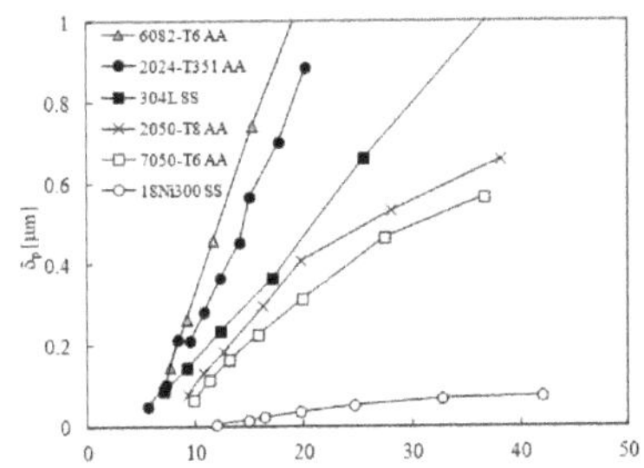

Figure 7. Evolution of plastic CTOD range, δ_p, with effective load range (R = 0.01).

The experimental results showed a relatively low influence of specimen thickness and stress on FCGR, which indicates that the crack closure phenomenon may have a low relevance. Figure 8 presents the evolution of crack closure with the crack length. The crack closure level was quantified by two parameters:

$$U_{clos} = \frac{F_{open} - F_{min}}{F_{max} - F_{min}} \times 100 \quad (5)$$

where F_{open} is the crack opening load, and

$$R_{eff} = \frac{F_{open}}{F_{max}} \quad (6)$$

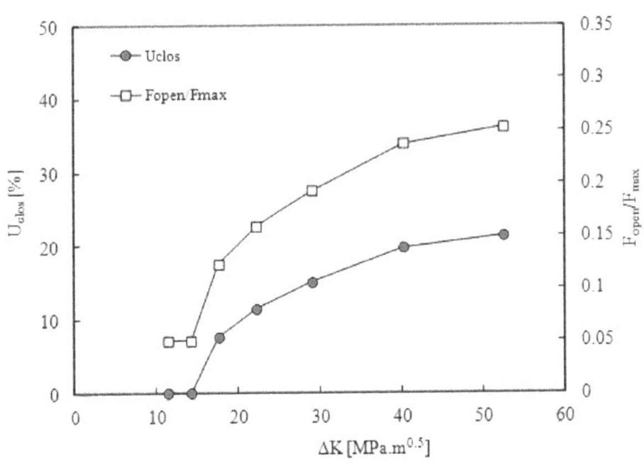

Figure 8. Crack closure level versus ΔK (18Ni300; R = 0.05).

The first parameter quantifies the percentage of load range during which the crack is closed, while the second one is the fraction of crack opening load divided by the maximum load. The crack closure level increases with crack length up to 21% (or 25% for F_{open}/F_{max}). In literature the crack closure level is usually quantified by R_{eff}, and values of 0.25, the maximum value obtained here for the highest crack length studied, are relatively low. This is explained by the relatively low plastic deformation obtained for this material. Values in the range of 0.2 to 0.3 are typical of plane strain state [25–29]. However, no crack closure was found for the SLM steel in the case of plane strain state.

Figure 9 shows the plastic CTOD range, δ_p, versus da/dN for different crack lengths and stress ratios. According to this figure, there is a progressive increase of FCGR with δ_p, as could be expected. The increase of stress ratio, R, moves the curve up. This variation, which should not exist, is relatively small and can be explained by errors of δ_p and da/dN. In fact, in a previous work of the authors [21] it was shown that the da/dN-δ_p curves are invariant to the stress ratio. In other words, the approach based on plastic CTOD range is able to remove the effect of R observed when ΔK is being used. This indicates that δ_p dictates fatigue crack growth. Vasco et al. [30] also observed independence relatively to stress ratio. In that study, δ_p was measured experimentally using Digital Image Correlation. A 2nd-order polynomial was fitted to the results obtained for R = 0.05:

$$\frac{da}{dN} = 95.987 \times \delta_p^2 + 1.052 \times \delta_p \tag{7}$$

where the units of da/dN and δ_p are μm/cycle and μm, respectively. The correlation coefficient is $R^2 = 0.9842$ and its validity is limited to δ_p in the range 0–0.8 μm. Figure 9 also presents results for other materials. The material models for the 304L stainless steel and the 18Ni300 steel were obtained using C(T) specimens, while the aluminum alloys were studied with M(T) specimens. For the same δ_p, the fatigue crack growth rate is significantly higher for the laser sintered material (18Ni300) than for the remaining materials.

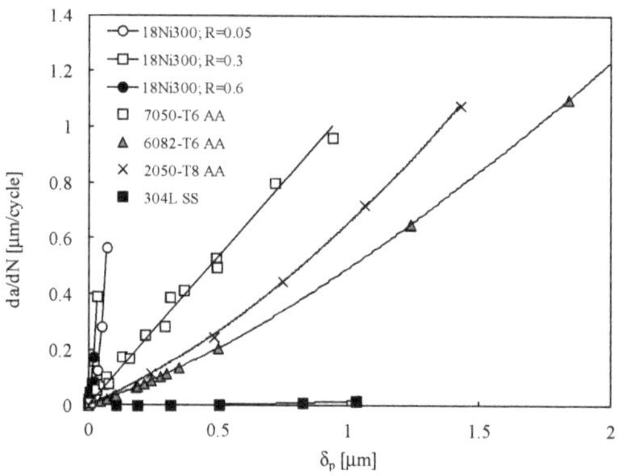

Figure 9. FCGR versus plastic CTOD, δ_p, for different materials.

ASTM E647 standard defines the experimental procedure to obtain fatigue threshold, ΔK_{th}. The load range is gradually reduced until there is no crack propagation. In each loading step, some crack propagation is required to eliminate the influence of crack closure from previous load. The load range reduction can be done setting the stress ratio, R, or the maximum load. In the constant R load reduction method, the maximum and minimum loads are successively reduced such that the stress ratio ($R = K_{min}/K_{max}$) remains constant. However, the measured values of ΔK_{th} are affected by stress ratio, being this effect linked to crack closure phenomenon. To overcome the effect of stress ratio, the constant K_{max} test procedure is proposed in the ASTM E647 standard. In this case, K_{min} is progressively increased in order to eliminate the effect of crack closure. The resulting effective threshold stress intensity, $\Delta K_{th,eff}$ is often referred to as an intrinsic measurement of fatigue crack growth resistance. The experimental approach is however laborious and time consuming. Two alternatives to predict ΔK_{th} were proposed here, based on the numerical analysis of plastic CTOD. In the first numerical approach, K_{max} was kept constant (= 23.4 MPa.m$^{0.5}$), while K_{min} increased. Figure 10a presents the plastic CTOD range versus ΔK. The decrease of ΔK decreased δ_p, as expected, and the extrapolation to the horizontal axis gave a threshold value ΔK_{th} = 11.2 MPa.m$^{0.5}$. It is assumed that fatigue threshold corresponds to the onset of zero plastic deformation. The second numerical approach uses the range of elastic regime to define the fatigue threshold. In Figure 6, the load range between points B and C defines the elastic regime of CTOD. The corresponding ΔK_{th} is obtained from this load range using the K solution for the C(T) specimen. Please note that this is also an effective load range, free of crack closure. Figure 10b presents the results obtained for two different load cases. The reduction of δ_p to zero defines the fatigue threshold. There is a convergence to a nearly constant value, which is a good indication of the robustness of the approach. The value obtained with the constant K_{max} approach is also presented, being slightly higher that the values obtained using the COTD-load curve. However, the experimental value is 5.2 MPa.m$^{0.5}$ [31] being significantly lower than the numerical predictions. This great difference can be explained by the effect of environment. In fact, the near-threshold FCGR is quite small, giving time for the action of environment.

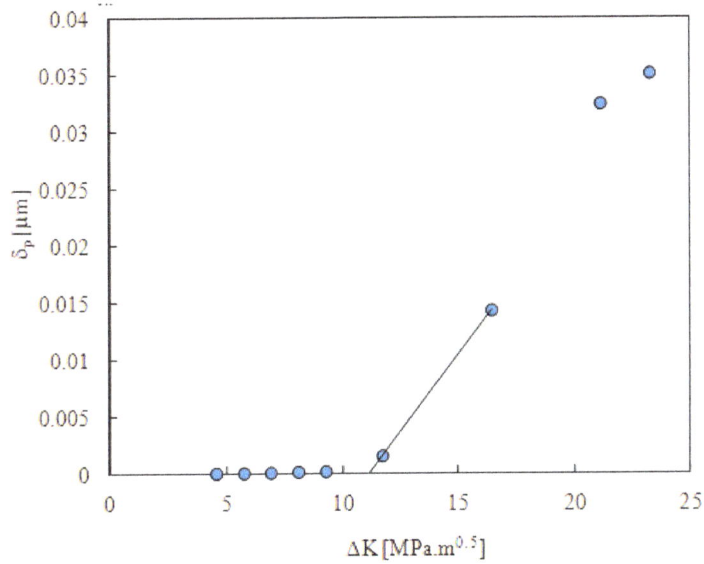

(a)

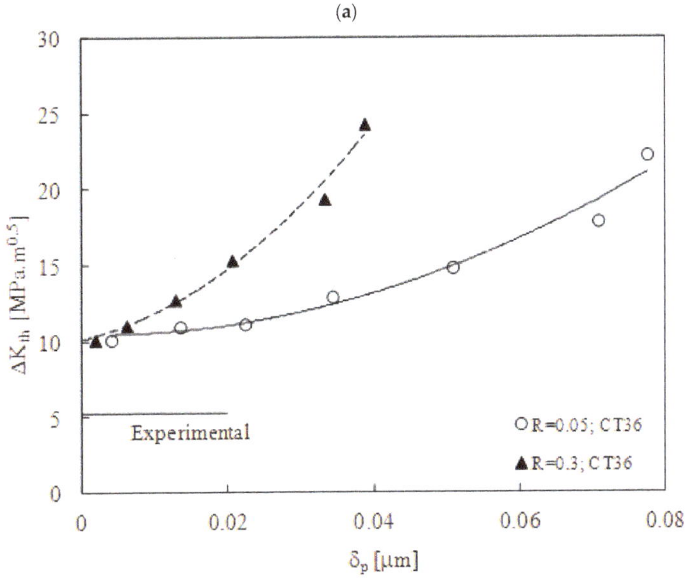

(b)

Figure 10. Numerical prediction of fatigue threshold. (**a**) Constant K_{max} approach (a = 17.272 mm; K_{max} = 23.4 MPa.m0.5). (**b**) Analysis based on the CTOD versus load plots.

4. Conclusions

The fatigue crack growth of the AISI 18Ni300 maraging steel made by Selective Laser Melting was studied numerically and experimentally. The experimental study considered ΔK as the crack driving force while the numerical work used the plastic CTOD. The main conclusions are:

The FCG rate, da/dN, plotted versus ΔK showed a classical Paris law regime with a typical value of the slope. This indicates that cyclic plastic deformation is probably the mechanism controlling fatigue crack growth. The influence of stress ratio and specimen thickness was found to be relatively small, indicating a reduced level of crack closure. The analysis of fracture surfaces by SEM showed propagation between deposition layers.

A complementary analysis was developed, based on plastic CTOD range, to better understand crack growth mechanisms. The level of plastic deformation at the crack tip was found to be relatively low, which also explains the relatively low level of crack closure. Consequently, the curve da/dN versus plastic CTOD range is clearly above the curves for other materials.

Author Contributions: Conceptualization, F.A., J.F. and C.C.; formal analysis, F.A. and P.P.; investigation, J.J., J.C. and L.S.; writing—original draft preparation, F.A.; writing—review and editing, F.A.; funding acquisition, J.F.

Funding: The authors acknowledge sponsorship under the project Hibsilaser, funded by the European Regional Development Fund (FEDER), through the Portugal-2020 program (PT2020), under the Regional Operational Program of the Center (CENTRO-01-0145-FEDER-028789) and the Foundation for Science and Technology IP/MCTES through national funds (PIDDAC).

Conflicts of Interest: The authors declare no conflict of interest. The funders had no role in the design of the study; in the collection, analyses, or interpretation of data; in the writing of the manuscript, or in the decision to publish the results.

References

1. Walker, K.F.; Liu, Q.; Brandt, M. Evaluation of fatigue crack propagation behaviour in Ti-6Al-4V manufactured by selective laser melting. *Int. J. Fatigue* **2017**, *104*, 302–308. [CrossRef]
2. Yadollahi, A.; Shamsaei, N.; Thompson, S.M.; Elwany, A.; Bian, L. Effects of building orientation and heat treatment on fatigue behavior of selective laser melted 17-4 PH stainless steel. *Int. J. Fatigue* **2017**, *94*, 218–235. [CrossRef]
3. Croccolo, D.; De Agostinis, M.; Fini, S.; Olmi, G.; Bogojevic, N.; Ciric-Kostic, S. Effects of build orientation and thickness of allowance on the fatigue behaviour of 15–5 PH stainless steel. *Fatigue Fract. Eng. Mater. Struct.* **2017**, *41*, 1–17. [CrossRef]
4. Spierings, A.B.; Starr, T.L.; Wegener, K. Fatigue performance of additive manufactured metallic parts. *Rapid Prototyp. J.* **2013**, *19*, 88–94. [CrossRef]
5. Mohammad, S.; Razavi, J.; Ferro, P.; Berto, F. Fatigue Assess-ment of Ti–6Al–4V Circular Notched Specimens Produced by Selective Laser Melting. *Metals* **2017**, *7*, 291.
6. Greitemeier, D.; Palm, F.; Syassen, F.; Melz, T. Fatigue performance of additive manufactured TiAl6V4 using electron and laser beam melting. *Int. J. Fatigue* **2017**, *94*, 211–217. [CrossRef]
7. Wang, Y.; Bergstrom, J.; Burman, C. Four-point bending fatigue behaviour of an iron-based laser sintered material. *Int. J. Fatigue* **2006**, *28*, 1705–1715. [CrossRef]
8. Ferreira, J.A.M.; Santos, L.M.S.; da Silva, J.; Costa, J.M.; Capela, C. Assessment of the fatigue life on functional hybrid laser sintering steel components. *Procedia Struct. Integr.* **2016**, *1*, 126–133. [CrossRef]
9. Laird, C.; Smith, G.L. Initial Stages of Damage in High Stress Fatigue in Some Pure Metals. *Phil. Mag.* **1963**, *95*, 1945–1963. [CrossRef]
10. Laird, C. The Influence of Metallurgical Structure on the Mechanisms of Fatigue Crack Propagation. *Fatigue Crack Propag.* **1967**, 131–180.
11. Edwards, P.; Ramulu, M. Fatigue performance evaluation of selective laser melted Ti-6Al-4V. *Mat. Sci. Eng. A-Struct.* **2014**, *598*, 327–337. [CrossRef]
12. Lanes, L.; Torres, Y.; Anglada, M. On the fatigue crack growth behaviour of WC-Co cemented carbides: Kinetics descrption, microstructural effects and fatigue sensitivity. *Acta Mater.* **2002**, *50*, 2381–2393. [CrossRef]

13. Antunes, F.V.; Ferreira, J.A.M.; Branco, C.M.; Byrne, J. Influence of stress state on high temperature fatigue crack growth in Inconel 718. *Fatigue Fract. Eng. Mat. Struct.* **2001**, *24*, 127–135. [CrossRef]
14. Tong, J.; Zhao, L.G.; Lin, B. Ratchetting strain as a driving force for fatigue crack growth. *Int. J. Fatigue* **2013**, *46*, 49–57. [CrossRef]
15. Antunes, F.V.; Branco, R.; Prates, P.A.; Borrego, L. Fatigue crack growth modelling based on CTOD for the 7050-T6 alloy. *Fatigue Fract. Eng. Mater. Struct.* **2017**, *40*, 1309–1320. [CrossRef]
16. ASTM E606. *Standard Test Method for Strain-Controlled Fatigue Testing*; American Society for Testing of Materials: West Conshohocken, PA, USA, 2012.
17. Voce, E. The relationship between stress and strain for homogeneous deformation. *J. Inst. Met.* **1948**, *74*, 537–562.
18. Chaboche, J.L. A review of some plasticity and viscoplasticity constitutive theories. *Int. J. Plast.* **2008**, *24*, 1642–1693. [CrossRef]
19. Lasdon, L.S.; Waren, A.D.; Jain, A.; Ratner, M.W. *Design and Testing of a Generalized Reduced Gradient Code for Non-Linear Optimization*; NTIS National Technical Information Service U. S. Department of Commerce: Cleveland, OH, USA, 1975.
20. Antunes, F.V.; Serrano, S.; Branco, R.; Prates, P.; Lorenzino, P. Fatigue crack growth in the 2050-T8 aluminium alloy. *Int. J. Fatigue* **2018**, *115*, 79–88. [CrossRef]
21. Antunes, F.V.; Rodrigues, S.M.; Branco, R.; Camas, D. A numerical analysis of CTOD in constant amplitude fatigue crack growth. *Theor. Appl. Fract. Mech.* **2016**, *85*, 45–55. [CrossRef]
22. Antunes, F.V.; Rodrigues, D.M. Numerical simulation of plasticity induced crack closure: Identification and discussion of parameters. *Eng. Fract. Mech.* **2008**, *75*, 3101–3120. [CrossRef]
23. Antunes, F.V.; Ferreira, M.S.C.; Branco, R.; Prates, P.; Gardin, C.; Sarrazin-Baudoux, C. Fatigue crack growth in the 304L stainless steel. *Eng. Fract. Mech.* **2019**, *214*, 487–503. [CrossRef]
24. Oliveira, M.C.; Alves, J.L.; Menezes, L.F. Algorithms and Strategies for Treatment of Large Deformation Frictional Contact in the Numerical Simulation of Deep Drawing Process. *Arch. Comput. Methods Eng.* **2008**, *15*, 113–162. [CrossRef]
25. McClung, R.C.; Thacker, B.H.; Roy, S. Finite element visualization of fatigue crack closure in plane stress and plane strain. *Int. J. Fracture* **1991**, *50*, 27–49.
26. Sehitoglu, H.; Sun, W. Modelling of plane strain fatigue crack closure. *ASME J. Eng. Mat. Technol.* **1991**, *113*, 31–40. [CrossRef]
27. LLorca, J.; Gálvez, V.S. Modelling plasticity-induced fatigue crack closure. *Eng. Fracture Mech.* **1990**, *37*, 185–196. [CrossRef]
28. Solanki, K.; Daniewicz, S.R.; Newman, J.C., Jr. Finite element modelling of plasticity-induced crack closure with emphasis on geometry and mesh refinement effects. *Eng. Fract. Mech.* **2003**, *70*, 1475–1489. [CrossRef]
29. Pokluda, J. Dislocation-based model of plasticity and roughness-induced crack closure. *Int. J. Fatigue* **2013**, *46*, 35–40. [CrossRef]
30. Vasco-Olmo, J.M.; Díaz, F.; Antunes, F.V.; James, M.N. Evaluación Experimental Del CTOD En El Crecimiento De Grieta A Fatiga A Partir De Los Campos De Desplazamientos. In Proceedings of the XXXV Encuentro del Grupo Español de Fractura, Málaga, Spain, 14–16 March 2018.
31. Santos, L.M.S.; Ferreira, J.A.M.; Jesus, J.S.; Costa, J.M.; Capela, C. Fatigue behaviour of selective laser melting steel components. *Appl. Fract. Mech.* **2016**, *85*, 9–15. [CrossRef]

applied sciences

MDPI

Communication

Cyclic Plasticity of the As-Built EOS Maraging Steel: Preliminary Experimental and Computational Results

Barry Mooney [1], **Dylan Agius** [2] **and Kyriakos I. Kourousis** [1,*]

[1] School of Engineering, University of Limerick, V94 T9PX Limerick, Ireland; barrymooney.bm@gmail.com
[2] Faculty of Engineering, University of Bristol, Bristol BS8 1TR, UK; dylan.agius@bristol.ac.uk
* Correspondence: kyriakos.kourousis@ul.ie; Tel.: +353-61-202217

Received: 21 January 2020; Accepted: 4 February 2020; Published: 12 February 2020

check for updates

Abstract: This short communication offers a preliminary view on ongoing research conducted on the as-built EOS maraging steel 300. The material's cyclic elastoplastic characteristics under strain-controlled loading have been investigated experimentally. Specimens fabricated under two primary orientations, horizontally and vertically to the build plate, have been tested. The obtained stress–strain hysteresis loops exhibited symmetry, with the vertical specimen showing a higher plastic strain energy dissipation capability than the horizontal specimen. Modelling of the material's elastoplastic behaviour was performed with a commonly used kinematic hardening rule, combined with both isotropic and anisotropic yield functions and elasticity moduli. The obtained simulations of the hysteresis loops, from the implementation of these two plasticity models, indicate the advantage of the anisotropic modelling approach over the isotropic approach. The anisotropic plasticity model describes in a more representative way the inherent elastic and plastic anisotropy of the as-built material. Further research is underway to explore the low cycle fatigue performance of this additively manufactured metal.

Keywords: maraging steel; plasticity; cyclic loading; low cycle fatigue; additive manufacturing

1. Introduction

Maraging steel 300 (MS300), also codified as American Iron and Steel Institute (AISI) 18Ni300, is an alloy typically used for structural components requiring very high strength, such as tooling, moulds, aircraft landing gear, and rocket casings. MS300 has been popular in laser powder bed fusion (LPBF) additive manufacturing (AM), mainly due to its good weldability. As with most AM alloys, there has been substantial research conducted on the characterisation of the MS300's mechanical properties under monotonic and cyclic loading. In the literature, a wide array of results can be found on its monotonic (tensile/compressive) performance, e.g., [1–6], while cyclic response results have mostly been reported for its high cycle fatigue (HCF) performance (in the range of 10^5–10^6 cycles, with the imposed stress/strain remaining within the elastic regime), e.g., [4,7–9]. However, for the full characterisation of the material's mechanical properties, it is necessary to investigate its performance under cyclic loading.

Published work on the performance of MS300 under low cycle fatigue (LCF) conditions (material loaded up to 10^3 cycles within the elastoplastic regime) is currently limited to a single study [10]. In particular, the work of Branco et al. [10] identified the strain-controlled LCF characteristics of as-built MS300 produced with a Concept Laser M3 AM system. This study was limited to the examination of test coupons that were fabricated by having their loading axis vertically to the built plate (vertical orientation).

The present short communication reports a preliminary set of experimental and simulation results obtained from an ongoing cyclic plasticity/LCF testing campaign on as-built MS300 produced with the EOS EOSINT M280 system. In particular, the experimental data from strain-controlled tests on vertically and horizontally fabricated test coupons are presented, analysed and discussed. This work offers an early look at experimental data examining the cyclic elastoplastic characteristics of this additively manufactured metal, complementing and extending the existing dataset of Branco et al. [10]. It also allows a comparison with a subset of the same previously published data and the evaluation of anisotropic plasticity modelling for simulation purposes. An isotropic plasticity modelling approach is commonly used by researchers and industry practitioners; however, it is a practice that may lead to higher computation error, which could, consequently, contribute to a substantial over- or under-estimation of the LCF life of this metal and parts produced with it.

2. Materials and Methods

The material used was MS1 (MS300) powder supplied by a metal AM equipment manufacturer (EOS). The powder consisted of spherical (36.69 μm median diameter) particles and had the chemical composition shown in Table 1 [5].

Table 1. Chemical composition of the MS1 (MS300) powder used.

	Ni	Mo	Co	Ti	Al	Cr	Mn	C	Fe
MS1 (MS300)	18.14	5.67	8.94	0.87	0.05–0.15	≤0.5	≤0.01	≤0.03	Balance

Rectangular test coupons (compliant with the ASTM E8M standard) were fabricated with EOS EOSINT M280 at the South Eastern Applied Materials (SEAM) Research Centre of the Waterford Institute of Technology. The geometry, dimensions and tolerances of the test coupons are illustrated in Figure 1. The EOS-recommended 'MS1 Performance 2.0' set of manufacturing parameters was used. The test coupons were fabricated at vertical and horizontal orientations relative to the build plate, with this communication reporting experimental data from a single set (one vertically and one horizontally produced test coupon). For simplicity, the test coupons are denoted as horizontal and vertical. Surface finishing was performed on the coupons via grinding and computer numerically controlled (CNC) wire electrical discharge machining (EDM) [5].

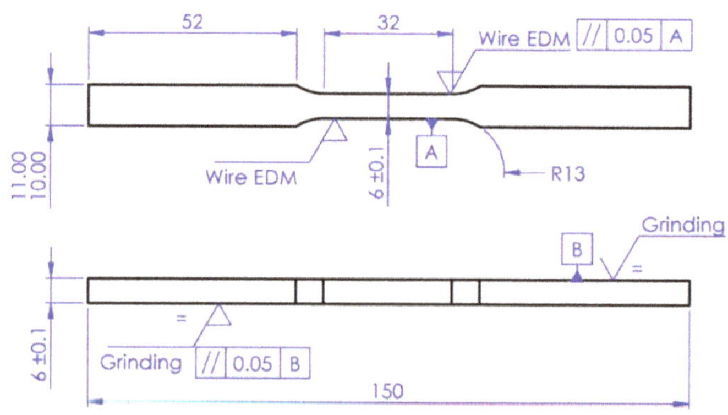

Figure 1. Test coupon geometry, dimensions and tolerances (in mm).

The mechanical tests were conducted at room temperature with a Zwick/Roell (Dartec) M1000/RK 100 kN servo-hydraulic closed-loop uniaxial testing machine fitted with ±100 kN fatigue rated hydraulic wedge grips. The two test coupons were subjected to fully reversed ($R = -1$, where R is the ratio of maximum to minimum strain) strain-controlled cyclic loading within a ± 0.9% strain range for the horizontal and ±0.95% for the vertical test coupon.

A microstructural analysis of the material of these specimens was not performed, as this has already been covered in our previously published studies for the same material [5,11].

3. Cyclic Plasticity Mathematical Modelling

The cyclic elastoplastic response of the material was mathematically modelled with a three term Armstrong–Frederick [12] kinematic hardening rule. The uniaxial formulation of this kinematic hardening rule is described by Equation (1), where X_i is the back-stress, c_i and γ_i are the material parameters and $d\varepsilon^p$ is the incremental plastic strain:

$$dX_i = c_i d\varepsilon^p - \gamma_i X_i \left| d\varepsilon^p \right|, \text{ for } i = 1,3 \quad (1)$$

Two types of yield surfaces (f) have been used for a separate implementation and analysis, the isotropic von Mises, combined with isotropic elasticity (a single value for the Elasticity Modulus, E), and the anisotropic Hill [13] (described by Equation (2)), combined with a Hooke's law orthotropic Elasticity Modulus (described by Equation (3)). For brevity, the two models are referred to in the paper as the isotropic and the anisotropic plasticity model.

$$f = F(\sigma_{22} - \sigma_{33})^2 + G(\sigma_{33} - \sigma_{11})^2 + H(\sigma_{11} - \sigma_{22})^2 + 2\left(L\sigma_{23}^2 + M\sigma_{31}^2 + N\sigma_{12}^2\right) = 1/2 \quad (2)$$

where σ_{ij} are the stress components and F, G, H, L, M and N are the coefficients of the Hill anisotropic yield surface.

$$\begin{bmatrix} \varepsilon_{11} \\ \varepsilon_{22} \\ \varepsilon_{33} \\ 2\varepsilon_{23} \\ 2\varepsilon_{31} \\ 2\varepsilon_{12} \end{bmatrix} = \begin{bmatrix} \frac{1}{E_1} & -\frac{v_{21}}{E_2} & -\frac{v_{31}}{E_3} & 0 & 0 & 0 \\ -\frac{v_{12}}{E_1} & \frac{1}{E_2} & -\frac{v_{32}}{E_3} & 0 & 0 & 0 \\ -\frac{v_{13}}{E_1} & -\frac{v_{23}}{E_2} & \frac{1}{E_3} & 0 & 0 & 0 \\ 0 & 0 & 0 & \frac{1}{G_{23}} & 0 & 0 \\ 0 & 0 & 0 & 0 & \frac{1}{G_{13}} & 0 \\ 0 & 0 & 0 & 0 & 0 & \frac{1}{G_{12}} \end{bmatrix} \begin{bmatrix} \sigma_{11} \\ \sigma_{22} \\ \sigma_{33} \\ \sigma_{23} \\ \sigma_{31} \\ \sigma_{12} \end{bmatrix} \quad (3)$$

where ε_{ij} are the components of strain, E_i are the Elasticity Modulus' tensor components in the direction of loading, v_{ij} are the Poisson ratios for extension in the i and j directions and G_{ij} are the Shear Modulus components in direction j on the plane whose normal is in direction i.

The anisotropic yield surface and orthotropic Hooke's law (anisotropic model) were selected to allow the implementation of previously identified and measured plastic and elastic anisotropic parameters of the MS300 material [5]. The isotropic model was used as a benchmark to the anisotropic plasticity model in order to evaluate the latter model's simulation accuracy.

4. Results and Discussion

4.1. Cyclic Mechanical Testing

The as-built EOS MS300 test coupons were strained to failure; however, only the data corresponding to the 28th cycle are reported and have been used in this analysis. This was a deliberate choice, since unplanned interruptions of the tests at the 29th cycle and the 516th cycle for the vertically and the horizontally built test coupons, respectively, may have compromised both the evolution of the hysteresis (stress–strain) loops and the LCF life of the material. Nevertheless, the 28th cycle data were sufficient to illustrate the cyclic elastoplastic characteristics of this material and to allow a comparison with the Branco et al. [10] mid-life (32nd cycle) results for the same strain range (±0.9%) or close to that (i.e., ±0.95% for the vertical coupon). The obtained stress–strain hysteresis loops along with the aforementioned Branco et al. [10] experimental data are illustrated in Figure 2.

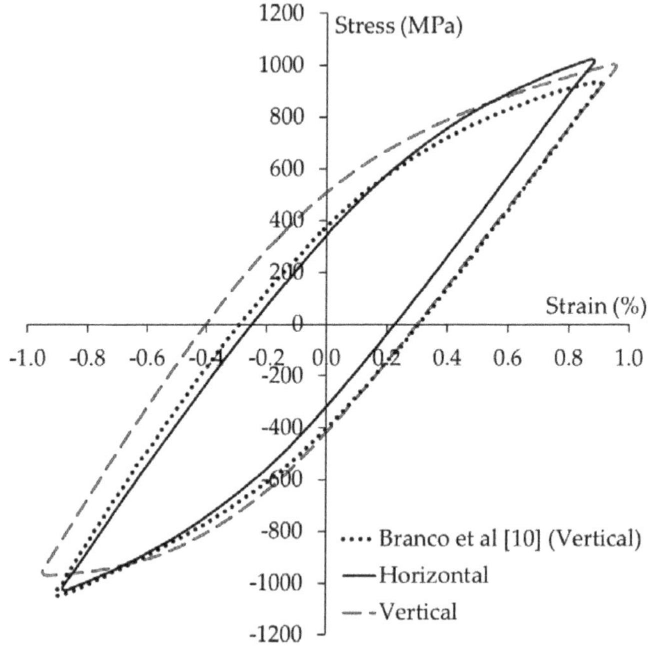

Figure 2. The hysteresis loops of the tested as-built EOS MS300 horizontal and vertical test coupons (at the 28th cycle) and the Branco et al. [10] published data for the as-built MS300 vertical coupon.

The hysteresis loops of the tested coupons (vertical, horizontal) appear to have symmetry, both in terms of their shape and in absolute stress values (approximately 900 and 1000 MPa, respectively). It can be observed that this is not the case for the Branco et al. [10] data, as the hysteresis loop has a negative stress asymmetry of approximately 100 MPa. Of note is that the area enclosed by the hysteresis loops, representing the plastic strain energy dissipation, is considerably smaller for the horizontal and the Branco et al. [10] vertical coupon compared to that of the vertical coupon. This is an indication of a higher LCF resistance of the vertical coupon; however, safe conclusions cannot be made unless further tests are performed for the estimation of the LCF life of as-built EOS MS300.

4.2. Cyclic Plasticity Simulation

The parameters of the kinematic hardening rule (Equation (1)), used in both the isotropic and anisotropic plasticity model, were calibrated manually (employing a successive trial and error continuous optimisation approach) from the experimental hysteresis loop data (for the vertical coupon). These parameters are listed in Table 2. Moreover, the yield stress was calibrated to $\sigma_y = 420$ MPa to account for the cyclic properties of the material (as opposed to using the yield stress measured from monotonic tensile tests).

Table 2. Parameters of the kinematic hardening rule for the as-built EOS MS300, used in both the isotropic and the anisotropic plasticity model.

Kinematic Hardening			
c_1	241,800 MPa	γ_1	930
c_2	98,280 MPa	γ_2	945
c_3	14,112 MPa	γ_3	96

For the isotropic hardening model, the Elasticity Modulus, *E*, selected was 160 GPa and the Poisson ratio, *v*, was 0.35 (based on the data reported in [5]).

The coefficients of the Hill anisotropic yield surface (Equation (2)) and the orthotropic Hooke's law (Equation (3)) (anisotropic plasticity model) were obtained from a previously published paper [11] on the same material (as-built EOS MS300). These coefficients are listed in Table 3. The methodology for the calculation of these coefficients is outside the scope of this paper; therefore, the reader is referred to the same publication for a detailed description [11].

Table 3. Coefficients of the Hill anisotropic yield surface and the orthotropic Hooke's law tensor for the as-built EOS MS300, used in the anisotropic plasticity model [12].

Hill		Orthotropic Hooke			
F	0.798	E_1	137 GPa	ν_{12}	0.43
G	0.638	E_2	161 GPa	ν_{21}	0.93
H	0.202	E_3	122 GPa	ν_{13}	0.26
L	1.167	G_{12}	122 GPa	ν_{31}	0.01
M	1.500	G_{12}	122 GPa	ν_{23}	0.24
N	1.500	G_{23}	122 GPa	ν_{32}	0.35

The two plasticity models were implemented for this study's experimental data (hysteresis loops stress–strain curves), derived from the two tested coupons (horizontal, vertical). Figure 3 presents the simulation results, where a good prediction of the experimental data can be observed. Of particular interest is that a closer prediction was obtained with the anisotropic plasticity model, which is attributed to the more representative description of the (elastic and plastic) anisotropy of the as-built EOS MS300 (measured in [5]). This is more clearly evidenced in the case of the horizontal test coupon data, where the hysteresis loop is simulated very well.

This, qualitative comparison is complemented by a calculation of the average error between experimental and simulated data (across the range of the stress–strain curve). The results of this calculation are included in Table 4, where the clear advantage of adopting the anisotropic plasticity model over the isotropic model is confirmed again.

Table 4. Average error (%) for the vertical and horizontal test coupon experimental and simulation data for the isotropic and the anisotropic plasticity model.

	Isotropic	Anisotropic
Vertical	43.5%	36.3%
Horizontal	58.6%	21.1%

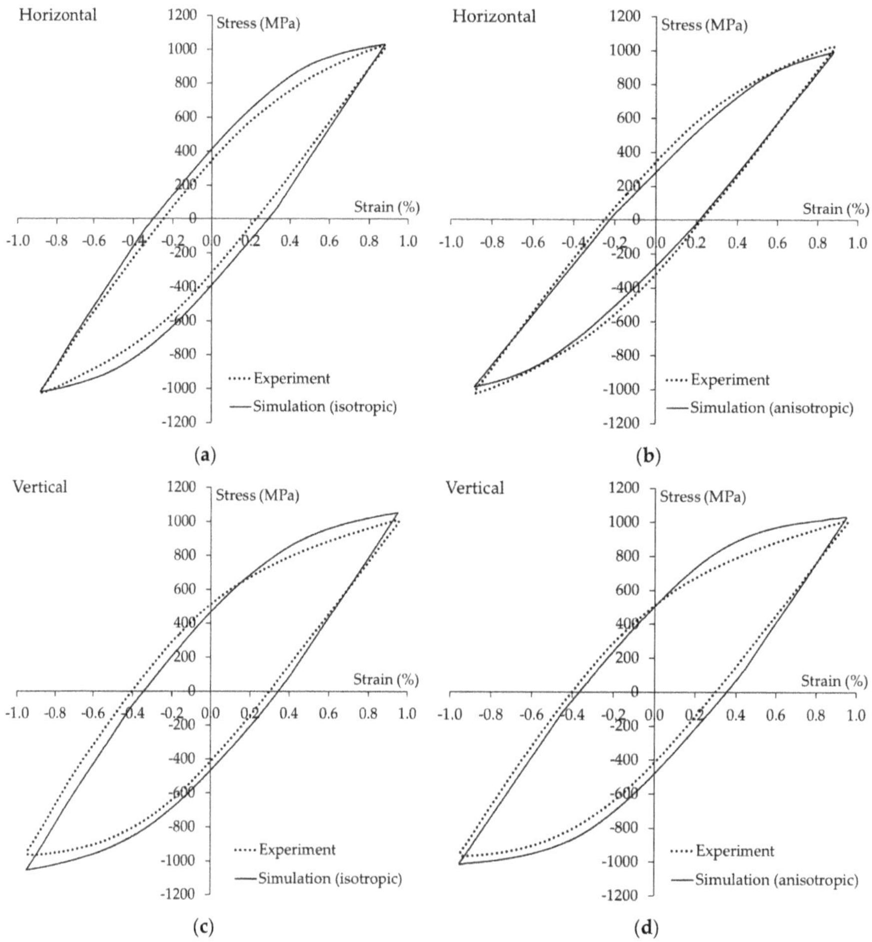

Figure 3. Experimental versus simulation results for the (**a**) isotropic plasticity model for the horizontal coupon; (**b**) anisotropic plasticity model for the horizontal coupon; (**c**) isotropic plasticity model for the vertical coupon; and (**d**) anisotropic plasticity model for the vertical coupon.

5. Conclusions

The obtained experimental and simulation results for the as-built EOS MS300 and the analysis have identified the following important features for this material:

- The stress–strain hysteresis loops produced from cyclic elastoplastic strain loading exhibited symmetrical characteristics, for both horizontal and vertical test coupons.
- Vertical test coupons appear to have a higher capacity for plastic strain energy dissipation than horizontal coupons, as indicated by their hysteresis loop area.
- The plasticity models used in this study were capable of simulating the hysteresis loops well.
- The anisotropic plasticity model offers a more representative description of the mechanical behaviour of this metal, resulting in higher accuracy simulations than the isotropic model.

A strain-controlled test campaign is ongoing, where coupons subjected to various strain ranges will be tested until failure. This will allow the estimation of the LCF life of the material, along with obtaining stress–strain hysteresis loops. The results and findings from this experimental investigation are expected to be reported in a subsequent full research paper.

Author Contributions: Conceptualization, K.I.K; methodology, B.M., D.A. and K.I.K.; software, D.A. and K.I.K.; validation, B.M., D.A. and K.I.K.; formal analysis, B.M., D.A. and K.I.K.; investigation, B.M., D.A. and K.I.K.; resources, B.M., D.A. and K.I.K.; data curation, B.M., D.A. and K.I.K.; writing—original draft preparation, B.M., D.A. and K.I.K.; writing—review and editing, B.M., D.A. and K.I.K.; visualization, B.M., D.A. and K.I.K.; supervision, K.I.K.; project administration, K.I.K.; funding acquisition, K.I.K and B.M. All authors have read and agreed to the published version of the manuscript.

Funding: This research was partially funded by the Irish Research Council, through the Government of Ireland Postgraduate Research Programme, and the Faculty of Science and Engineering of the University of Limerick.

Acknowledgments: The support of the South Eastern Applied Materials (SEAM) Research Centre, of the Waterford Institute of Technology, in manufacturing the test coupons is acknowledged.

Conflicts of Interest: The authors declare no conflict of interest.

References

1. Casati, R.; Lemke, J.N.; Tuissi, A.; Vedani, M. Aging behaviour and mechanical performance of 18-Ni 300 steel processed by selective laser melting. *Metals* **2016**, *6*, 218. [CrossRef]
2. Becker, T.H.; Dimitrov, D. The achievable mechanical properties of SLM produced Maraging Steel 300 components. *Rapid Prototyp. J.* **2016**, *22*, 487–494. [CrossRef]
3. Tan, C.; Zhou, K.; Ma, W.; Zhang, P.; Liu, M.; Kuang, T. Microstructural evolution, nanoprecipitation behavior and mechanical properties of selective laser melted high-performance grade 300 maraging steel. *Mater. Des.* **2017**, *134*, 23–34. [CrossRef]
4. Suryawanshi, J.; Prashanth, K.G.; Ramamurty, U. Tensile, fracture, and fatigue crack growth properties of a 3D printed maraging steel through selective laser melting. *J. Alloys Compd.* **2017**, *725*, 355–364. [CrossRef]
5. Mooney, B.; Kourousis, K.I.; Raghavendra, R. Plastic anisotropy of additively manufactured maraging steel: Influence of the build orientation and heat treatments. *Addit. Manuf.* **2019**, *25*, 19–31. [CrossRef]
6. Song, J.; Tang, Q.; Feng, Q.; Ma, S.; Setchi, R.; Liu, Y.; Han, Q.; Fan, X.; Zhang, M. Effect of heat treatment on microstructure and mechanical behaviours of 18Ni-300 maraging steel manufactured by selective laser melting. *Opt. Laser Technol.* **2019**, *120*, 105725. [CrossRef]
7. Croccolo, D.; De Agostinis, M.; Fini, S.; Olmi, G.; Robusto, F.; Kostic, S.C.; Vranic, A.; Bogojevic, N. Fatigue response of as-built DMLS maraging steel and effects of aging, machining, and peening treatments. *Metals* **2018**, *8*, 505. [CrossRef]
8. Damon, J.; Hanemann, T.; Dietrich, S.; Graf, G.; Lang, K.-H.; Schulze, V. Orientation dependent fatigue performance and mechanisms of selective laser melted maraging steel X3NiCoMoTi18-9-5. *Int. J. Fatigue* **2019**, *127*, 395–402. [CrossRef]
9. Croccolo, D.; De Agostinis, M.; Fini, S.; Olmi, G.; Robusto, F.; Ciric-Kostic, S.; Moraca, S.; Bogojevic, N. Sensitivity of direct metal laser sintering Maraging steel fatigue strength to build orientation and allowance for machining. *Fatigue Fract. Eng. Mater. Struct.* **2019**, *42*, 374–386. [CrossRef]
10. Branco, R.; Costa, J.D.M.; Berto, F.; Razavi, S.M.J.; Ferreira, J.A.M.; Capela, C.; Santos, L.; Antunes, F. Low-cycle fatigue behaviour of AISI 18Ni300 maraging steel produced by selective laser melting. *Metals* **2018**, *8*, 32. [CrossRef]
11. Mooney, B.; Kourousis, K.I.; Raghavendra, R.; Agius, D. Process phenomena influencing the tensile and anisotropic characteristics of additively manufactured maraging steel. *Mater. Sci. Eng. A* **2019**, *745*, 115–125. [CrossRef]
12. Armstrong, P.J.; Frederick, C.O. *A Mathematical Representation of the Multiaxial Bauschinger Effect*; G.E.G.B. Report RD/B/N 731; Berkeley, CA, USA, 1996.
13. Hill, R. A theory of the yielding and plastic flow of anisotropic metals. *Proc. R. Soc Lond. Ser. A Math. Phys. Sci.* **1948**, *193*, 281–297.

applied sciences

MDPI

Article

Crack Initiation and Propagation Fatigue Life of Ultra High-Strength Steel Butt Joints

Markus J. Ottersböck [1,*], Martin Leitner [1], Michael Stoschka [1] and Wilhelm Maurer [2]

[1] Chair of Mechanical Engineering, Montanuniversität Leoben, Franz-Josef-Straße 18, 8700 Leoben, Austria; martin.leitner@unileoben.ac.at (M.L.); michael.stoschka@unileoben.ac.at (M.S.)
[2] voestalpine Stahl GmbH, voestalpine-Straße 3, 4020 Linz, Austria; wilhelm.maurer@voestalpine.com
* Correspondence: ottersboeck@stud.unileoben.ac.at

Received: 31 August 2019; Accepted: 13 October 2019; Published: 29 October 2019

check for updates

Abstract: The division of the total fatigue life into different stages such as crack initiation and propagation is an important issue in regard to an improved fatigue assessment especially for high-strength welded joints. The transition between these stages is fluent, whereas the threshold between the two phases is referred to as technical crack initiation. This work presents a procedure to track crack initiation and propagation during fatigue tests of ultra high-strength steel welded joints. The method utilizes digital image correlation to calculate a distortion field of the specimens' surface enabling the identification and measurement of cracks along the weld toe arising during the fatigue test. Hence, technical crack initiation of each specimen can be derived. An evaluation for ten ultra high-strength steel butt joints reveals, that for this superior strength steel grade more than 50% of fatigue life is spent up to a crack depth of 0.5 mm, which can be defined as initial crack. Furthermore, a notch-stress based fatigue assessment of these specimens considering the actual weld topography and crack initiation and propagation phase is performed. The results point out that two phase models considering both phases enable an increased accuracy of service life assessment.

Keywords: fatigue strength; ultra high-strength steel; butt joints; weld topography; crack detection and tracking; crack propagation

1. Introduction

1.1. *Fatigue Assessment of Welds Considering Crack Initiation and Propagation*

The fatigue life of a component is a very complex physical process which could be divided to five stages [1]: dislocation movement, crack nucleation, micro- and macro crack propagation and the final fracture. For fatigue assessments of welded joints, the individual consideration of each phase is usually too complicated. Most procedures focus on the stages up to a crack of detectable size. Fracture mechanical approaches, on the other hand, neglect all stages before macro crack propagation. In order to increase the accuracy of technical fatigue assessments, two-phase models considering both crack initiation and propagation are introduced [2]. Hereby, the first three stages are treated as the crack initiation phase (index init) followed by a crack propagation phase (index cp). Thus, the total fatigue life up to burst fracture N_f sums up the cycles of both phases N_{init} and N_{cp}, respectively. The transition between these two phases is referred to as technical crack initiation at a crack depth a_{th}; its value varies depending on the selected model [3–6]. In the literature, various approaches exist for an assessment of the crack initiation phase, such as the fatigue notch factor K_f [7–9], strain life approaches [10,11] or the notch stress intensity factor [12,13]. The crack propagation phase, on the other hand, is generally assessed using linear elastic fracture mechanics.

1.2. Detection of the Crack Initiation Point

The result of a standard load controlled fatigue test is the number of load cycles until burst fracture at a certain load level, including the crack initiation and propagation phase. For cylindrical small-scale specimens without considerable imperfections, such as pores, the crack propagation phase is rather small compared to crack initiation. Therefore, the S-N curves obtained by such standard tests could be applied in fatigue life assessment concepts to estimate the time to crack initiation [14]. For welded components, on the other hand, the specimens should exhibit a certain width/sheet thickness ratio in order to ensure a steady state residual stress state and crack initiation from a plain weld rather than the specimen's edges [15]. In practice, a minimum width of three to five times the sheet thickness is suggested to avoid such effects [16,17] causing welded specimens to be comparably large even in the case of thin sheets. Therefore, the crack propagation life of welds significantly contributes to the specimen's total fatigue life [18–20]. Hence, the knowledge about the technical crack initiation enables a split of the fatigue assessment to crack initiation and crack propagation phases. This is of great interest even for fatigue tests of welded specimens, as it makes a more precise assessment of a structure's fatigue life possible.

A number of different methods exist for monitoring crack propagation. For welded structures, local strain measurement at hot spots [21,22] is widely used due to its flexibility and easy handling. This procedure is expedient if the specimen has a low number of distinct hot spots, such as a longitudinal stiffener, where crack initiation will occur. Otherwise, a high number of strain gauges is necessary to cover the whole highly-stressed weld toe area.

Another common technique is the application of a mixture of zinc oxide and glycerine on the weld toe [21,23] giving a colour change from white to black when a crack initiates. The main disadvantage of this method lies in the fact that the weld is covered by a viscous fluid preventing any exchange of oxygen leading to a possible change of crack initiation and propagation behaviour.

A purely optical tracking of crack propagation is performed in [24,25]. Hereby, an industrial camera tracks the crack's surface extension using its contrast to the surrounding surface. This procedure achieves high accuracy but it requires a fine surface quality and special lighting.

In [24], the optical tracking is compared to the crack propagation detected via crack gauges. For this procedure, a crack gauge is applied on the surface region where the crack is expected. During the fatigue test, constant current flows through the crack and the resistance based on voltage drop is measured. As the crack opens, the crack gauge is torn apart, thereby increasing the resistance of the gauge, which can be linked to the surface crack length. The comparison between optical measurement and crack gauges in [24] reveals a sound accordance.

The eddy current method applied in [26,27] enables non-contact measurement of surface cracks even below paint coatings. Thereby, eddy currents proportional to the conductivity of the material are induced by placing an AC-energized probe coil near the surface of the specimen. Discontinuities, such as cracks, affect the magnitude and phase of the induced current. The consequential magnetic field is then sensed by the probe indicating the presence and size of the surface crack.

Digital image correlation (DIC) is used for estimating stress intensity factors in [28] and the consequential identification of crack propagation law parameters in [29]. In [30], DIC and crack opening displacement techniques are combined to monitor crack propagation tests. Therewith, the fatigue behaviour of thin cracked plates made of aluminium alloy in crack propagation mode I are analysed.

1.3. Effect of Local Weld Geometry on Fatigue Performance

Modern welding technologies involving highly automated processes enable the production of high-quality welds exhibiting a very smooth transition at the weld toe. Nevertheless, variations in weld geometry due to the dynamic welding processes cannot be eliminated entirely. These very local discontinuities often govern the fatigue performance of welds due to its locally pronounced notch effects. Due to the increasing demands on quality of the welds of high-strength structural components, the effect of local weld geometry on the fatigue performance is of particular relevancy. Therefore, an assessment of such variations is essential in order to determine the maximum allowable size and extension.

In [31–33], the linear elastic stress concentration factor is utilized to take weld geometry into account. Thereby, Ref. [31,32] determine the stress concentration factor by the aid of linear elastic FE-analysis, whereas [33] uses analytical equations. In general, this procedure is prone to overestimating the effect of the actual geometry, as support effects are not considered. However, a local fatigue notch factor can be obtained by evaluating the local notch stress course, the stress gradient and, finally, the corresponding supporting factor [34].

Radaj proposed to replace the actual weld toe transition radius by a fictitious radius ρ_f in order to take support effects into account [9] (see Equation (1)). In [33], this procedure leads to better matching results than the use of the actual transition radius.

$$\rho_f = \rho + 1\text{mm} \tag{1}$$

The effective notch stress concept [9,35–37] proposes a reference notch radius of ρ_{ref} = 1 mm independent of the actual geometry. Thereby, a crack-like weld toe exhibiting an actual weld toe radius of ρ = 0 mm is assumed leading to a fictitious radius $\rho_f = \rho_{ref}$ = 1 mm in Equation (1). This concept is part of the International Institute of Welding (IIW) guideline [38], verified and consistent for structural steels and aluminium alloys. However, it does not consider the actual weld toe geometry. Therefore, it tends to deliver conservative results especially for high-quality high-strength welds with extremely shallow weld toes.

In [39], the critical distance approach developed by Lawrence [10] is applied. Thereby, the fatigue notch factor K_f is calculated from the stress concentration factor K_t using the Peterson formula. The required parameter a^*, the critical distance, is related to the ultimate tensile strength of the material.

Neuber proposed in [8] an evaluation method based on stress averaging in order to consider microstructural support. Hereby, the fatigue assessment is performed using an effective stress σ_{eff} obtained by averaging of the stress distribution in depth $\sigma(x)$ over the microstructural support length ρ^* at the surface layer instead of the linear elastic stress maximum (see Equation (2)). The fraction of σ_{eff} and the nominal stress σ_n gives the fatigue relevant effective stress concentration factor K_f (see Equation (3)).

$$\sigma_{\text{eff}}(\rho^*) = \frac{1}{\rho^*} \int_{y_0}^{y_0+\rho^*} \sigma(y)dy \tag{2}$$

$$K_{\text{f}}(\rho^*) = \frac{\sigma_{\text{eff}}(\rho^*)}{\sigma_{\text{n}}} \tag{3}$$

Values for the key parameter ρ^* are provided by Neuber himself for various materials in [8]. For welded structural steels exhibiting a microstructure similar to steel castings a microstructural support length of ρ^* = 0.4 mm is suggested leading to the effective notch stress concept with a fictitious transition radius of ρ_{ref} = 1 mm, which is mentioned above in Radaj's concept. In [40], comprehensive investigations on the effect of weld geometry on the fatigue strength under consideration of the microstructural support length are published. Here, a significantly smaller microstructural support length of around ρ = 0.1 mm reveals the lowest scatter band of all fatigue test results. This is traced back to the high hardness of the weld, where Neuber's initial suggestion for ferritic steel is more appropriate instead of cast steel.

In many cases, the effects of the weld geometry regarding to fatigue strength is assessed using fracture mechanics neglecting the crack initiation phase. In [41], the effects of variations of weld geometry parameters are studied systematically. A comparison to test results reveals the best agreement using a semi-elliptical crack with an initial depth of a_{init} = 0.05 mm. The work published in [42] presents a holistic study on the effects of the weld geometry applying the IBESS-model [43], a fracture mechanics-based prediction of the fatigue strength of welded joints. Mashiri et al. [44] carried out crack propagation analysis using the boundary element analysis. Their investigations on the effect of weld profile and undercuts on fatigue life shows the best fit at an initial crack length of a_{init} = 0.1 mm. In [45,46], the Frost diagram is utilized to assess undercuts from a fracture mechanical perspective. The focus hereby lies in the establishment of undercut tolerances for industry.

Concluding, numerous local approaches are currently available for fatigue assessment of welded joints, but for calculation one has to apply proper local weld toe geometry values especially in the case of high-quality welds.

2. Experimental Investigations

The experimental work of which results are utilized in this paper, especially specimen manufacturing and details regarding the fatigue testing setup, have been already published in [47]. This chapter provides a short overview on the invoked test procedure for completion. In addition, investigations on the crack propagation parameters of the weld's base material are performed and described in this section.

2.1. Fatigue Tests

The two-layer butt-welded specimens are manufactured using S1100 base material of 6 mm sheet thickness, T89 metal core filler wire and standard M21 shielding gas. The mechanical properties of the base and filler metal are listed in Table 1. An optical inspection of the specimens after the manufacturing process reveals undercuts at the weld toe of several specimens (see Figure 1). These specimens undergo a surface topography scan prior to fatigue testing in order to investigate the effect of these imperfections. The tumescent tensile fatigue tests were carried out under standard atmosphere at a stress ratio of R = 0.1 and a test frequency of f = 10 Hz up to burst fracture or a run-out level of 10 million load cycles. The statistical evaluation in finite life regime is performed according to [48] using arbitrary slope values, whereas the run-out level is assessed by the arcsin$\sqrt{P}$-procedure [49] with a decrease of 10% per decade in fatigue strength corresponding to a slope of k = 22, as suggested in [50]. Figure 2 shows the fatigue test result, where the effect of the weld undercuts regarding to fatigue strength is clearly visible. However, in some cases the reduction in fatigue life is far less pronounced than expected based on the evaluation of each weld's stress concentration factor (see Section 4.1 and [47,51]). Figure 3 shows the fracture surface of a defective specimen clearly identifying the crack initiation point at the undercut.

Table 1. Mechanical properties of the base and filler material.

Material	Yield Strength σ_y (MPa)	Tensile Strength σ_u (MPa)	Elongation A (%)	Impact Work ISO-V KV (J)
Base, S1100	≥1100	≥1140	≥8	≥ 27 @ −20 °C
Filler, T89	≥890	≥940	≥15	≥ 47 @ −40 °C

Figure 1. Specimen with weld toe exhibiting a local undercut [47].

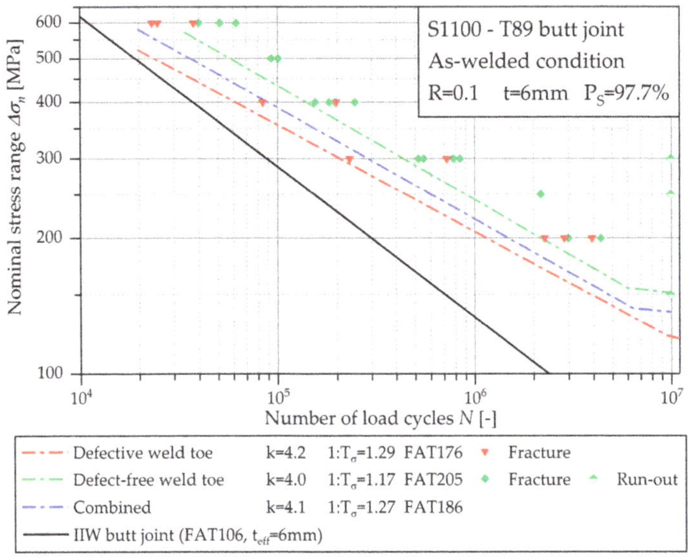

Figure 2. Fatigue test results according to the nominal stress evaluation [47].

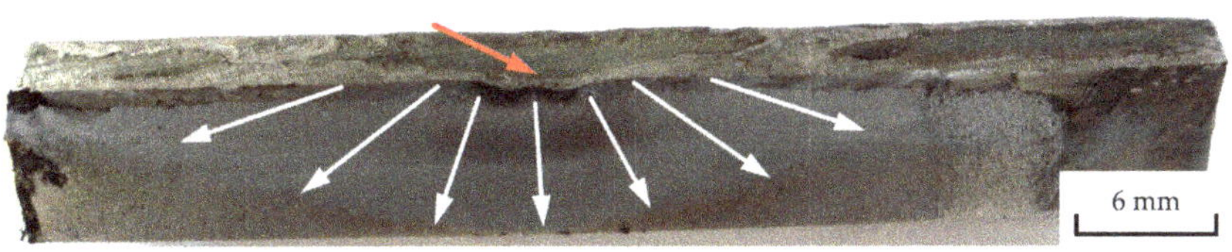

Figure 3. Fracture surface clearly showing the crack initiation point at the undercut (red arrow) [47] ($\Delta\sigma_n$ = 200 MPa, N_f = 2.87×10^6, R = 0.1).

2.2. Determination of Crack Propagation Parameters

The assessment of crack propagation life requires the knowledge of suitable material parameters describing the crack propagation behaviour. Therefore, the determination procedure utilizing crack gauges presented in [24] is applied to measure the crack propagation of S1100 base material exhibiting 6 mm sheet thickness. Hereby, four single edge notch tension (SENT) specimens are tested under constant amplitudes at a stress ratio of R = 0.1 recording the respective crack propagation. The subsequent evaluation of the fatigue crack growth rate is performed according to ASTM E647 [52]; the according geometry factor to determine the specimens stress intensity factor is based on [53]. Figure 4 shows the resulting long-crack propagation behaviour of all four specimens.

In the literature, many different models are provided describing the materials' behaviour in the da/dN—ΔK regime; for example, in [54,55]. For this work, the crack propagation model according to Paris-Erdogan [56] is applied requiring only two parameters C_P and m_P (see Equation (4)). A non-linear fitting procedure was used to determine these parameters for the present experimental data, whose result is plotted in Figure 4 as well. The consequential values are C_P = 8.3509×10^{-10} and m_P = 1.721 for units of mm/cycle and MPa$\sqrt{\text{mm}}$, respectively.

$$da/dN = C_P \cdot \Delta K^{m_P} \tag{4}$$

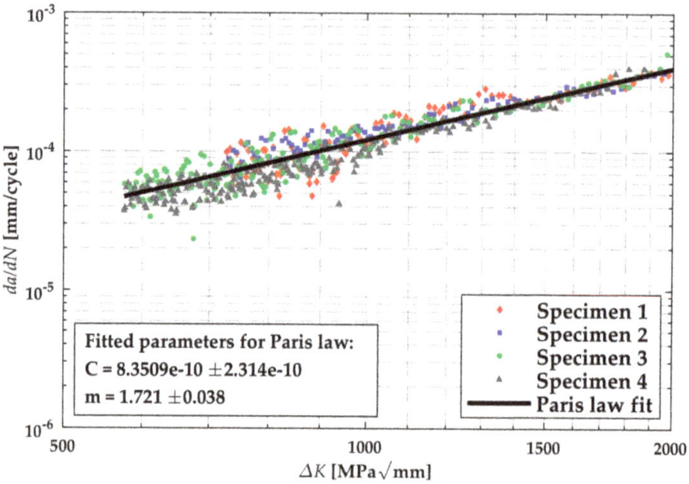

Figure 4. Stable crack propagation curve of the investigated S1100 base material at R = 0.1.

3. Optical Detection of Crack Initiation and Propagation

The following section presents an engineering feasible method for detecting crack initiation and tracking its propagation of welded specimens. This quite straightforward procedure is based on digital image correlation of pictures of the specimens acquired during the experiment using a common single-lense reflex (SLR) camera system.

3.1. Experimental Setup

The setup for this method is very simple and does not require any changes on the specimen itself (see Figure 5). A common digital SLR camera is placed around 20–30 cm off the specimen's surface. The exact distance is figured out empirically for each specimen type in order to fit the region of interest to the camera's captured area and its focus settings. In order to trigger the camera during the fatigue test, the cameras trigger button is connected to the control computer for the hydraulic test rig. The applied camera is a Canon (Ōta, Tokyo, Japan) EOS 650D featuring a comparably high resolution of 18 megapixels. The lens selected for the pictures is a Canon EF-S 60mm f/2.8 Macro USM macro lens enabling a detailed view of the region of interest without the risk of collision with the clamping devices. A lighting system consisting of an LED lamp is installed to ensure lighting independent of the time of day.

The image acquisition is performed at the specimens' upper load F_u causing possible cracks to open. Before running the test, a reference image is taken at maximum loading as a basis of comparison, which is required by the subsequent post-testing digital image correlation procedure. At first, the possibility of synchronized image acquisition during fatigue testing was checked. However, as the image has to be taken exactly at the maximum load, a very short shutter time is required in order to ensure a clear picture with adverse impact on the image quality. Hence, the run cycle of the fatigue test stops at maximum load for the image acquisition for a few seconds (see Figure 6). Due to the short retention time and standard conditions around the specimen, holding the specimen at maximum load does not affect the crack initiation and propagation behaviour.

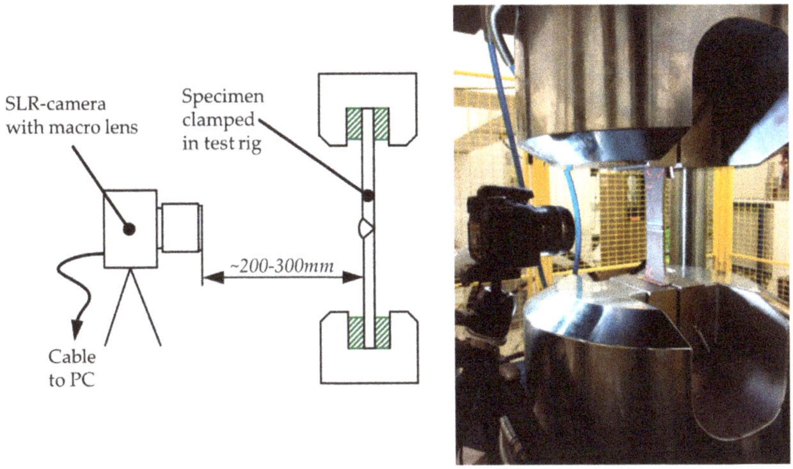

Figure 5. Setup for image acquisition: (**a**) Schematic layout; (**b**) setup at test rig.

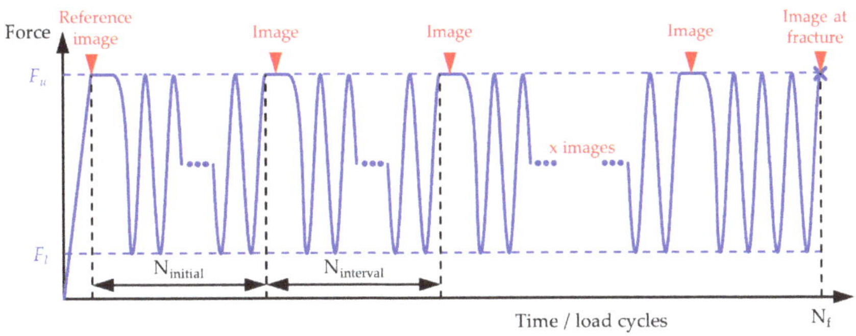

Figure 6. Schematic block testing programme for image acquisition.

The image acquisition starts with a reference image before test start and continues after a certain number of load cycles $N_{initial}$. This initial cycle number is based on previous tests and strongly depends on the load level of the respective test. The a priori fatigue testing reference image and the image taken at the initial cycle number show no distinct difference, implying early-stage crack propagation. After the initial image acquisition, the fatigue test is continued block by block. The number of load cycles between two images $N_{interval}$ increases with the load cycle number according to Table 2. Both measures reduce the number of images to be taken without a significant decrease of the procedure's accuracy. A final image of the broken specimen at N_f provides insight on the failure-tripping crack in case of several initiated cracks.

Table 2. Definition of intervals for image acquisition.

Current Load Cycle Number N (–)			Interval between Image Acquisition $N_{interval}$ (–)
0≤	N	<100,000	2000
100,000≤	N	<400,000	5000
400,000≤	N	<800,000	10,000
800,000≤	N	<1500,000	25,000
1500,000 ≤	N		50,000

3.2. Crack Detection and Tracking Procedure

A Matlab©-based procedure (R2017b, The Mathworks, Inc.; Natick, Massachusetts, USA) is developed for detection of crack initiation and subsequent tracking of crack propagation. Essentially, it is divided into two sections. At first, digital image correlation is applied to the acquired images in order to get a local distortion field of the specimen's surface. Therefore, an algorithm submitted to Matlab Central File Exchange by Elizabeth Jones [57] (File ID #43073) provides the basic functions. This code is evaluated using standard test images by the Society for Experimental Mechanics (SEM) in the *DIC-Challenge*, see [58] and the website https://sem.org/dic-challenge/. The second section detects crack initiation and tracks crack propagation using the distortion field gathered in the first step.

The procedure's structure and operation is described in the following subsection using exemplarily a specimen tested at a nominal stress range of $\Delta\sigma_n = 400$ MPa, an image acquisition after $N_{initial} =$ 60,000 load cycles and $N_f =$ 85,039 load cycles until burst fracture.

3.2.1. Local Distortion Field of the Specimen's Surface

A crack in a structure leads to a local reduction of the stiffness along with a significant increase of local displacement in case of a load perpendicular to the crack surface. Therefore, the change in local displacement is a suitable parameter for the detection of surface cracks. Hereby, digital image correlation is an appropriate tool in order to measure this parameter using the images acquired during the fatigue tests. Digital image correlation is a camera-based toolbox to acquire full field displacements by cross-correlation. Usually, speckle patterns are applied to the measured surface in order to increase the surface contrast. However, in the present case the texture of the specimens' surface itself turned out to be sufficient for correlation in the desired regions. This is a great advantage compared to other crack detection methods as any surface modification could lead to a change in the weld's crack initiation and propagation behaviour.

The image correlation procedure compares each image at a certain number of load cycles with the reference image acquired at the beginning of the fatigue test. In order to exclude any effects coming from optical distortion due to the camera lens, the Matlab© built-in *Camera Calibrator* is applied to estimate the geometric parameters of the utilized lens. Subsequently, the distortion effects are removed of the acquired images. Figure 7 shows the grayscale images prepared for image correlation of Reference (a) and last image (b) of the exemplary specimen. Here, the x-axis corresponds to the horizontal and the y-axis to vertical direction. The last image already shows a large crack at the upper weld toe leading close to burst fracture. However, the actual size of the crack is not visible. In order to increase the correlation speed, parts of the image are excluded from correlation which do not contain relevant information, such as the areas to the left and right of the specimen. Here, Figure 7a shows the processed net surface as a red rectangle. The correlation procedure requires a set of parameters affecting its quality and accuracy as well as the calculation time which are described in the documentation of the code (see [59]). Hereby, a suitable set of parameters for the current images is invoked using the documented approach and further empirical investigations. One of the most sensitive parameters is the grid spacing defining the distance in pixels between two adjacent control points. At these points, the output distortion is computed. Therefore, the grid spacing represents the resolution quality of the result directly linked to computation time. For example, an image with a size of 6000 × 4000 pixels and a grid spacing of 15 pixels results in a distortion field of 400 × 267 entries. Here, a large grid spacing gives a rather coarse distortion field whereas a fine grid spacing leads to a disproportional long calculation time of several hours per image and increases the risk of poor correlation results. Thus, a grid spacing between 15 and 20 pixels is recommended. A more detailed insight on the correlation procedure and its parameters is given in [60].

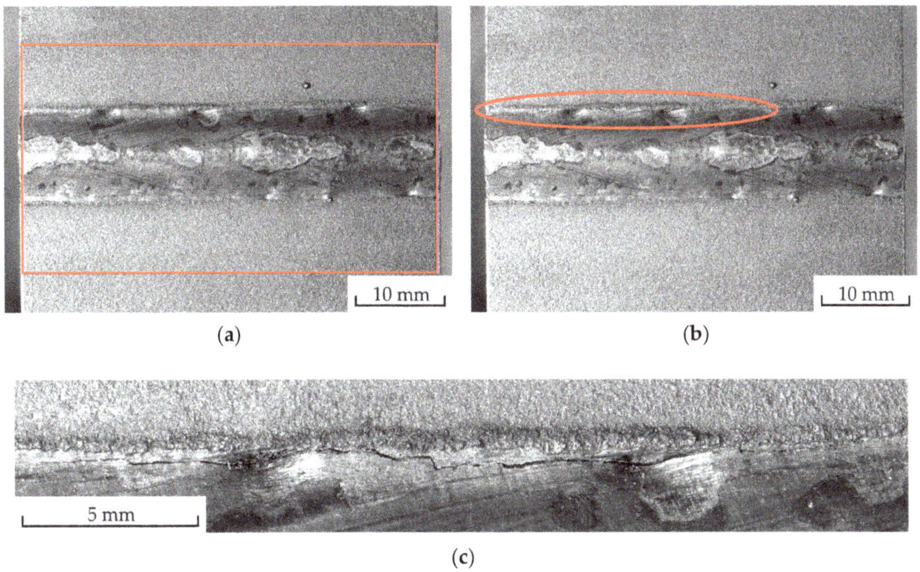

Figure 7. Reference and last image of an exemplary specimen, $\Delta\sigma_n = 400$ MPa, $N_f = 85{,}039$: (**a**) Reference image before fatigue test; (**b**) image at N = 84,000 with crack at upper weld toe; (**c**) magnification of the crack region of (**b**).

The result of the image correlation procedure for N = 84,000 is depicted in Figure 8a as the vertical displacement field s_y in the load direction of the specimen. This outcome includes the local displacement as well as a potential global movement of the specimen due to a small amount of sliding in the clamping area. Therefore, the vertical gradient of the vertical displacement s'_y is introduced in order to highlight local changes in vertical displacement and simultaneously exclude possible global movement. The corresponding result of the displacement gradient is depicted in Figure 8b explicitly indicating the present cracks only.

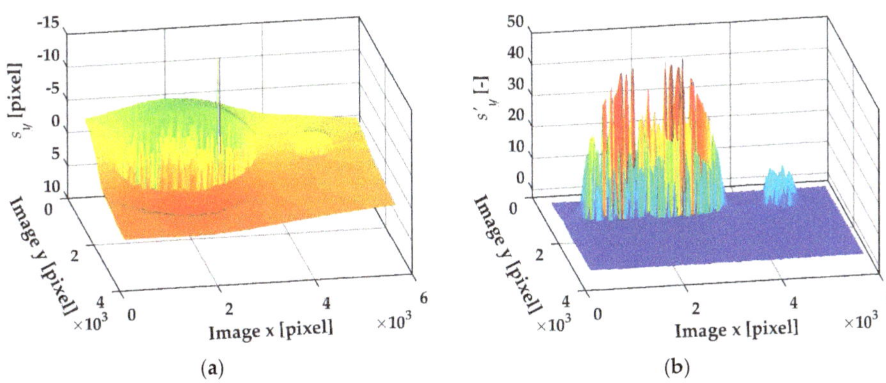

Figure 8. Result of image correlation at N = 84,000: (**a**) Displacement field s_y in vertical (y)-direction; (**b**) vertical gradient of vertical displacement field s'_y.

3.2.2. Crack Detection and Tracking

This section deals with the identification of a crack and its tracing over the load cycle number based on the results of the digital image correlation procedure. This includes recognition of a newly initiated crack to an existing one, as well as the merging of two small cracks into a larger one. This part of the procedure is once again divided into two subsections; the first covers the detection of all cracks for each image separately. The second traces the growth with increasing load cycle number and determines possible merges, as well as initiation of new cracks.

The identification of cracks in an image is performed utilizing the Matlab implemented image processing toolbox. Therefore, the gradient field is converted to a binary black-white image with an initial threshold value $s'_{y,thres}$ of two pixels/pixel. If the maximum gradient value of an image exceeds forty pixels/pixel, the threshold value is adapted to five percent of the maximum value. The reason for this lies in the enlargement of the plastic zone at the crack tip with increasing crack length [55], which is considered by this adaptive threshold setting. It has to be noted that the definition and the course of the threshold value is, besides the parameters of the image correlation, the primary influencing factor regarding the finally-measured crack length. Therefore, it plays a decisive role in the procedure's verification process (see Section 3.3.). Figure 9 shows the black-white image related to Figure 8b with a threshold value of $s'_{y,thres} = 2.39$. Figure 10 plots the result of crack identification process in the gradient field of Figure 8b.

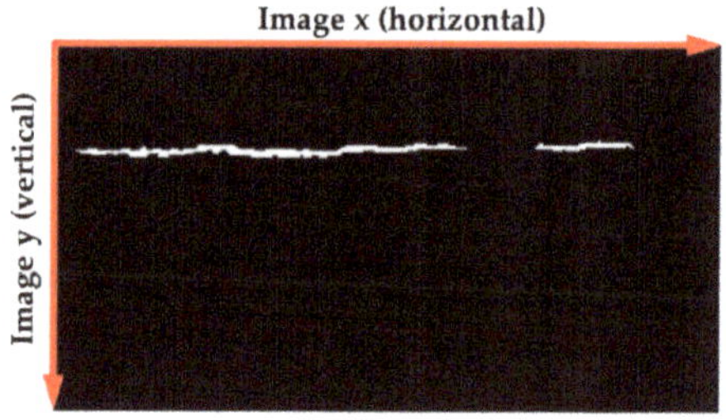

Figure 9. Binarized matrix of Figure 8 indicating crack regions.

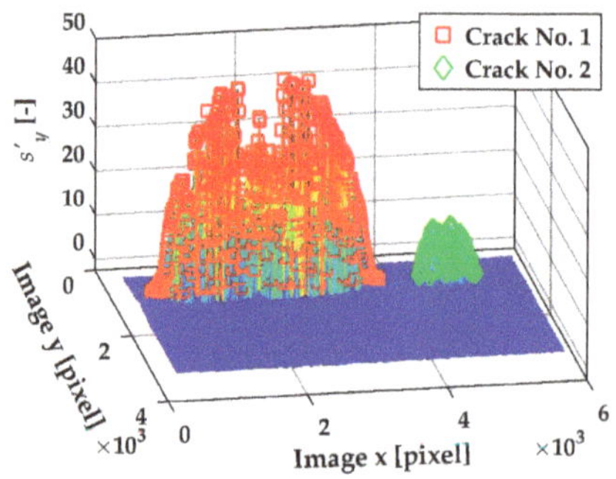

Figure 10. Gradient field s'_y of Figure 8 with marked results of the crack identification process.

Based on the results of each image, crack tracking is performed over all of the images. At first, the last image before fracture is considered by identifying the number and positions of the final cracks before specimen failure. This information is essential, as all crack initiations will occur in these areas. However, due to this action, the procedure is not suitable for application during the test without any modification. Subsequently, a loop through all images assigns the respective cracks to a final crack.

Additionally, the merging of cracks is noted at this point by a comparison to the previous image's state. Finally, the determined surface crack length is converted from pixels to millimetres.

Figure 11 shows the procedures result, the surface crack length *2c* in mm with respect to the load cycle number. Crack number one actually initiated at two different positions and merged just a few thousand load cycles before final fracture. Figure 12a shows the image of the first step at 60,000 load cycles with marked crack number one in red and two in green dye. In this picture, the two separate initiations of crack one are visible, which will merge within further crack growth. Figure 12b depicts the image of the last step before fracture at 84,000 load cycles where crack one is already merged. The respective fracture surface after 85,039 load cycles in Figure 12c gives a good opportunity to verify the result. Here, the two initiation points within the left crack are clearly visible. Furthermore, all three crack initiation points are located at local undercuts, which are indicated in the figure.

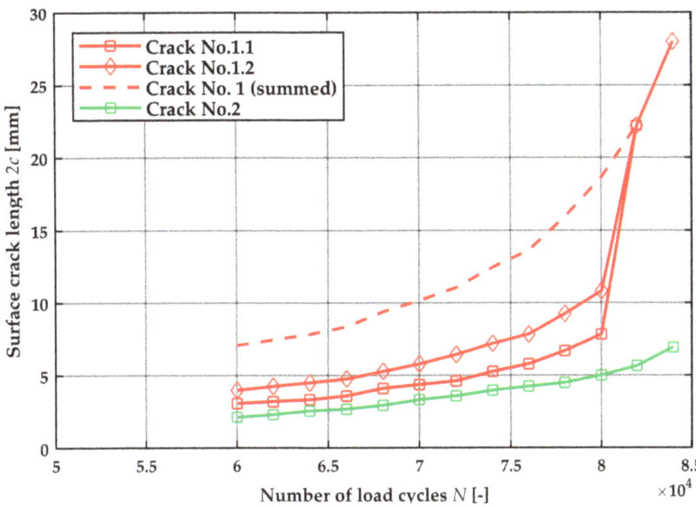

Figure 11. Surface crack length *2c* over load cycle number *N*.

3.3. Validation of the Procedure

The procedure for crack detection and tracking presented in this section is based on estimations and parameters, which are assessed empirically. Therefore, the procedure requires a validation process in order to prove, that the assessed crack length is properly evaluated. This is of importance, as the result not only includes the presence of a crack but also provides the absolute surface crack length value in millimetres.

Two different approaches are used for validation. The first, more comprehensive test includes block load testing of a notched specimen in order to create beach marks at the fracture surface, which can be easily compared to the procedures result. The second approach is based on fracture surfaces of welded specimens exhibiting a second not through-crack, whose measurements can be compared to the procedure's results as well.

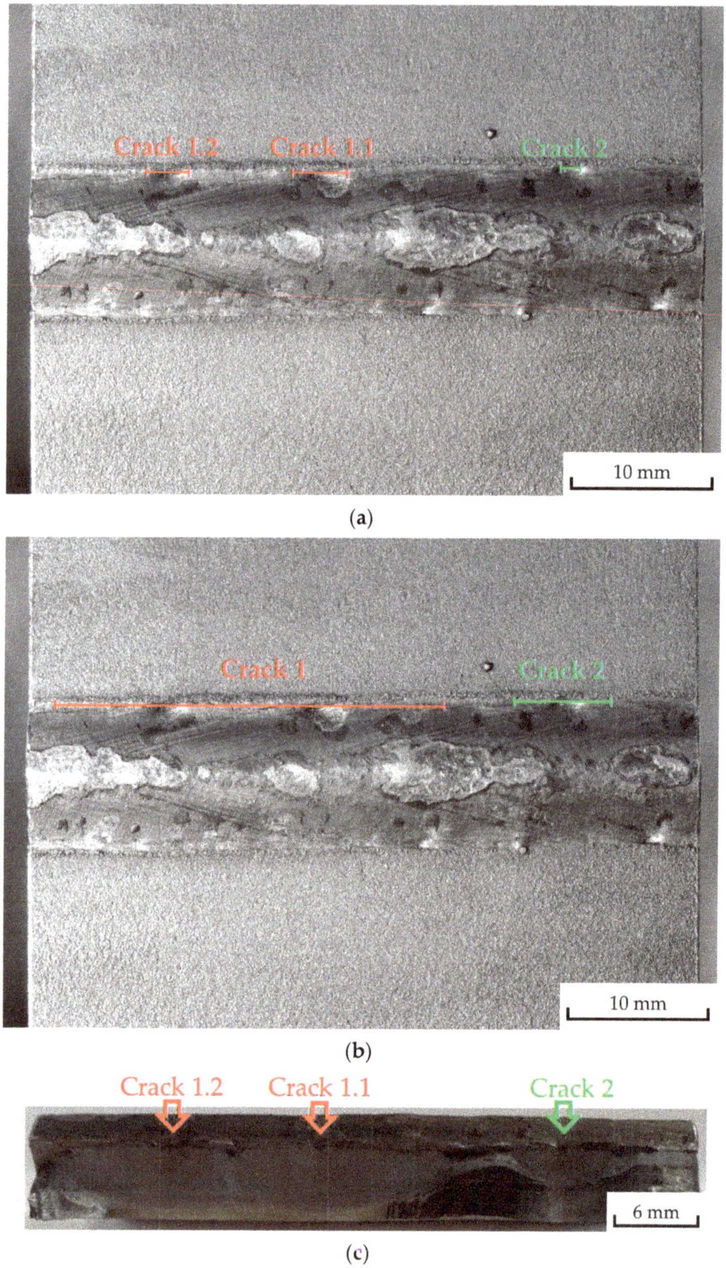

Figure 12. Position and extension of determined cracks marked at specimen surface and comparison of the crack identification procedures result with the final fracture surface: (**a**) $N = 60{,}000$; (**b**) $N = 84{,}000$; (**c**) fracture surface at $N_f = 85{,}039$ with three clearly visible crack initiation points.

3.3.1. Beach Marks

The first validation approach utilizes beach marks at a fracture surface as basis for comparison. Therefore, a block load fatigue test is performed using a notched specimen made of 10 mm S355 construction steel. The spark-eroded semi-elliptical notch placed at the centre of the 40 mm wide specimen exhibits a depth of 1 mm and a surface length of 2.5 mm and acts as initial crack. The block load test consists of alternating sequences of 20,000 load cycles at an upper level $\Delta\sigma = 250$ MPa and a lower level of $\Delta\sigma = 150$ MPa at a constant stress ratio of $R = 0.1$. Therefore, crack propagation rates vary for each block, which appears at the fracture surface. Further details on specimen geometry and experimental procedure are given in [61]. During fatigue testing, an image is acquired every 5000 load cycles corresponding to a number of four images per block.

After the fatigue test, a high-resolution image of the fracture surface is acquired. In total, six beach marks have enough contrast to ensure a reliable measurement. Figure 13a shows the fracture surface of the block load test including the measurement results. The result of the crack detection and tracking procedure is plotted in Figure 13b; additionally, the measurement results are shown as horizontal lines. Table 3 shows the according numbers including the percentage difference. The comparison between measurement and assessment leads to very satisfactory results, the deviation for all assessed beach marks is less than two percent. This confirms the set of parameters used for the image correlation and crack detection procedure.

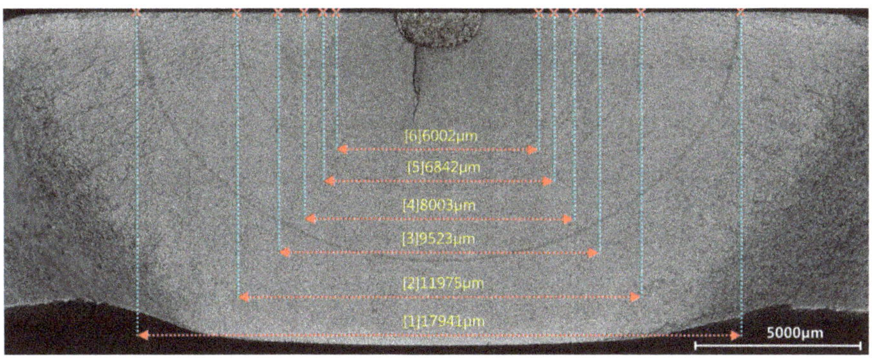

(**a**)

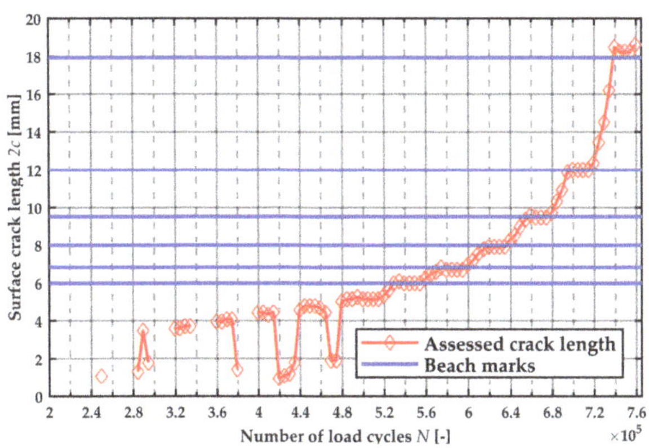

(**b**)

Figure 13. Comparison of beach marks length and assessed crack length: (**a**) Fracture surface with the surface length measurement of visible bench marks; (**b**) development of the crack length over the load cycle number.

Table 3. Comparison between the measured and assessed surface crack length.

Beach Mark		Measured Crack Length 2*c* (mm)	Assessed Crack Length 2*c* (mm)	Deviation (%)
No.	N (–)			
1	740,000	17.94	18.23	1,62
2	700,000	11.98	11.97	−0,08
3	660,000	9.52	9.52	0
4	620,000	8.00	7.92	−1.00
5	580,000	6.84	6.72	−1.75
6	540,000	6.00	6.00	0

At this point it should be noted that the crack propagation of this specimen starts at both ends of the semi-elliptical notch nearly at the same load cycle number. Especially during low load blocks, the procedure detects these two cracks separately, neglecting the presence of the notch in between. Therefore, the assessed surface crack length in such cases is about the surface length of the notch smaller than the total tip to tip length. This explains the behaviour of the surface crack length trace in Figure 13b jumping up to a load cycle number of 480,000.

3.3.2. Fracture Surfaces

The second part of the method's validation utilizes separate, not-through base plate thickness cracks in the fracture surface. The fracture-causing crack itself is not suitable for a comparison with the calculated value, as the transition zone from stable crack propagation to rupture is vague. This inhibits a precise measurement of the final surface crack length. The number of such separated, not-through cracks is limited and takes place preferably in high load levels, where the tendency for multi-crack initiation is higher. However, two suitable fracture surfaces are available for validation within the present test series (see Figure 14).

(**a**)

(**b**)

Figure 14. Fracture surfaces used for validation of the calculated crack length: (**a**) Specimen 1; $\Delta\sigma_n$ = 400 MPa, N_f = 85,039; (**b**) specimen 2; $\Delta\sigma_n$ = 600 MPa, N_f = 30,871.

The measurement of the actual crack length in the fracture surface is performed utilizing a Matlab©-tool introduced in [25]. The script fits a semi-ellipse to a number of points manually selected in the image using a non-linear fitting algorithm. Figure 15 displays the result of the fitting procedure for both specimens including the crack measures. Hereby, the shape of the crack in Figure 15a fits the semi-ellipse quite well, whereas the smaller crack in Figure 15b deviates from the semi-elliptical shape considerably. However, the surface length is assessed properly.

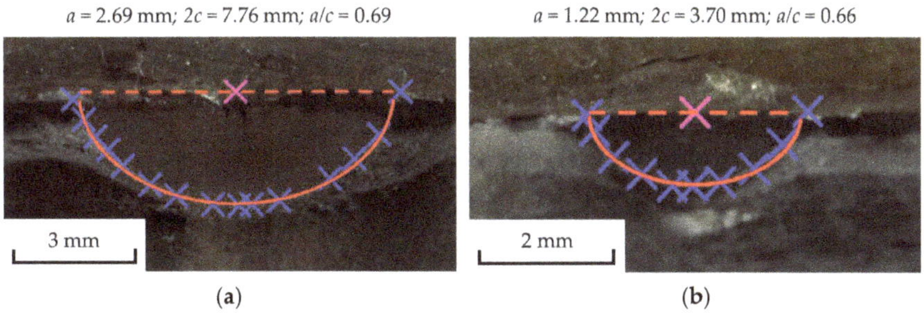

Figure 15. Measurement of not-through crack size by fitting a semi-ellipse: (**a**) Specimen 1; (**b**) specimen 2.

Due to the strategy of the presented crack detection procedure, the measurement of the surface crack length cannot be performed at burst fracture. Thus, quadratic extrapolation from the last four measurement results up to the number of load cycles at burst fracture is performed. This assumption could lead to an underestimation of the actual crack length. However, it delivers sufficient accuracy and does not require any local material parameters affecting the extrapolation.

Table 4 presents the calculated surface crack length with respect to load cycle numbers for both investigated cracks including the extrapolated values at burst fracture. The measured values originating from the fracture surface are listed as well for direct comparison. Hereby, the actual surface crack length exceeds the extrapolated values of about 2% in both cases. This, once more, validates the approach and the determined parameters of the crack detection and tracking procedure.

Table 4. Crack propagation of selected not-through cracks and comparison to fracture surface measurement.

Specimen 1			**Specimen 2**		
Number of Load Cycles N (–)	**Assessed Crack Length $2c$ (mm)**	**Measured Crack Length $2c$ after Final Rupture (mm)**	**Number of Load Cycles N (–)**	**Assessed Crack Length $2c$ (mm)**	**Measured Crack Length $2c$ after Final Rupture (mm)**
60,000	1.41		20,000	-	
62,000	2.31		22,000	-	
64,000	2.56		24,000	1.19	
66,000	2.70		26,000	1.70	3.70
68,000	2.95		28,000	2.39	
70,000	3.34		30,000	3.24	
72,000	3.59	7.76	30,871	3.67 *	
74,000	3.98				
76,000	4.24				
78,000	4.49				
80,000	5.01				
82,000	5.65				
84,000	6.93				
85,039	7.66 *				

* Quadratic extrapolation.

4. Fatigue Assessment

This section assesses the fatigue life of welded ultra high-strength steel butt joints and evaluates the effect of undercuts.

4.1. Neuber's Stress Averaging Method

The stress averaging method proposed by Neuber utilizes the microstructural support length ρ^* in regard to microstructural support. As already stated in Section 1, suggestions for this parameter often deviate from determinations using the lowest scatter-band of the resultant S-N curve as a key factor, especially in the case of welded specimens.

The local stress distribution at the weld toe and its trace in depth is well known due to surface topology scans and subsequent numerical analysis of this geometry of all ten specimens (see [47,51]). The values for the effective stress $\sigma_{eff}(\rho^*)$ and the corresponding fatigue notch factor $K_f(\rho^*)$ are calculated for each specimen at the respective points of crack initiation. The required stress distribution in depth $\sigma(x)$ is taken from the numerical analysis of the actual specimen's weld topography. Hereby, the value of ρ^* is varied in order to obtain the S-N curve with minimum scatter-band, where a minimum value of ρ^* = 0.1 mm must be maintained due to the mesh seed of about 30 μm in depth. Figure 16 shows the development of the consequential S-N curves for specimen fracture with varying ρ^* including the P_S = 97.7% line of the statistical evaluation. Additionally, the curves for nominal stress $\Delta\sigma_n$ and notch stress $\Delta\sigma_{notch}$ using the individual stress concentration factors K_t are displayed. The result of the statistical evaluation for crack initiation and total fatigue life for effective notch stress using selected values of ρ^* as well as nominal and notch stress are listed in Table 5; Table 6 provides the corresponding data for each specimen.

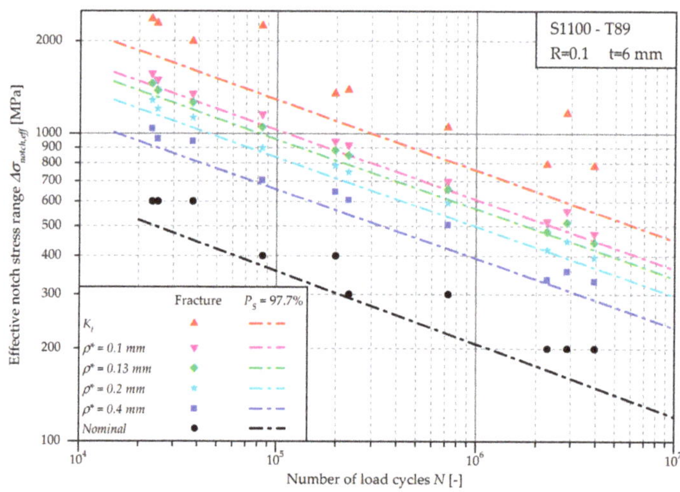

Figure 16. S-N curves for specimen fracture depending upon the microstructural support length ρ^*.

Table 5. Statistical evaluation for crack initiation and total fatigue life.

Evaluation Method	Crack Initiation			Burst Fracture		
	Slope k (–)	FAT Value (MPa)	Scatter Index $1{:}T_\sigma$ (–)	Slope k (–)	FAT Value (MPa)	Scatter Index $1{:}T_\sigma$ (–)
Nominal stress	4.43	160	1.30	4.20	176	1.29
Eff. notch stress (ρ^* = 0.4 mm)	4.68	307	1.16	4.46	337	1.15
Eff. notch stress (ρ^* = 0.3 mm)	4.68	340	1.14	4.45	372	1.13
Eff. notch stress (ρ^* = 0.2 mm)	4.66	392	1.11	4.44	427	1.10
Eff. notch stress (ρ^* = 0.13 mm)	4.64	452	1.09	4.43	489	1.09
Eff. notch stress (ρ^* = 0.1 mm)	4.62	489	1.08	4.43	523	1.10
Notch stress (K_t)	4.37	626	1.41	4.38	650	1.51

Table 6. Overview on local stress concentration and fatigue test data for each crack initiation point.

Specimen No.	K_t (–)	K_f (–)		$\Delta\sigma_{eff}$ (MPa)	N_{th} (–)	N_f (–)	N_{th}/N_f (%)
		ρ = 0.10 mm	ρ = 0.13 mm				
1	3.93	2.60	2.42	1453.5	10,000	23,497	43
	4.50	2.69	2.49	1492.2	10,520	-	-
2	3.37	2.36	2.21	884.5	142,840	199,292	71
3	3.97	2.60	2.40	479.8	1601,800	2281,981	70
	5.61	2.89	2.63	1052.6	52,650	85,039	53
4	4.75	2.83	2.60	1039.5	58,530	-	-
	4.45	2.71	2.50	1001.2	67,580	-	-
5	5.82	2.80	2.57	513.6	1596,300	2873,617	52
6	3.50	2.33	2.20	660.0	635,190	728,340	87
7	3.92	2.36	2.21	441.5	2610,560	3958,013	67
8	3.33	2.24	2.10	1259.4	24,000	37,839	66
	4.60	2.40	2.20	1322.7	19,330	-	-
9	3.81	2.49	2.30	1378.7	13,790	25,060	52
10	4.61	3.06	2.84	852.9	155,070	232,717	67

Figure 17 illustrates the development of the S-N curves slope k, scatter-band $1{:}T_\sigma$ and the fatigue class FAT of crack initiation fatigue life (a) and total fatigue life (b) with respect to ρ^*. Thereby, the statistical evaluation of the resulting S-N curves according to [48] is performed separately for each selected ρ^*. In the case of technical crack initiation, the least scattering is obtained by the minimum evaluated microstructural support length of ρ^* = 0.1 mm. The least scattering for total fatigue life is reached by ρ^* = 0.13 mm. In both cases, the consequential scatter-band shows very low values of $1{:}T_\sigma$ = 1.081 and $1{:}T_\sigma$ = 1.088. This result corresponds to the findings in [40], where a similar procedure yields in a stress averaging length of ρ^* = 0.05–0.1 mm for the minimum scatter-band. Whereas the slope of the evaluated S-N curve s remains rather constant, the FAT value naturally increases with decreasing averaging length. In Figure 18, the resulting effective S-N curve is pictured showing a significant decrease in scattering compared to the nominal stress results in Figure 2.

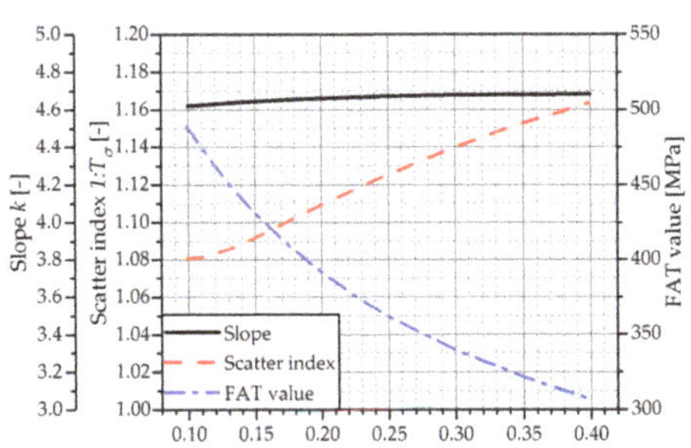

(**a**)

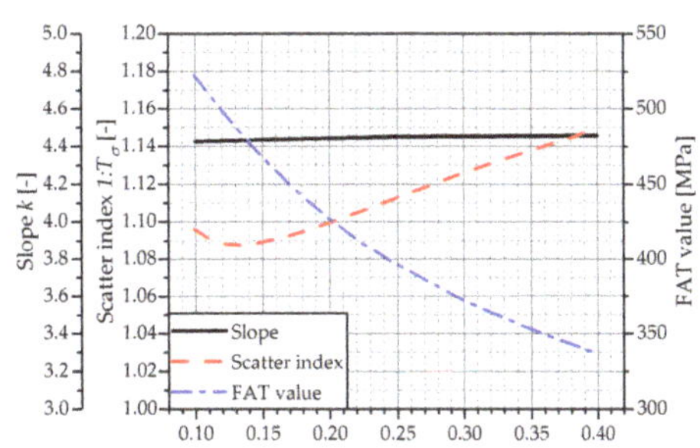

(**b**)

Figure 17. Development of slope, scatter index and FAT value with microstructural support length; (**a**) Crack initiation; (**b**) burst fracture.

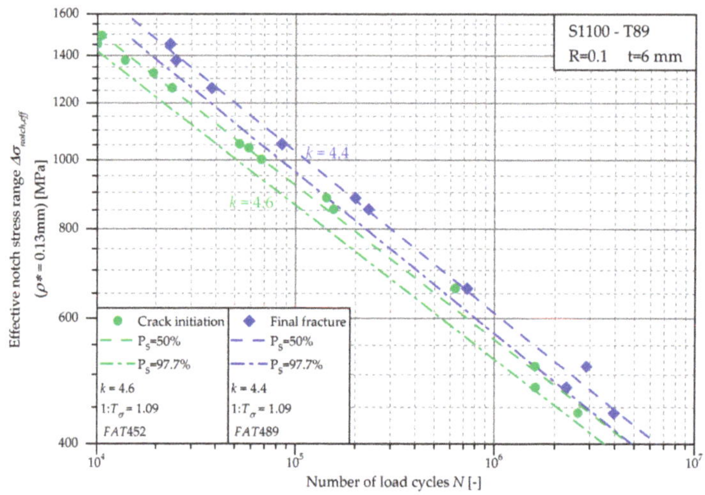

Figure 18. S-N curve including crack initiation life and cycles to final rupture of all investigated specimens.

4.2. Crack Initiation and Propagation Life

The procedure for optical crack detection and tracking presented in Section 3 allows a separation of the crack initiation life from the total fatigue life until burst fracture. However, the method's main result is the surface crack length *2c*, whereas the crack depth *a* is of importance for specimen failure. Therefore, a formula for the crack aspect ratio *a/c* given in [62] is applied (see Equation (5)). This expression was derived from fillet weld toes by fitting of experimental data, but according to the authors, it can be conservatively used for butt welds as well.

$$2c = -0.27 + 6.34 \cdot a \quad \text{for } 0.1 \text{ mm} < a < 3 \text{ mm} \tag{5}$$

The threshold crack depth defining the line between crack initiation and crack propagation life is set to a_{th} = 0.5 mm for all investigated specimens [3,4,63]. According to Equation (5), the corresponding surface crack length is *2c* = 2.9 mm, giving a rather low crack aspect ratio of *a/c* = 0.345. This value is quite comprehensible, as the local stress concentration at the weld toe decreases rapidly in the sheet thickness direction and therefore encourages the crack to grow in the surface direction first [64,65]. The fracture surface in Figure 14b exhibits a small crack on the left hand side marked by a yellow arrow exhibiting a very low aspect ratio. A detailed measurement reveals an aspect ratio of *a/c = 0.35* validating the result of Equation (5) for the present case. The crack detection procedure delivers the surface crack length in quite coarse steps of load cycle numbers. In order to determine the according threshold load cycle number N_{init} for a_{th}, linear interpolation is applied.

Figure 18 shows the effective notch S-N curve using ρ^* = 0.13 mm of specimen failure in blue dye and crack initiation with a_{th} = 0.5 mm in green dye. Table 6 summarizes the effective notch stress fatigue test data. Furthermore, the statistical evaluation including lines of *50%* and *97.7%* probability of survival are depicted. Some specimens exhibit more than one propagating crack leading to a number of fourteen crack initiation points of ten specimens. In such cases, the effective notch stress values are evaluated at the respective position of the crack initiation points of the particular specimen.

Figure 19 displays the portion of crack initiation life for the ten through cracks with respect to the effective notch stress range. Hereby, the portion of crack propagation life goes along with increasing stress range although large scatter is observed. However, the technical crack initiation phase seems to be the major portion of the fatigue life. In this context, Equation (6) presents a formula for the trace of the crack propagation fraction based on the statistical evaluation of the test results for crack initiation (index *init*) and burst fracture (index *f*). The consequential equation is a power law with basis $\Delta\sigma$ and the difference of the slopes of the S-N curves as the exponent are applicable for varying the probabilities of survival. However, the scatter indices in the present case are nearly identical causing the equation to be almost independent of the employed probability of survival P_S-value. The result of Equation (6) is depicted in Figure 19 for P_S = 97.7%.

$$\frac{N_{init}}{N_f}(\Delta\sigma) = \frac{N_{\mathrm{FAT}}\cdot\left(\frac{FAT_{init}}{\Delta\sigma}\right)^{k_{init}}}{N_{\mathrm{FAT}}\cdot\left(\frac{FAT_f}{\Delta\sigma}\right)^{k_f}} = \frac{(FAT_{init})^{k_{init}}}{\left(FAT_f\right)^{k_f}}\cdot(\Delta\sigma)^{k_f-k_{init}} = const\cdot(\Delta\sigma)^{k_f-k_{init}} \tag{6}$$

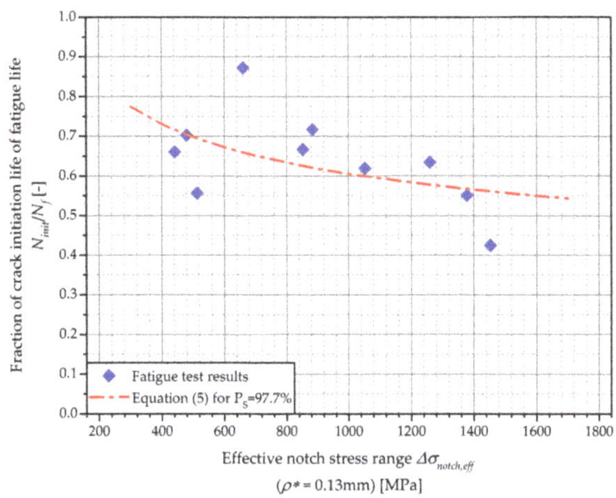

Figure 19. The fraction of the crack initiation life in total fatigue life over the effective notch stress range.

4.3. Assessment of Crack Propagation Life by Fracture Mechanics

The concluding part of this work deals with the calculation of the stable crack propagation life for all specimens using linear fracture mechanics according to the Paris law. Thereby, each crack initiation point of every specimen is processed using the data presented within this work. Subsequently, the results are compared to the actual fatigue test results.

Two approaches are pursued in this section:

1. Start of calculation at the previously determined threshold load cycle number N_{th} with an initial crack length of $a_{init} = a_{th}$ = 0.5 mm.
2. Calculation from the test start with an initial crack length of $a_{init} = u$ + 0.1 mm as recommended in [38].

For each of these two approaches three different sets of crack propagation parameters for the Paris law are utilized:

1. S1100 base material determined in Section 2.2.;
2. IIW parameters for welds suggested in [38]; and
3. Best fit parameters resulting by a least square fit of the fatigue test results.

The following three subsections give a comprehensive insight in the applied crack propagation methodology invoking the weight function approach and local geometry-dependent stress distribution approximation formulae.

4.3.1. Crack Propagation Analysis

If the crack initiation phase of a component is significantly exceeded, the fatigue assessment of a structure cannot be continued with stress-based approaches. Thereby, the fatigue action at a crack tip is described by the stress intensity factor (SIF) range ΔK (see Equation (7)). The crack propagation rate for each load cycle is then calculated by crack propagation laws, such as the Paris law (see Equation (4) and [56]).

$$\Delta K = \Delta\sigma \cdot \sqrt{\pi \cdot a} \cdot Y(a, c) \tag{7}$$

The stress intensity factor strongly depends on crack configuration and geometrical shapes which are considered by a correction function $Y(a,c)$. For standardized crack configurations and load cases, parametric formulae have been published (see, e.g., [66,67]).

4.3.2. Weight Functions Approach

The idea behind this technique [68–70] is to split a given random stress distribution $\sigma(x)$ into differential pairs of unit forces opening the crack. The weight function $m(x, a)$ then describes the effect of each differential force on the crack. The stress intensity factor for a given crack length a is, thus, an integration over the crack length (see Equations (8) and (9)). This procedure can be applied for the deepest point 'a' and the surface points 'c' of a crack by application of the appropriate weight function. In [69], general forms for the weight functions for semi-elliptical surface cracks in plates with finite thickness are published (see Equations (10) and (11)). Figure 20 provides an overview on the crack geometry parameters and the stress distribution at the weld toe due to axial loading.

$$K_a = \int_0^a \sigma(x) \cdot m_a(x, a) dx \quad \textit{for the deepest point 'a' of a crack} \tag{8}$$

$$K_c = \int_0^a \sigma(x) \cdot m_c(x, a) dx \quad \textit{for the surface point 'c' of a crack} \tag{9}$$

$$m_a(x, a) = \frac{2}{\sqrt{2\pi\,(a - x)}}\left[1 + M_{1a}\left(1 - \frac{x}{a}\right)^{\frac{1}{2}} + M_{2a}\left(1 - \frac{x}{a}\right) + M_{3a}\left(1 - \frac{x}{a}\right)^{\frac{3}{2}}\right] \tag{10}$$

$$m_c(x, a) = \frac{2}{\sqrt{\pi x}}\left[1 + M_{1c}\left(\frac{x}{a}\right)^{\frac{1}{2}} + M_{2c}\left(\frac{x}{a}\right) + M_{3c}\left(\frac{x}{a}\right)^{\frac{3}{2}}\right] \tag{11}$$

The determination of the parameters M_{1a}–M_{3a} and M_{1c}–M_{3c} is covered by many publications for various crack shapes. A sound approach is provided in [69]. However, [71] offers a refined set of formulas enabling a wider range of crack proportions; therefore, these are selected for the present work.

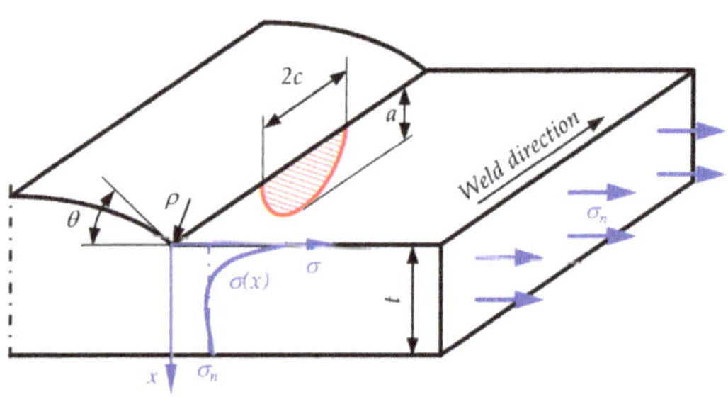

Figure 20. Sketch of the weld geometry including weld and crack geometry parameters.

4.3.3. Stress Distribution

A precise formulation of the local stress distribution in depth $\sigma(x)$ constitutes an important aspect concerning the proper assessment of the individual stress intensity factors. In [72] an approximation for the stress concentration factor is presented based on the local weld toe radius ρ, the flank angle θ and the sheet thickness t (see Equation (12) and Figure 20). Furthermore, a topology-dependent solution for the stress distribution in depth is shown in [73]. Equation (13) presents the main part of this approximation for pure tension loading. Details for parameter G_m, a correction for the stress distribution away from the stress peak, can be found in the cited work. This set of equations is also referred to and listed in the IIW recommendations [38].

$$K_{tHall}(\rho,\theta,t) = 1 + 0.388 \cdot \theta^{0.37} \cdot \left(\frac{\rho}{t}\right)^{-0.454} \tag{12}$$

$$\sigma(x) = \frac{K_{tHall}(\rho,\theta,t) \cdot \sigma_0}{2\sqrt{2}} \cdot \left[\left(\frac{x}{\rho} + \frac{1}{2}\right)^{-\frac{1}{2}} + \frac{1}{2}\left(\frac{x}{\rho} + \frac{1}{2}\right)^{-\frac{3}{2}}\right] \cdot \frac{1}{G_m(\rho,\theta,t)} \tag{13}$$

4.3.4. Calculation of Crack Propagation

The methods and approaches presented previously in this section allow an individual calculation of the crack propagation for each point of the specimen under consideration of the local weld toe topography. Thereby, the progress of crack length with load cycle numbers $a(N)$ is calculated by numerical integration of Equation (4) with step size ΔN (see Equation (14)). Within this procedure, the stress intensity factor value ΔK has to be recalculated after each step upon the current crack length value $a(N)$. The value of ΔN is made dependent on the respective number of N_f to ensure at least fifty steps to a final crack length $a_f = 0.9t$ where burst fracture is assumed.

$$a(N + \Delta N) = a(N) + C_P \cdot \Delta K[a(N)]^{m_P} \cdot \Delta N \tag{14}$$

4.3.5. Determination of "Best Fit" Parameters for the Paris Power Law

In addition to the existing material parameters for the S1100 base material and the suggestions according to IIW, an additional set of parameters is determined by best fit to the fatigue test results. This set is characterised by the minimum sum of squared errors, SSE (see Equation (15)). Thereby, the difference of the log values of the calculated and experimental load cycle numbers is used, otherwise high load cycle numbers would be overweighted. The linear elastic crack propagation parameters C_P and m_P are varied within a meaningful range of $1 \times 10^{-13} \leq C_P \leq 1 \times 10^{-7}$ and $1.5 \leq m_P \leq 3$.

$$SSE = \sum \left(\log_{10} N_{f\_calc} - \log_{10} N_f\right)^2 \tag{15}$$

4.3.6. Crack Propagation Results

The outcome for cases 1–3, where the crack propagation calculation starts at the determined crack initiation point at a crack length of a_{init} = 0.5 mm, is illustrated in Figure 21. Case 1 one employs the crack propagation parameters of the S1100 base material. The IIW parameters for welded joints in case 2 deliver a conservative approach across all considered specimens. In general, the use of these parameters leads to a very short crack propagation phase. The application of the S1100 base material parameters in case 1 leads to a non-conservative assessment for specimens with burst fracture below 1×10^5 load cycles. Naturally, the best fit parameters show the best accordance, as expected. Hereby, the exponent m_P is similar, but C_P is about three times smaller compared to the IIW values. Interestingly, all three parameter sets underestimate the crack propagation phase of the three specimens with the burst fracture above one million load cycles, significantly.

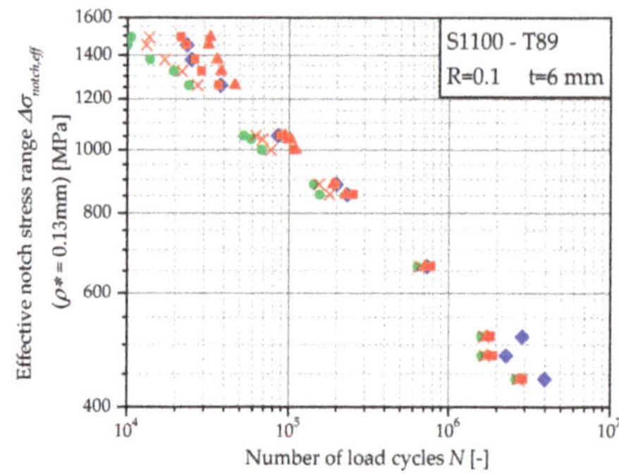

		SSE	*k*	$1:T_\sigma$	*FAT*
Crack initiation	●		4.6	1.09	452
Fracture	◆		4.4	1.09	489
Case 1	▲	0.141	3.8	1.08	453
Case 2	×	0.259	4.5	1.08	458
Case 3	■	0.065	4.2	1.08	467

Figure 21. S-N curve of crack propagation calculation starting at crack initiation.

The results of cases 4–6, where the crack initiation phase is neglected, are shown in Figure 22. As already expected based on the previous results, the IIW recommendation in case 5 leads to a very conservative assessment. The use of the S1100 base material parameters in case 4 results in a very steep S-N curve with a slope of just $k = 1.7$. Here, the specimens exposed to the highest effective notch stress are assessed quite well, but the assessment becomes more and more conservative with the decreasing effective notch stress near to the level of case 5. The best fit parameters reveal once more the best accordance, naturally. However, the assessment is non-conservative for specimens with a burst fracture below 2×10^5 load cycles.

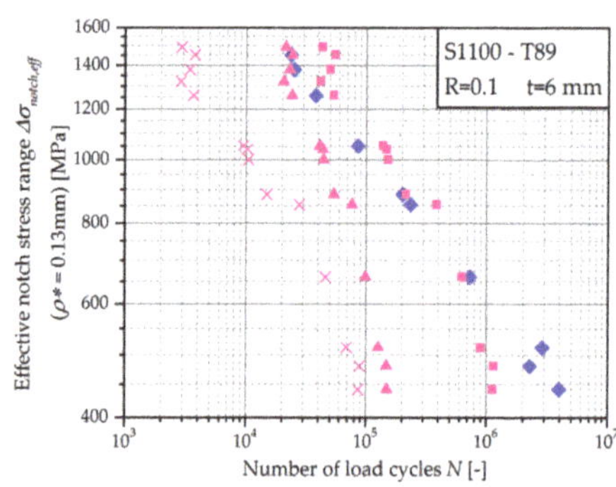

		SSE	*k*	$1:T_\sigma$	*FAT*
Fracture	◆		4.4	1.09	489
Case 4	▲	6.710	1.7	1.24	92
Case 5	×	14.251	3.0	1.24	147
Case 6	■	0.965	2.9	1.26	339

Figure 22. S-N curve of crack propagation calculation from the start.

Table 7 provides an overview on the parameters of the investigated cases. Furthermore, the resulting SSE as well as the result of the statistical evaluation of the consequential S-N curve for each case is shown.

Table 7. Summary of the investigated crack propagation cases including the statistical evaluation of the consequential S-N curves.

	Start Crack Length a_{init} (mm)	Start Load Cycle Number N_{start} (–)	Parameters for Paris Law			SSE (–)	Statistical Evaluation of S-N Curve		
			Material	C_P	m_P		Slope k (–)	FAT Value (MPa)	Scatter Index $1{:}T_\sigma$ (–)
Case 1	0.5	N_{th}	S1100	8.35×10^{-10}	1.72	0.141	3.80	453	1.077
Case 2	0.5	N_{th}	IIW	5.21×10^{-13}	3.00	0.259	4.47	458	1.076
Case 3	0.5	N_{th}	Best fit	1.78×10^{-13}	2.90	0.065	4.17	467	1.084
Case 4	u + 0.1	0	S1100	8.35×10^{-10}	1.72	6.710	1.73	92	1.239
Case 5	u + 0.1	0	IIW	5.21×10^{-13}	3.00	14.251	3.03	147	1.243
Case 6	u + 0.1	0	Best fit	1.00×10^{-13}	2.85	0.985	2.92	339	1.255

5. Summary and Conclusions

This study deals with the determination of the crack initiation fatigue life and the subsequent macroscopic stable crack propagation life during fatigue tests of ultra high-strength steel S1100 butt joints exhibiting noticeable local undercuts. The underlying experiments show a drop of about 15% in fatigue strength if undercuts are present whereas the inverse slope of about k = 4 remains almost unchanged. Furthermore, a large scattering of the fatigue strength with the size and shape of the undercut is observed. However, the decrease is not as pronounced as one would expect based on the evaluated stress concentration factors using finite element analysis. Here, previous work in [47,51] revealed a K_t of about 1.9–2.5 for regions without undercut compared to values for K_t from 3.5–5.4 for local undercuts.

One main part of this work is the introduction of a simple and cost-effective procedure to determine the crack size during specimen fatigue testing. Therefore, a DSLR-camera is utilized to take pictures of the weld surface in specified intervals. In order to ensure comparable situations for each photo, the fatigue test is interrupted and the upper force is applied to the specimen. An implemented digital image correlation procedure compares each picture with the initial state and calculates the consequential distortion field. Excessive distortions are then identified as crack and monitored until specimen burst rupture. The procedure subsequently provides a load cycle versus surface crack length for each crack within the defined area. The related crack depth is additionally calculated using equations from the literature enabling the determination of a crack initiation load cycle number at a specified crack depth.

Neuber's stress averaging method is utilized to calculate the effective notch stress range for each specimen. Therefore, the results of FE-analyses for each specimen based on the measured weld geometry are taken. Hereby, values of the microstructural support length of $\rho^* = 0.1$ mm for technical crack initiation and $\rho^* = 0.13$ mm for burst fracture are obtained for minimum scattering regression analysis. These results show good accordance with the published results for similar steel grades.

The second part deals with the application of the presented crack detection procedure on the fatigue tests. It reveals a comparably high portion of crack initiation phase of more than 50% fatigue life even though only specimens with undercuts are evaluated. Basically, the cycles to crack initiation tend to decrease with the increasing effective notch stress range, but quite high scattering is observed.

Finally, a linear elastic crack propagation calculation is performed. The base material featured parameters for the Paris law by SENT-testing. As expected, these show a significant deviation from the suggestions according to IIW for welded joints. However, the evaluated base material S1100 parameters match published results for similar steel types and grades. The crack propagation assessment is started either at the number of cycles to crack initiation, determined with the presented optical detection procedure, or the assessment is carried out from the start of the fatigue test. Furthermore, three different sets of material parameters are used for the case study. The main conclusion of this investigation is that the crack initiation phase is not neglectable. Otherwise, the assessment tends to deliver very conservative results, especially for welds in the high-cycle fatigue domain above one million load cycles. In detail, the material parameters obtained by the IIW recommendation consistently leads to conservative results. If the crack initiation phase is considered, the derived FAT value is just 6 MPa, or 1% higher than the FAT value of the crack initiation S-N curve. Compared to the fracture FAT value, this corresponds to a deviation of −7%. In the case of crack calculation from the start, the result is far more conservative, with a deviation of about −70%. The usage of the base material parameter set delivers non-conservative results for very high effective notch stresses and conservative results for the high-cycle domain. The derived FAT values show similar deviation to the fatigue test result as the evaluations using the IIW-parameters.

Summing up, the provided methodology gives additional insight into the crack initiation and propagation behaviour of high-quality welded butt joints and can act as the basis to compare other steel grades in regard to quality measures, such as undercuts on fatigue testing and crack growth.

Author Contributions: Conceptualization: M.J.O., M.L., M.S. and W.M.; methodology: M.J.O., M.L. and M.S.; software, M.J.O.; validation: M.J.O., M.L. and M.S.; formal analysis: M.J.O.; investigation: M.J.O.; resources: M.L. and W.M.; data curation: M.J.O.; writing—original draft preparation: M.J.O.; writing—review and editing: M.L. and M.S.; visualization: M.J.O.; supervision: M.L. and M.S.; project administration: M.L. and W.M.; funding acquisition: M.L. and W.M.

Funding: Financial support was given by the Austrian Research Promotion Agency (FFG).

Acknowledgments: Special thanks are given to the Austrian Research Promotion Agency (FFG), who founded the research project by funds of the Austrian Ministry for Transport, Innovation and Technology (bmvit) and the Federal Ministry of Science, Research and Economy (bmwfw).

Conflicts of Interest: The authors declare no conflict of interest.

Nomenclature

a	Crack depth (mm)
a_f	Final crack depth for crack propagation calculation (mm)
a_{init}	Initial crack depth for crack propagation calculation (mm)
a_{th}	Threshold crack depth between crack initiation and propagation (mm)
a/c	Crack aspect ratio (–)
c	Half surface crack length, half width of elliptical crack (mm)
C_P	Crack growth rate coefficient according to Paris law (ΔK in MPa$\sqrt{}$mm; da/dN in mm/cycle)
E	Young's modulus (MPa)
FAT	Fatigue class according to IIW, stress range $\Delta\sigma$ at $N = 2{\cdot}10^6$ load cycles and $P_S = 97.7\%$ (MPa)
K	Stress intensity factor (MPa$\sqrt{}$mm)
K_f	Fatigue notch factor (–)
K_t	Stress concentration factor (–)
k	Inverse slope of S/N-curve (–)
M_a	Weight function parameters for deepest point of a surface crack
$m_a(x,a)$	Weight function for deepest point of surface crack
M_c	Weight function parameters for surface point of a surface crack
$m_c(x,a)$	Weight function for surface point of surface crack
m_p	Slope of crack growth rate curve according to Paris law (–)
N	Load cycle number (–)
N_f	Load cycle number at specimen burst fracture (–)
$N_{initial}$	Load cycle number before start of image acquisition (–)
$N_{interval}$	Interval load cycle number between two image acquisitions (–)
N_{init}	Load cycle number at crack length a_{th}, threshold between crack initiation and propagation (–)
N_{cp}	Load cycle number of crack propagation until burst fracture (–)
N_k	Transition knee point of S/N-curve (–)
P_S	Probability of survival (–)
R	Load stress ratio (–)
s_y	Displacement in y-direction (specimen loading direction) (pixel)
s'_y	Gradient of displacement in y-direction (–)
$s'_{y,\mathrm{thres}}$	Threshold for gradient of displacement in y-direction (–)

SSE	Sum of squared errors (–)
t	Sheet thickness (mm)
$1 : T_\sigma$	Scatter index of S/N-curve, ratio of stress range $\Delta\sigma$ at $P_S = 10\%$ and $P_S = 90\%$ (–)
ρ	Weld toe radius (mm)
ρ^*	Microstructural support length (mm)
θ	Weld flank angle (°)
σ	Stress (MPa)
Recurring indices	
eff	Effective value
n	Nominal value
notch	Notch value
Δ	Range, difference of upper and lower value

References

1. Radaj, D. Review of fatigue strength assessment of nonwelded and welded structures based on local parameters. *Int. J. Fatigue* **1996**, *18*, 153–170. [CrossRef]
2. Lassen, T.; Recho, N. *Fatigue Life Analyses of Welded Structures*; ISTE: London, UK; Newport Beach, CA, USA, 2006.
3. Chattopadhyay, A.; Glinka, G.; El-Zein, M.; Qian, J.; Formas, R. Stress analysis and fatigue of welded structures. *Weld. World* **2011**, *55*, 2–21. [CrossRef]
4. Hou, C.-Y.; Charng, J.-J. Models for the estimation of weldment fatigue crack initiation life. *Int. J. Fatigue* **1997**, *19*, 537–541. [CrossRef]
5. Remes, H. Strain-based Approach to Fatigue Strength Assessment of Laser-welded Joints. Ph.D. Thesis, Helsinki University of Technology, Espoo, Finland, 2008.
6. Lihavainen, V.-M. A Novel Approach for Assessing the Fatigue Strenght of Ultrasonic Impact Treated Welded Structures. Ph.D. Thesis, Lappeenranta University of Technology, Lappeenranta, Finland, 2006.
7. Peterson, R.E. Notch sensitivity. In *Metal Fatigue*; Sines, G., Waisman, J.L., Dolan, T.J., Eds.; McGraw-Hill: New York, NY, USA, 1959; pp. 293–306.
8. Neuber, H. Über die Berücksichtigung der Spannungskonzentration bei Festigkeitsberechnungen. *Konstruktion* **1968**, *20*, 245–251.
9. Radaj, D. *Design and Analysis of Fatigue Resistant Welded Structures*; Abington: Cambridge, UK, 1990.
10. Lawrence, F.V.; Ho, N.J.; Mazumdar, P.K. Predicting the fatigue resistance of welds. *Ann. Rev. Mater. Sci.* **1981**, *11*, 401–425. [CrossRef]
11. Seeger, T. *Grundlagen für Betriebsfestigkeitsnachweise. Stahlbau-Handbuch*, 3rd ed.; Stahlbau-Verl.-Ges: Köln, Germany, 1996; pp. 5–123.
12. Atzori, B.; Lazzarin, P. Notch sensitivity and defect sensitivity under fatigue loading: Two sides of the same medal. *Int. J. Fract.* **2001**, *107*, 1–8. [CrossRef]
13. Glinka, G. Energy density approach to calculation of inelastic strain-stress near notches and cracks. *Eng. Fract. Mech.* **1985**, *22*, 485–508. [CrossRef]
14. Haibach, E. *Betriebsfestigkeit. Verfahren und Daten zur Bauteilberechnung*; Springer: Berlin, Germany, 2006.
15. Krebs, J.; Hübner, P.; Kassner, M. *Eigenspannungseinfluss auf Schwingfestigkeit und Bewertung in geschweißten Bauteilen*; DVS-Verlag: Düsseldorf, Germany, 2004.
16. Lieurade, H.-P.; Huther, I.; Maddox, S.J. *Recommendations on the Fatigue Testing of Welded Components*; LETS Global: Rotterdam, The Netherlands, 2006.
17. Stoschka, M.; Leitner, M.; Fössl, T.; Posch, G. Effect of high-strength filler metals on fatigue. *Weld. World* **2012**, *56*, 20–29. [CrossRef]
18. Maddox, S.J. The effect of mean stress on fatigue crack propagation a literature review. *Int. J. Fract.* **1975**, *11*, 389–408.
19. Verreman, Y.; Nie, B. Short-crack growth and coalescence along the toe of a manual fillet weld. *Fatigue Frac. Eng. Mat. Struct.* **1991**, *14*, 337–349. [CrossRef]
20. Fricke, W. Fatigue analysis of welded joints: State of development. *Mar. Struct.* **2003**, *16*, 185–200. [CrossRef]
21. Baumgartner, J.; Bruder, T. Influence of weld geometry and residual stresses on the fatigue strength of longitudinal stiffeners. *Weld. World* **2013**, *57*, 841–855. [CrossRef]

22. Leitner, M.; Barsoum, Z.; Schäfers, F. Crack propagation analysis and rehabilitation by HFMI of pre-fatigued welded structures. *Weld. World* **2016**, *60*, 581–592. [CrossRef]
23. Baumgartner, J.; Waterkotte, R. Crack initiation and propagation analysis at welds—Assessing the total fatigue life of complex structures. *Mat. Wiss. Werkstofftech.* **2015**, *46*, 123–135. [CrossRef]
24. Simunek, D.; Leitner, M.; Grün, F. In-situ crack propagation measurement of high-strength steels including overload effects. *Proc. Eng.* **2018**, *213*, 335–345. [CrossRef]
25. Simunek, D.; Leitner, M.; Maierhofer, J.; Gänser, H.-P. Crack growth under constant amplitude loading and overload effects in 1:3 scale specimens. *Proc. Struct. Integr.* **2017**, *4*, 27–34. [CrossRef]
26. Todorov, E.I.; Mohr, W.C.; Lozev, M.G.; Thompson, D.O.; Chimenti, D.E. Detection and Sizing of Fatigue Cracks in Steel Welds with Advanced Eddycurrent techniques. In Proceedings of the AIP Conference—34th Annual Review of Progress in Quantitative Nondestructive Evaluation, Golden, CO, USA, 22–27 July 2007; pp. 1058–1065.
27. Lamtenzan, D.; Glenn, W.; Lozev, M.G. *Detection and Sizing of Cracks in Structural Steel Using the Eddy Current Method*; US Department of Transportation Federal Highway Administration FHWA-RD-00-018; Turner-Fairbank Highway Research Center: McLean, VA, USA, 2000.
28. Roux, S.; Réthoré, J.; Hild, F. Digital image correlation and fracture: An advanced technique for estimating stress intensity factors of 2D and 3D cracks. *J. Phys. D Appl. Phys.* **2009**, *42*, 214004. [CrossRef]
29. Mathieu, F.; Hild, F.; Roux, S. Identification of a crack propagation law by digital image correlation. *Int. J. Fatigue* **2012**, *36*, 146–154. [CrossRef]
30. Ozelo, R.R.M.; Sollero, P.; Sato, M.; Barros, R.S.V. Monitoring crack propagation using digital image correlation and cod technique. In Proceedings of the COBEM 2009 20th International Congress of Mechanical Engineering, ABCM, Gramado, Brazil, 15–20 November 2009.
31. Alam, M.M.; Barsoum, Z.; Jonsén, P.; Kaplan, A.F.H.; Häggblad, H.Å. The influence of surface geometry and topography on the fatigue cracking behaviour of laser hybrid welded eccentric fillet joints. *Appl. Surf. Sci.* **2010**, *256*, 1936–1945. [CrossRef]
32. Caccese, V.; Blomquist, P.A.; Berube, K.A.; Webber, S.R.; Orozco, N.J. Effect of weld geometric profile on fatigue life of cruciform welds made by laser/GMAW processes. *Mar. Struct.* **2006**, *19*, 1–22. [CrossRef]
33. Lillemäe, I.; Remes, H.; Liinalampi, S.; Itävuo, A. Influence of weld quality on the fatigue strength of thin normal and high strength steel butt joints. *Weld. World* **2016**, *60*, 731–740. [CrossRef]
34. Rennert, R.; Kullig, E.; Vormwald, M.; Esderts, A.; Siegele, D. *Analytical Strength Assessment of Components Made of Steel, Cast Iron and Aluminium Materials in Mechanical Engineering: FKM Guideline*, 6th ed.; VDMA-Verl: Frankfurt, Germany, 2012.
35. Olivier, R.; Köttgen, V.B.; Seeger, T. *Welded Joints I—Fatigue Strength Assessment Method for Welded Joints Based on Local Stresses*; Forschungshefte No. 143; Forschungskuratorium Maschinenbau: Frankfurt, Germany, 1989.
36. Olivier, R.; Köttgen, V.B.; Seeger, T. *Welded Joints II—Investigation of Inclusion into Codes of a Novel Fatigue Strength Assessment Method for Welded Joints in Steel*; No. 180; Forschungskuratorium Maschinenbau: Frankfurt, Germany, 1994.
37. Morgenstern, C.; Sonsino, C.M.; Hobbacher, A.; Sorbo, F. Fatigue design of aluminium welded joints by the local stress concept with the fictitious notch radius of rf = 1 mm. *Int. J. Fatigue* **2006**, *28*, 881–890. [CrossRef]
38. Hobbacher, A. *Recommendations for Fatigue Design of Welded Joints and Components*, 2nd ed.; Springer: New York, NY, USA, 2016.
39. Nykänen, T.; Björk, T.; Laitinen, R. Fatigue strength prediction of ultra high strength steel butt-welded joints. *Fatigue Frac. Eng. Mater. Struct.* **2013**, *36*, 469–482. [CrossRef]
40. Liinalampi, S.; Remes, H.; Lehto, P.; Lillemäe, I.; Romanoff, J.; Porter, D. Fatigue strength analysis of laser-hybrid welds in thin plate considering weld geometry in microscale. *Int. J. Fatigue* **2016**, *87*, 143–152. [CrossRef]
41. Nykänen, T.; Marquis, G.; Björk, T. Effect of weld geometry on the fatigue strength of fillet welded cruciform joints. In *Proceedings of the International Symposium on Integrated Design and Manufacturing of Welded Structures*; Lappeenranta University of Technology: Lappeenranta, Finland, 2007.
42. Schork, B.; Kucharczyk, P.; Madia, M.; Zerbst, U.; Hensel, J.; Bernhard, J.; Tchuindjang, D.; Kaffenberger, M.; Oechsner, M.; Zerbst, U. The effect of the local and global weld geometry as well as material defects on crack initiation and fatigue strength. *Eng. Fract. Mech.* **2018**, *198*, 103–122. [CrossRef]

43. Madia, M.; Zerbst, U.; Th. Beier, H.; Schork, B. The IBESS model—Elements, realisation and validation. *Eng. Fract. Mech.* **2018**, *198*, 171–208. [CrossRef]
44. Mashiri, F.R.; Zhao, X.L.; Grundy, P. Effects of weld profile and undercut on fatigue crack propagation life of thin-walled cruciform joint. *Thin Walled Struct.* **2001**, *39*, 261–285. [CrossRef]
45. Steimbreger, C.; Chapetti, M.D. Fatigue strength assessment of butt-welded joints with undercuts. *Int. J. Fatigue* **2017**, *105*, 296–304. [CrossRef]
46. Steimbreger, C.; Chapetti, M.; Hénaff, G. Undercut tolerances in industry from a fracture mechanic perspective. *MATEC Web Conf.* **2018**, *165*, 21009. [CrossRef]
47. Ottersböck, M.J.; Leitner, M.; Stoschka, M.; Maurer, W. Effect of Weld Defects on the Fatigue Strength of Ultra High-strength Steels. *Proc. Eng.* **2016**, *160*, 214–222. [CrossRef]
48. ASTM. *Practice for Statistical Analysis of Linear or Linearized Stress-Life (S-N) and Strain-Life (-N) Fatigue Data*; E739; ASTM International: West Conshohocken, PA, USA, 2015.
49. Dengel, D.; Harig, H. Estimation of the fatigue limit by progressively-increasing load tests. *Fatigue Frac. Eng. Mat. Struct.* **1980**, *3*, 113–128. [CrossRef]
50. Sonsino, C.M. Course of SN-curves especially in the high-cycle fatigue regime with regard to component design and safety. *Int. J. Fatigue* **2007**, *29*, 2246–2258. [CrossRef]
51. Ottersböck, M.J.; Leitner, M.; Stoschka, M. Characterisation of actual weld geometry and stress concentration of butt welds exhibiting local undercuts. *Eng. Fail. Anal.* **2019**. in review.
52. ASTM. *Standard Test Method for Measurement of Fatigue Crack Growth Rates*; E647; ASTM International: West Conshohocken, PA, USA, 2000.
53. Tada, H.; Paris, P.C.; Irwin, G.R. *The Stress Analysis of Cracks Handbook*, 3rd ed.; ASME: New York, NY, USA, 2000.
54. Beden, S.M.; Abdullah, S.; Ariffin, A.K. Review of fatigue crack propagation models for metallic components. *Eur. J. Sci. Res.* **2009**, *28*, 364–397.
55. Richard, H.A.; Sander, M. Ermüdungsrisse. In *Erkennen, Sicher Beurteilen, Vermeiden*, 3rd ed.; Springer Vieweg: Wiesbaden, Germany, 2012.
56. Paris, P.; Erdogan, F. A critical analysis of crack propagation laws. *J. Basic Eng.* **1963**, *85*, 528. [CrossRef]
57. Jones, E.M.C. *Improved Digital Image Correlation*; Version 4; University of Illinois: Champaign, IL, USA, 2015.
58. Reu, P.L.; Toussaint, E.; Jones, E.M.C.; Bruck, H.A.; Iadicola, M.; Balcaen, R.; Turner, D.Z.; Siebert, T.; Lava, P.; Simonsen, M. DIC Challenge: Developing images and guidelines for evaluating accuracy and resolution of 2D analyses. *Exp. Mech.* **2018**, *58*, 1067–1099. [CrossRef]
59. Jones, E.M.C. *Documentation for Matlab-Based DIC Code*; Version 4; University of Illinois: Champaign, IL, USA, 2015.
60. Ottersböck, M.J. Einfluss von Imperfektionen auf die Schwingfestigkeit hochfester Stahlschweißverbindungen. Ph.D. Thesis, Montanuniversität Leoben, Leoben, Austria, 2019.
61. Simunek, D.; Leitner, M.; Maierhofer, J.; Gänser, H.-P. Fatigue crack growth under constant and variable amplitude loading at semi-elliptical and V-notched steel specimens. *Proc. Eng.* **2015**, *133*, 348–361. [CrossRef]
62. Engesvik, K.M. Analysis of Uncertainties in the Fatigue Capacity of Welded Joints. Ph.D. Thesis, University of Trondheim, Trondheim, Norway, 1981.
63. Baumgartner, J. Enhancement of the fatigue strength assessment of welded components by consideration of mean and residual stresses in the crack initiation and propagation phases. *Weld. World* **2016**, *60*, 547–558. [CrossRef]
64. Radaj, D.; Sonsino, C.M.; Fricke, W. *Fatigue Assessment of Welded Joints by Local Approaches*, 2nd ed.; Woodhead Publishing: Sawston, UK, 2006.
65. Goyal, R.; Glinka, G. Fracture mechanics-based estimation of fatigue lives of welded joints. *Weld. World* **2013**, *57*, 625–634. [CrossRef]
66. Newman, J.C.; Raju, I.S. Stress-intensity factor equations for cracks in three-dimensional finite bodies, ASTM STP 791. In *Fracture Mechanics: Fourteenth Symposium—Volume I: Theory and Analysis*; Lewis, J.C., Sines, G., Eds.; ASTM International: West Conshohocken, PA, USA, 1983; pp. I238–I265.
67. Murakami, Y.; Murakami, Y. *Stress Intensity Factors Handbook*; Pergamon Press: Oxford, UK, 1990.
68. Bueckner, H.F. Novel principle for the computation of stress intensity factors. *J. Appl. Math. Mech.* **1970**, *50*.
69. Shen, G.; Glinka, G. Weight functions for a surface semi-elliptical crack in a finite thickness plate. *Theor. Appl. Fract. Mech.* **1991**, *15*, 247–255. [CrossRef]

70. Glinka, G.; Shen, G. Universal features of weight functions for cracks in mode I. *Eng. Fract. Mech.* **1991**, *40*, 1135–1146. [CrossRef]
71. Wang, X.; Lambert, S.B. Stress intensity factors for low aspect ratio semi-elliptical surface cracks in finite-thickness plates subjected to nonuniform stresses. *Eng. Fract. Mech.* **1995**, *51*, 517–532. [CrossRef]
72. Hall, M.S.; Topp, D.A.; Dover, W.D. *Parametric Equations for Stress Intensity Factors in Weldments*; Project Report TSC/MSH/0244; Technical Software Consultants Ltd.: Milton Keynes, UK, 1990.
73. Monahan, C.C. *Early Fatigue Crack Growth at Welds*; Computational Mechanics: Southampton, UK, 1995.

applied sciences

MDPI

Article

Fracture Toughness of Hybrid Components with Selective Laser Melting 18Ni300 Steel Parts

Luis M. S. Santos [1], Joel de Jesus [1,*], José M. Ferreira [1], José D. Costa [1] and Carlos Capela [1,2]

[1] CEMMPRE, Department of Mechanical Engineering, Rua Luís Reis Santos, Pinhal de Marrocos, 3030788 Coimbra, Portugal; luis_lms@sapo.pt (L.M.S.S.); martins.ferreira@dem.uc.pt (J.M.F.); jose.domingos@dem.uc.pt (J.D.C.); ccapela@ipleiria.pt (C.C.)

[2] Instituto Politécnico de Leiria, ESTG, Department of Mechanical Engineering, Morro do Lena–Alto Vieiro, 2400-901 Leiria, Portugal

* Correspondence: joel.jesus@uc.pt; Tel.: +351-239-790-736; Fax: +351-239-790-701

Received: 27 July 2018; Accepted: 2 October 2018; Published: 11 October 2018

check for updates

Abstract: Selective Laser Melting (SLM) is currently one of the more advanced manufacturing and prototyping processes, allowing the 3D-printing of complex parts through the layer-by-layer deposition of powder materials melted by laser. This work concerns the study of the fracture toughness of maraging AISI 18Ni300 steel implants by SLM built over two different conventional steels, AISI H13 and AISI 420, ranging the scan rate between 200 mm/s and 400 mm/s. The SLM process creates an interface zone between the conventional steel and the laser melted implant in the final form of compact tension (CT) samples, where the hardness is higher than the 3D-printed material but lower than the conventional steel. Both fully 3D-printed series and 3D-printed implants series produced at 200 mm/s of scan rate showed higher fracture toughness than the other series built at 400 mm/s of scan rate due to a lower level of internal defects. An inexpressive variation of fracture toughness was observed between the implanted series with the same parameters. The crack growth path for all samples occurred in the limit of interface/3D-printed material zone and occurred between laser melted layers.

Keywords: selective laser melting (SLM); maraging steel 300; fracture toughness; hybrid parts; implanted parts

1. Introduction

Selective Laser Melting (SLM) refers to layer material addition techniques that allow the generation of complex 3D-printed parts by selectively joining successive layers of powder material on top of each other, using thermal energy supplied by a focused, computer-controlled laser beam. This technology is a very high-energy process, as each layer of metal powder must be heated above the melting point of the metal [1,2]. Moreover, this additive manufacturing process can provide for complex geometry components, repair defects (as an alternative to laser welding [3]) and reconstruction in the moulds industry. The SLM process is largely affected by parameters such as laser power, scan speed rate, protective atmosphere, material used in the process and materials to joint, among others. The wrong choice of these parameters can lead to a high porosity, bad adhesion between layers and a materials interface which can decrease its mechanical behavior.

The literature reports an abundance of recent research about the morphologic and mechanical properties of SLM steels. Kempen et al. [4] studied the microstructure and mechanical properties of SLM 18Ni-300 steel and observed that a higher layer thickness and/or scan speed causes a decrease in density. A cellular-dendritic growth mechanism in the solidification was observed where intercellular spacing is less than 1 μm which leads to an excellent strength and hardness in both as-produced and age hardened conditions.

Casalino et al. [5] realized an experimental investigation and statistical optimization of the SLM process in 18Ni300 maraging steel, where the correlation between the mechanical and density was positive, and the surface roughness decreases with the energy density. Bai et al. [6] studied the influence of parameters on SLM process and mechanical properties evolution of maraging steel 300. Like previous authors, they prove the considerable influence of the scan rate, laser power and scan spacing as regards the density of a part. Low power laser and high scan rate lead to a low density produced part and high power laser and slow scan rate can achieve 99% of density in a produced part. Casati et al. [7] also obtained 99.9% of relative density in cubic samples produced by SLM using gas atomized H11 grade steel powder.

The presence of defects and microstructural heterogeneities creates uncertainty in mechanical properties of SLM produced parts, which complexify their generalized use [8]. Moreover, high SLM mechanical properties are required for structural materials applications. As is often reported in the literature, SLM process affects the material properties due to residual stresses, defects or inherent roughness of the surface [5,9–11].

The authors of Reference [12] studied the mechanical behavior of selective laser melting 18Ni300 steel specimens, concluding that a very high scan rate (400 or 600 mm/s) causes a high percentage of porosity and a consequent drastic reduction of the tensile strength, stiffness and elongation at failure, compared with the results obtained for a 200 mm/s scan rate. The fatigue crack propagation path occurred predominantly in a transgranular direction, crossing the laser layer.

The growing direction and grain aspect ratio can lead to anisotropic behavior of SLM parts. The loading direction, in respect to layer deposition plane and grain aspect ratio, plays an important role in its mechanical properties. The effect of building orientation on tensile properties and fatigue crack growth was studied by Suryawanshi et al. [13] and observed a lack of anisotropy for selective laser melting 18Ni300 steel. Strong intra-layer bonding and the absence of dominant texture were the reasons ascribed to this behavior. The same authors concluded that mechanical characteristics of aged SLM are broadly similar to those reported for the aged maraging steel of the same grade.

On the other hand, Mazur et al. [14] verified that the tensile and fatigue behavior in mechanical tests of SLM manufactured H13 specimens compared with the results to those conventionally produced showed tensile strength of as-built specimens to be up to 30% lower than conventional material. Fatigue strength was also significantly lower than reference data. These results were related to the high roughness of the 3D-printed surface and the presence of high tensile residual stresses. Ackermann et al. [15] realized impact tests in a three different conditions: 3D-printed, 3D-printed heat-treated and conventional heat-treated specimens in H13 tool steel. These authors observed that the H13 steel is so brittle and that it substantially limits direct use of the printed part. Proper heat treatment must be carried out after SLM process in order to obtain desired properties.

Studies regarding hybrid parts are not common. Cyr et al. [16] studied the fracture behavior of additively manufactured 18Ni300/H13 hybrid hard steels. These authors observed a presence of an interface between the different steels. On the other hand, tensile tests of hybrid as-printed sample show that the fracture occurs in the H13 steel, away from the interface. The authors of this research [17] also studied the morphologic and fatigue strength of additively manufactured 18Ni300/H13 and 18Ni300/AISI420 hybrid samples. The presence of an interface between the different steels was also observed. The fatigue strength has a lower influence due to the substrate (H13 or AISI420) and the hybrid samples show a lower fatigue life when compared with the as-printed samples. Moreover, the fatigue crack initiation, propagation and final fracture were localized in the as-printed material, being different of the Cyr et al. [16] results because these authors used an untreated H13 steel in the as-printed samples.

Campanelli et al. [18] studied 18Ni300 marage steel clads deposited onto an AISI 304 substrate. These authors observed that the dilution with the substrate is as low as 10 to 28%, which warrants the adhesion of marage to AISI 304 steel, and they developed a statistical analysis and mathematical model that can be used to select the process parameters to produce high-density parts, reaching a average density value of 98.2%.

No publications about the fracture toughness of conventional steels with 18Ni300 steel implants built by SLM were found in the literature, justifying the importance of the current investigation in order to understand the influence on material fracture of the geometric shape, the presence of defects and load mode. The results of this study can lead to a better understanding of mechanical design and reliability of mechanical projects of conventional steels with 18Ni300 steel implants built by SLM. Therefore, the main objective of this work is to study the fracture toughness of both fully 3D-printed specimens and hybrid specimens composed of a conventional steel substrate and a steel implant, both 3D-printed by SLM.

2. Material and Methods

2.1. Materials and Specimens Manufacturing

The specimens used in this research were the Compact Tension type (CT), manufactured with SLM parts growing in load direction. Figure 1 shows the geometry and main dimensions of the specimens and a schematic view of the hybrid samples.

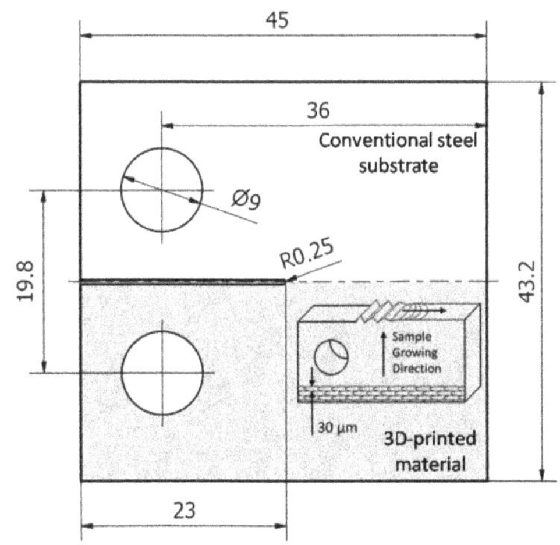

Figure 1. Specimen geometry, dimensions and schematic view of the hybrid parts (dimensions in mm).

Three different kind of specimens were manufactured for this study: fully 3D-printed sample using recycled powder of AISI 18Ni300 steel (SM) of which the average particle diameter ranges between 0.01–0.056 mm and two hybrid samples, for which half specimen was a conventional machined steel and the remaining was a steel implant built by SLM using the maraging AISI 18Ni300 previously referred. The two steel substrates were the tempered AISI H13 steel (HS/SM) and the stainless steel AISI 420, both machined by conventional processes (SS/SM). Specimens with 6 and 3 mm thickness were produced with notches open by wire EDM (Electrical discharge machining) centered in the interface zone. Table 1 presents the main chemical composition according to the manufacturer, for the three steels. Table 2 presents tensile test results for SS/SM, HS/SM and SM samples, obtained by the authors [17] using round specimens with 4 mm of diameter according to ASTM E8/E8M-16a [19].

Table 1. Chemical composition of steels (CM-Conventional Manufacturing). Reproduced with the permission from [17], copyright Elsevier, 2016.

Steel	C	Ni	Co	V	Mo	Ti	Al	Cr	P	Si	Mn	Fe
18Ni300 (powder)	0.01	18.2	9.0	-	5.0	0.6	0.05	0.3	0.01	0.1	0.04	Balance
AISI H13 (CM)	0.4	-	-	0.94	1.30	-	-	5.29	0.017	1.05	0.036	Balance
AISI 420 (CM)	0.37	-	-	0.17	-	-	-	14.22	0.021	0.64	0.037	Balance

Table 2. Tensile properties for SS/SM, HS/SM and SM specimens. Reproduced with the permission from [17], copyright Elsevier, 2016.

Specimen	SM		SS/SM		HS/SM	
Scan rate (mm/s)	200	400	200	400	200	400
Ultimate Tensile Stress [MPa]	1147	1032	1135	1000	1146	1000
Yield stress [MPa]	1000	910	1115	900	1125	900
Extension at break [%]	6	2.2	3.6	1.5	4.2	1.7
Young's modulus [GPa]	180	180	190	180	190	180

The 3D-printing was done by Lasercusing® process, in a "Concept Laser" equipment model "M3 linear" using as an energy source a laser type Nd:YAG with a maximum power of 100 W in continuous wave mode, a wavelength of 1064 nm and 0.2 mm of spot laser diameter. The samples were manufactured using two scan rates, 200 and 400 mm/s, adding layers of 30 μm thickness with hatch spacing of 100 μm and 25% overlapping, growing towards the direction corresponding to the application of load in the mechanical tests. After manufacturing the specimens were mechanically polished.

2.2. Fracture Toughness Tests

The fracture toughness tests were performed according to the BS 7448-3 standard [20] using CT specimens in loading Mode I. Before the fracture toughness tests a fatigue pre-crack was created at room temperature in an Instron Electroplus machine under constant amplitude tensile loading with sinusoidal load wave and stress ratio R = 0.05. In accordance with this standard, a pre-crack of 4 mm length was created from the notch, in order to comply with the two standard requirements: the pre-crack has to be higher than 2.5% of the W parameter and greater than 1.3 mm from the notch. The crack length was measured using a travelling microscope (45× of magnification) with an accuracy of 10 μm, as shown in Figure 2.

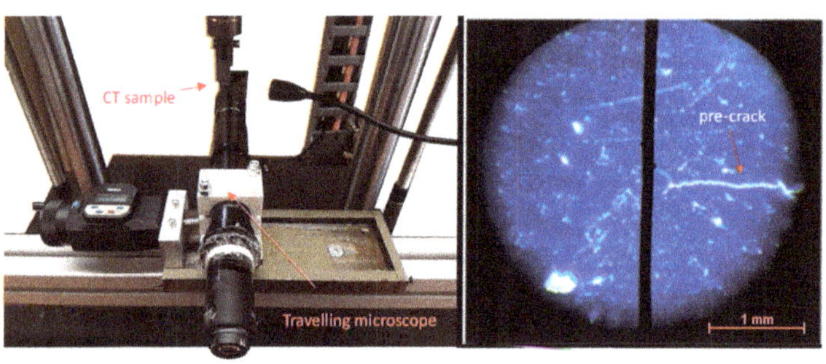

Figure 2. Pre-crack length measurement apparatus.

The fracture toughness tests were carried out in an Instron machine, model 4206, using a displacement rate of 2 mm/min. The crack opening displacement was measured using an MTS clip gauge model 632.02C-221 (MTS Systems, MN, USA) (range: −2.5 mm to 2.5 mm) fixed with an apparatus especially created for this purpose. Figure 3 shows the CT sample and extensometer assembly.

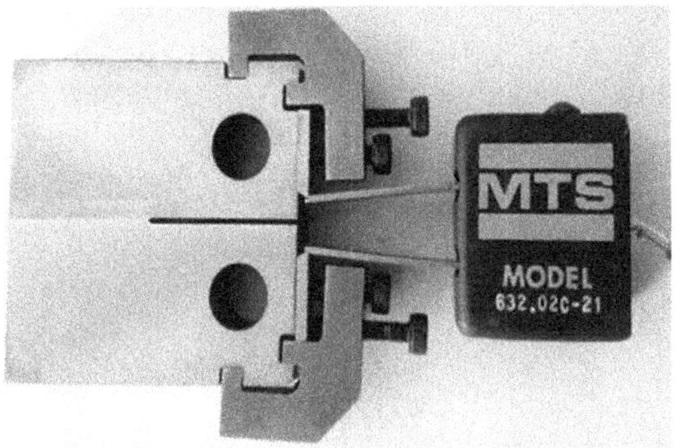

Figure 3. Sample-extensometer assembly.

The fracture toughness K_{Ic}, and the critical integral, J_c, were calculated according the BS 7448-3 [20] standard through the Equations (1)–(4), respectively. Equation (5) was used for the validation of the critical stress intensity factor values in Mode I loading (K_{Ic}) by the criteria of excessive plastic deformation at the crack tip.

$$K_Q = \frac{F_Q}{BW^{0.5}} \times f'\left(\frac{a_0}{W}\right) \quad (1)$$

$$f'\left(\frac{a_0}{W}\right) = \frac{\left(2+\frac{a_0}{W}\right)\left[0.886+4.64\frac{a_0}{W}-13.32\frac{a_0^2}{W^2}+14.72\frac{a_0^3}{W^3}-5.6\frac{a_0^4}{W^4}\right]}{\left(1-\frac{a_0}{W}\right)^{1.5}} \quad (2)$$

$$J = \left[\frac{F}{BW^{0.5}} \times f'\left(\frac{a_0}{W}\right)\right]^2 \frac{\left(1-\vartheta^2\right)}{E} + \frac{\eta_p U_p}{B(W-a_0)} \quad (3)$$

$$\eta_p = 2 + 0.522\left(1-\frac{a_0}{W}\right) \quad (4)$$

$$2.5\left(\frac{K_Q}{\sigma_{YS}}\right)^2 \geq (W-a_0) \quad (5)$$

where a_0 = 14 mm is the crack length, W the distance indicated in Figure 1, B is the sample thick, F_Q is a particular load value determined from the fracture test chart, U_p correspond to the plastic component of the area under load vs. elongation curve, E is the Young modulus, ν the Poisson coefficient and F is the maximum load.

Complementary analyses were performed to characterize the microstructure and the fracture mechanisms. The metallographic and crack morphologic analysis before and after the fracture tests were performed according to the ASTM E3 standard metallographic practice [21], being used for etching a solution composed of 4 g of picric acid, 1 mL of hydrochloric acid (HCL) and 95 mL ethyl alcohol. Afterwards, the samples were observed through an optical microscope Leica DM4000 M LED. The hardness tests were performed using a Struers Type Duramin-1 microhardness tester, with a load of 10 N during 15 s, according to the ASTM E 384 standard [22]. The local chemical compositions were obtained through the energy dispersive X-ray spectroscopy analysis. Finally, fractured surfaces of the broken surfaces and the interfaces of the specimens were analyzed in a scanning electron microscope (SEM) Philips® XL30 TMP.

3. Results and Discussion

3.1. Metallographic Analysis

Figure 4 shows the main microstructures in different regions for an SS/SM sample. Figure 4a shows the microstructure for the AISI 420 steel being a ferritic structure with M_{23} C_6 spherical carbides [23]. The interface between 3D-printed and conventional steel parts can be observed in Figure 4b,c (marked by an arrow) showing decarbonisation, and a thickness between 40–50 μm. Figure 4d reveals a martensitic microstructure for the 3D-printed half [24].

Similar analysis was carried out for HS/SM samples, as shown in Figure 5. The AISI H13 shows a tempered martensitic structure [25], as can be seen in Figure 5a. In this sample, the interface is not clearly identified, but is possible to note a different appearance between the 3D-printed material near the AISI H13 steel and the 3D-printed material distant from the interface (both zones are marked by arrows in Figure 5b). This variation in appearance is due to some decarbonisation, as observed in the previous sample.

The analysis by EDS (Energy-dispersive X-ray spectroscopy) of the chemical composition in the interface reveals that this is identical for both kinds of samples. Figure 6 presents an exemplary plot of the chemical composition for the interface of hybrid sample HS/SM 3D-printed with 200 mm/s of scan rate, where the Ni choice can be identified, which is characteristic for the 18Ni300 3D-printed steel, and other common choices for steels, namely C and Fe, with low carbon contents.

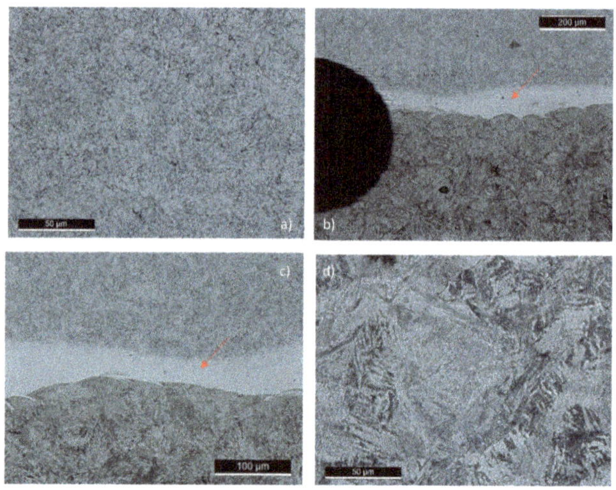

Figure 4. Metallographic analysis for SS/SM sample.

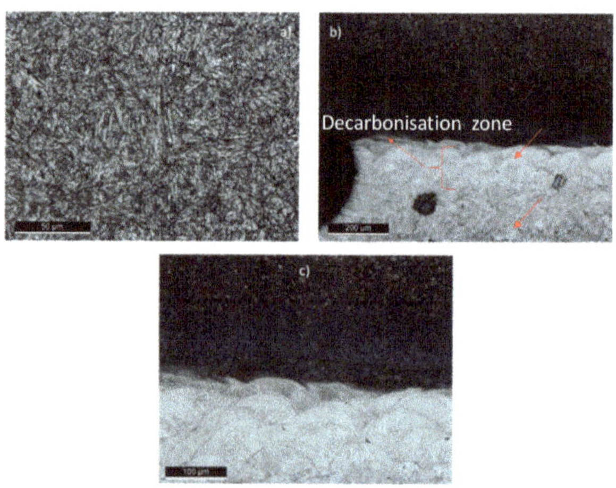

Figure 5. Metallographic analysis for HS/SM sample.

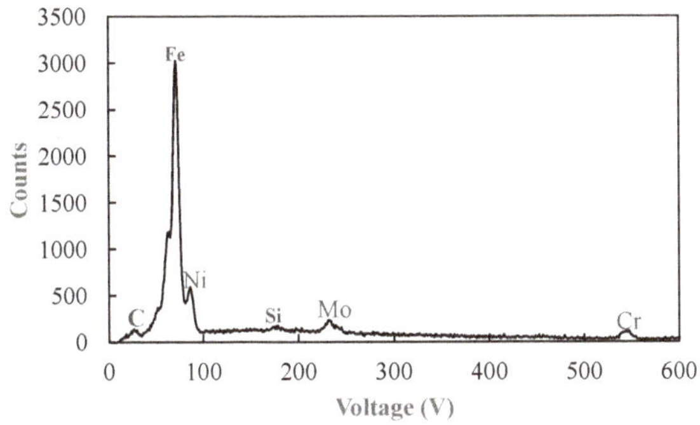

Figure 6. Chemical composition of the interface for hybrid sample HS/SM 3D-printed with 200 mm/s scan rate.

3.2. Microhardness

Figure 7 shows the hardness profiles for the three series analyzed. Similar hardness values were obtained for the 3D-printed half of about 330 HV_1. The average hardness value of ten measurements in the interface of about 450 HV_1 is quite similar for both samples, being 452 HV_1 for SS/SM and 446 HV_1 for HS/SM series. This is an expected result since the chemical composition and cooling rate during solidification in the interface of both series is similar. Finally, the AISI 420 steel presents a higher hardness than AISI H13, 532 HV_1 and 506 HV_1, respectively.

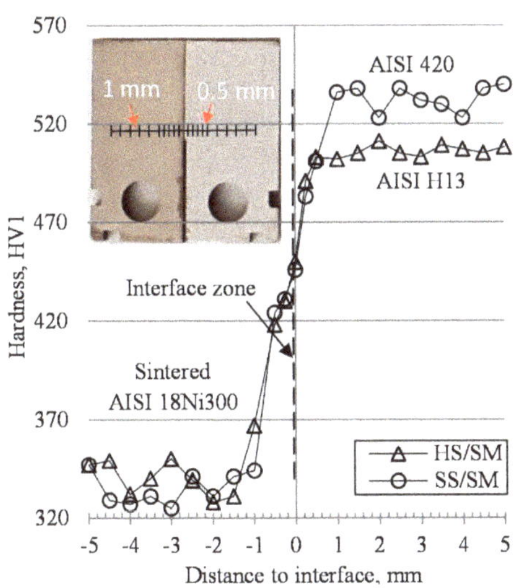

Figure 7. Hardness profile for both hybrid sample series.

3.3. Fracture Tests: Load vs. Displacement Curves

Figure 8 shows exemplary load-displacement curves of fracture tests on fully 3D-printed specimens. The identification of the curves is made according to the following code: the first number correspond to scan rate (2 = 200 mm/s; 4 = 400 mm/s), the letters identify the material composition of the sample, the subsequent two digits identifies the number of the specimen and the last digit is the thickness in mm.

The analysis of Figure 8 shows, as expected, that the maximum loads are reached for the samples with the greatest thickness (6 mm). On the other hand, 3D-printed specimens with a scan speed of

400 mm/s have both lower maximum load and lower maximum displacement than the 3D-printed specimens produced with a scan speed of 200 mm/s, which can be explained by the increase in internal defects (porosities, unmelted powder, unmelted zones, among others) and a consequent decrease of strength and ductility as was referred by References [11,12]. In addition, as is known, the thickness has an important role in the material behavior. The lower stiffness associated with the reduction of the thickness of the sample approximates the tests to a plane stress condition leading to a greater plastic deformation, which may invalidate the determination of the fracture toughness in Mode I (K_{Ic}). Therefore, for the 3 mm thick specimens it is necessary to use the critical value of the J Integral parameter (J_{Ic}).

Figure 9 shows some of the curves obtained for HS/SM and SS/SM hybrids specimens, all processed with a scan rate of 200 mm/s. From the analysis of the figure stands out a behavior similar to the one observed in the previous figure, since the fracture occurs predominantly on the 3D-printed side of the samples. However, in this case, the stiffness of HS/SM and SS/SM hybrids specimens with 6 mm thickness are very similar, leading to the conclusion that the fracture is mainly controlled by the 3D-printed part of the specimen. On the other hand, there is a greater dispersion of the maximum loads for the HS/SM hybrids specimens. The interface of this specimens presents greater microstructural heterogeneities between the 3D-printed material next to the interface with the AISI H13 steel and the 3D-printed material distant from the interface (both zones are marked by arrows in Figure 5b). As referred in the metallographic analysis (Section 3.1), this variation of microstructure is due to some decarbonisation.

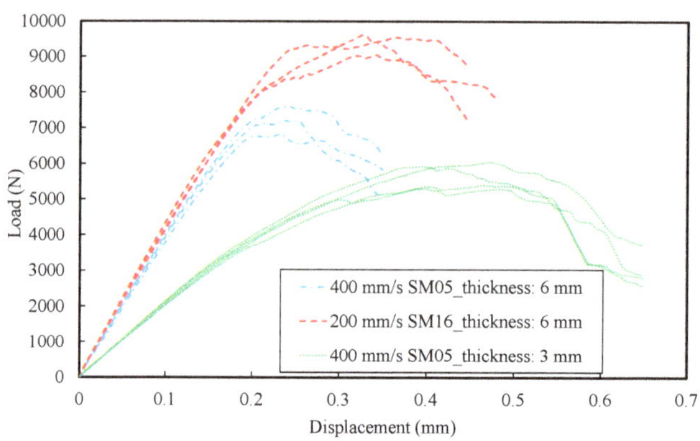

Figure 8. Fracture tests: Load vs. displacement curves for fully 3D-printed specimens.

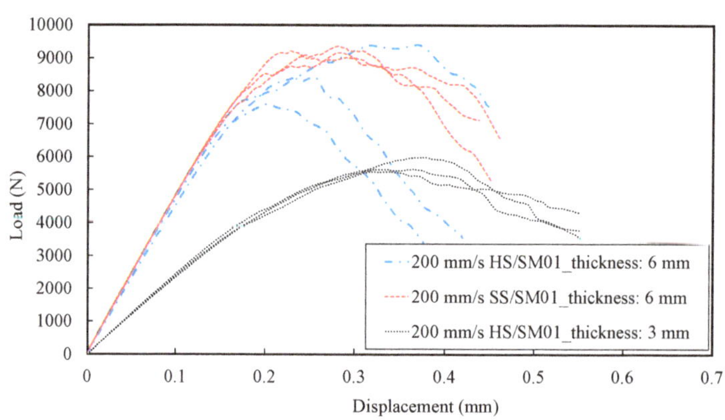

Figure 9. Fracture tests: Load vs. displacement curves for hybrid specimens.

3.4. Fracture Toughness

Tables 3–5 present the fracture toughness K_Q, the validity condition for $K_Q = K_{Ic}$ and J_c, for entirely 3D-printed (SM) samples, and hybrid SS/SM and HS/SM samples, for the two different scan rates (200 and 400 mm/s) and the two thickness values (3 and 6 mm). The results show that the majority of tests lead to invalid fracture toughness tests, because the geometry of samples do not guarantee a plain strain state, leading to an excessive plastic deformation at the crack tip. However, K_Q increases at least 10% when thickness decreases from 6 to 3 mm and the increase of the scan speed from 200 to 400 mm/s reduces K_Q above 20%. Hybrid samples show a slightly lower toughness than fully 3D-printed specimens, reaching almost 10% for HS/SM specimens, which is caused by microstructural variations at interface. The values of fracture toughness obtained in this work for the samples SM, SS/SM and HS/SM 3D-printed at 200 mm/s scan speed rate and 3 mm of thickness, are similar to those K_Q values obtained for the steel H13 by J.M. Costa et al. [3] in laser deposit welding Nd-YAG with 4 mm thick specimens ($K_Q = 79.08 \pm 31.76$ MPa.m$^{1/2}$).

The effects of thickness, substrate material and scan speed on the J_c parameter is similar to that observed for K_Q, but with more pronounced trends. For the scan speed of 200 mm/s, Jc on 3 mm thickness samples decreases about 15% and 30% for SS/SM and HS/SM samples, respectively, when compared with entirely 3D-printed specimens. The same comparison on J_c for 6 mm thickness leads to toughness reductions of about 7% and 45% for SS/SM and HS/SM samples, respectively. The increasing of the scan speed from 200 to 400 mm/s reduces about 50% the J_c values for both 3D-printed and hybrid samples.

Table 3. Fracture toughness tests results for SM specimens.

Sample Reference	Valid $K_Q = K_{Ic}$	K_Q (MPa.m$^{1/2}$)	Average K_Q (MPa.m$^{1/2}$)	J_c (N/mm)	Average J_c (N/mm)	Critical Crack, a_c (mm)
2SM01_3	No	72.99	80.17 ± 6.71	59.23	65.68 ± 5.35	31.26
2SM02_3	No	80.44		67.86		31.31
2SM03_3	Yes	90.85		73.26		31.11
2SM04_3	No	76.40		62.4		31.96
2SM13_6	No	76.38	72.85 ± 4.26	42.9	42.4 ± 2.5	29.39
2SM14_6	No	68.93		40.1		30.62
2SM15_6	No	77.77		46.3		31.03
2SM16_6	No	68.32		40.3		29.11
4SM05_6	No	49.22	55.42 ± 4.71	19.71	18.65 ± 4.10	29.65
4SM06_6	No	52.56		12.27		28.89
4SM07_6	No	60.71		23.71		30.23
4SM08_6	No	59.18		18.93		30.06

Table 4. Fracture toughness tests results for SS/SM specimens.

Sample Reference	Valid $K_Q = K_{Ic}$	K_Q (MPa.m$^{1/2}$)	Average K_Q (MPa.m$^{1/2}$)	J_c (N/mm)	Average J_c (N/mm)	Crack to Fracture, a_c (mm)
2SS/SM01_3	No	83.22	83.9 ± 1.1	48.9	55.82 ± 7.13	32.03
2SS/SM02_3	No	82.46		48.5		31.98
2SS/SM03_3	No	85.15		62.7		31.45
2SS/SM04_3	No	84.75		63.21		31.67
2SS/SM01_6	No	67.71	70.97 ± 2.55	41.14	39.02 ± 1.52	30.58
2SS/SM03_6	No	71.24		37.59		29.91
2SS/SM04_6	No	73.95		38.33		31.06
4SS/SM01_6	No	54.76	56.47 ± 2.36	17.08	19.56 ± 2.5	28.98
4SS/SM02_6	No	57.84		19.53		29.05
4SS/SM03_6	No	59.59		23.64		29.86
4SS/SM04_6	No	53.69		18.02		30.14

Table 5. Fracture toughness tests results for HS/SM specimens.

Sample Reference	Valid $K_Q = K_{Ic}$	K_Q ($MPa.m^{1/2}$)	Average K_Q ($MPa.m^{1/2}$)	J_c (N/mm)	Average J_c (N/mm)	Crack to Fracture, a_c (mm)
2HS/SM01_3	No	73.93	73.66 ± 5.77	40.7	44.78 ± 16.17	32.13
2HS/SM02_3	No	81.7		71.55		32.98
2HS/SM03_3	No	73.03		38.71		30.47
2HS/SM04_3	No	65.99		28.17		30.51
2HS/SM01_6	No	61.54	65.41 ± 3.45	21.05	22.84 ± 4.1	28.58
2HS/SM03_6	No	64.77		24.9		29.71
2HS/SM04_6	No	69.91		17.24		30.03

The estimated K_Q for the specimens built with 200 mm/s of scan rate and 6 mm thick showed a decrease about 10% from the K_{Ic} value of the commercial steel grade 300 [18]. On the other hand, the 3 mm thick samples presented similar values between the estimated K_Q and the K_{Ic} value of the commercial steel grade 300 [26] around 80 $MPa.m^{1/2}$.

Santos et al. [12] show the level of defects observed in two samples produced with 200 mm/s and 400 mm/s of speed rate scan, respectively. It can be observed that the level of defects (area) increases from 0.74% to 7.37% when the speed increases from 200 to 400 mm/s. Besides, the fracture toughness decreases with the increase of scan speed rate, as previously mentioned (Table 5). Therefore, it can lead to a conclusion that the fracture toughness of SLM maraging steel grade 300 is lower that the commercial steel grade 300 produced by conventional processes, due to the internal defects (porosities, unmelted powder, unmelted zones, among others) and their effects.

3.5. Fracture Surface Analysis

For a better understanding of the previous results, a microscopic analysis was carried out to observe the fracture surfaces and crack growth path of the SM, SS/SM and HS/SM samples with 6 mm of thickness 3D-printed at 200 mm/s. Figure 10 illustrates the crack growth path for SS/SM sample. Figure 10a shows that the fracture crack initiated and grew in the 3D-printed material near the interface zone. Therefore, the samples tend to fail near the midplane of the notch plane at greater microstructural heterogeneities and where the stresses are higher. The crack propagated in different laser layers' interfaces through internal defects as shown in Figure 10b. Figure 11 is a SEM image of a SS/SM specimen showing the different laser directions between layers (marked by red arrows) and the porosities defects (circumscribed by red circles); this figure confirms the failure at the laser layers interfaces.

(**a**)

(**b**)

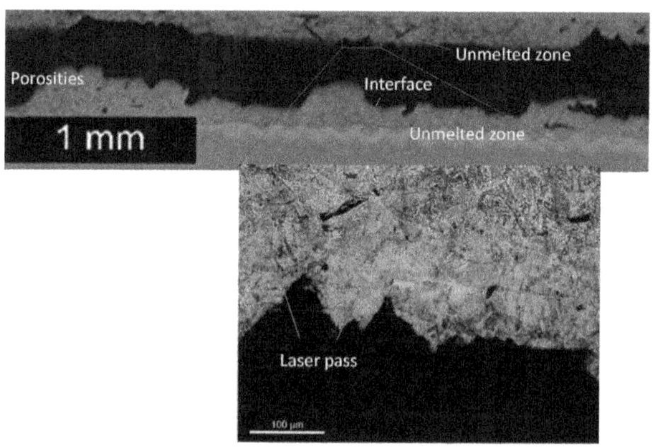

Figure 10. Crack growth path for SS/SM sample 3D-printed at 200 mm/s.

Figure 12 shows the crack growth path for a HS/ST sample 6 mm thick 3D-printed at 200 mm/s. Failure mechanisms are quite similar to those observed for SS/SM samples, but in this case the crack grew along the limit between the interface material and the 3D-printed material (Figure 12a). The crack path is determined by the variation of direction between laser layers, as illustrated in Figure 12b.

Figure 13 illustrates a SEM image of the fracture surface of HS/SM sample 6 mm tick 3D-printed at 200 mm/s. This image shows: the directions of laser passes marked by arrows for different layers, the porosity and the interstice between two different passes in the same directions (unmelted zone), both marked by red ellipses forms. Comparing Figures 11 and 13, it can be concluded that the main failure mechanisms are quite similar in both hybrid series, characterized by an interface failure between adjacent laser layers, since the same 3D-printed material was deposed and the same parameters were used in the SLM process.

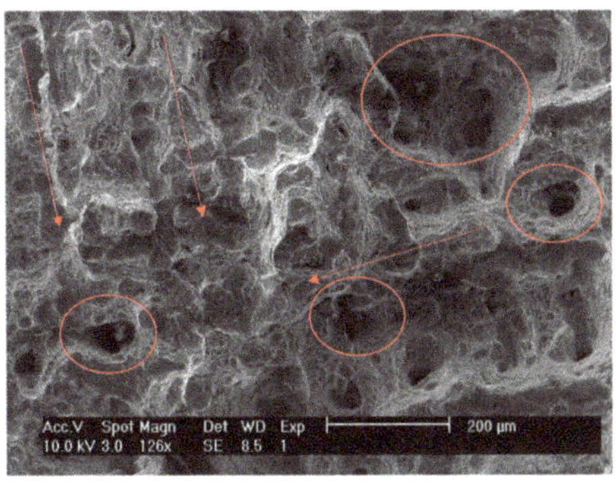

Figure 11. Fracture surface of SS/SM sample 3D-printed at 200 mm/s.

(a)

(b)

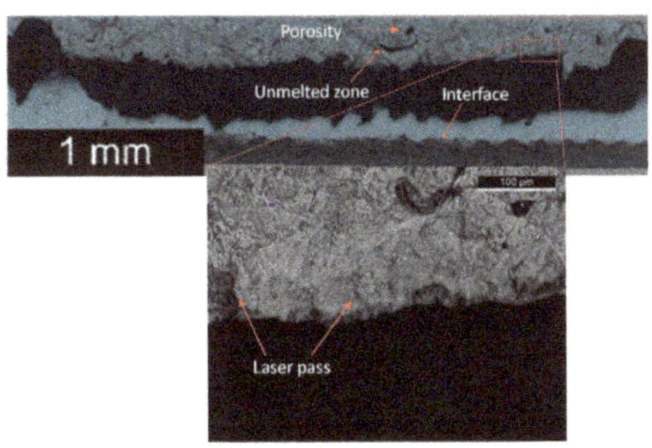

Figure 12. Crack growth path for HS/SM sample 3D-printed at 200 mm/s.

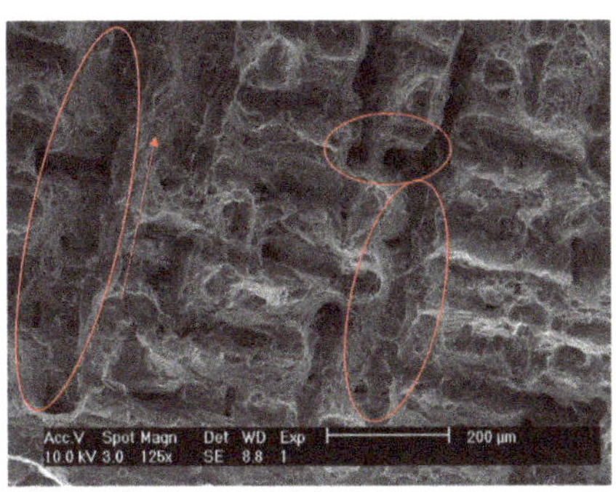

Figure 13. Fracture surface of HS/SM sample 3D-printed at 200 mm/s.

4. Conclusions

This work analyzed the fracture toughness performance for three different SLM 3D-printed compositions using CT specimens with 3 and 6 mm of thickness: samples of fully 3D-printed steel (maraging AISI 18Ni300), samples of AISI H13 with implant of 3D-printed maraging steel AISI 18Ni300 and samples of AISI 420 steel implant of 3D-printed maraging AISI 18Ni300. Two different scan rates (200 mm/s and 400 mm/s) were used to build the different samples through the SLM process. It was observed that a clear interface zone was created between 3D-printed and conventional steels by the SLM process. This interface presented the same chemical composition and hardness values for both ASI H13 and AISI 420 conventional steels. However, the interface showed a higher hardness than the 3D-printed material but lower than ASI H13 or AISI 420 steels. Additionally, the fracture toughness values did not exhibit significant variations, when comparing the different series for the same scan rate and thickness. Fracture toughness decreases when the sample thickness increases and fracture toughness values presented a great agreement with other methods, namely laser welding with modern Nd-YAG process. The SLM 18Ni300 maraging steel fracture toughness value is slightly lower when compared to the commercial steel grade 300 because the internal defects produced by SLM have an important role in decreasing the mechanical performance. Finally, the crack growth path for all hybrid specimens was along the boundary between the interface zone and the 3D-printed steel, showing a brittle mechanics fracture due to the defects produced by the SLM process.

Author Contributions: Investigation, Data Curation, L.M.S.S.; Investigation, Data Curation, Writing-Original Draft Preparation, Writing-Review & Editing, J.D.J.; Supervision, Writing-Review & Editing, J.M.F.; Investigation, Writing-Review & Editing, J.D.C.; Investigation, Resources, C.C.

Funding: Project number 016713 (PTDC/EMS-PRO/1356/2014) financed by Project 3599 "Promover a Produção Científica e Desenvolvimento Tecnológico e a Constituição de Redes Temáticas" (3599-PPCDT) and FEDER funds.

Acknowledgments: EROFIO S.A. industry for the supply of the testing samples.

Conflicts of Interest: The authors declare no conflicts of interest.

References

1. Fayazfar, H.; Salarian, M.; Rogalsky, A.; Sarker, D.; Russo, P.; Paserin, V.; Toyserkan, E. A critical review of powder-based additive manufacturing of ferrous alloys: Process parameters, microstructure and mechanical properties. *Mater. Des.* **2018**, *144*, 98–128. [CrossRef]
2. Balc, N.C.; Berce, P.; Pacurar, R. Comparison between SLM and SLS in producing complex metal parts. In *Annals of DAAAM for 2010 & Proceeding of the 21st International DAAAM Symposium, Volume 21, No. 1, ISSN 1726-9679*; DAAAM International: Vienna, Austria, 2010.
3. Costa, J.M.; Ferreira, J.M.; Capela, C. Fracture toughness of the heat affected zone on Nd-YAG laser welded joints. *Eng. Fail. Anal.* **2009**, *16*, 1245–1251. [CrossRef]
4. Kempena, K.; Yasaa, E.; Thijsb, L.; Krutha, J.-P.; Van Humbeeckb, J. Microstructure and mechanical properties of Selective Laser Melted 18Ni-300 steel. *J. Alloys Compd.* **2017**, *725*, 355–364. [CrossRef]
5. Casalino, G.; Campanelli, S.L.; Contuzzi, N.; Ludovico, A.D. Experimental investigation and statistical optimization of the selective laser melting process of a maraging steel. *Opt. Laser Technol.* **2015**, *65*, 151–158. [CrossRef]
6. Bai, Y.; Yang, Y.; Wang, D.; Zhang, M. Influence mechanism of parameters process and mechanical properties evolution mechanism of maraging steel 300 by selective laser melting. *Mater. Sci. Eng. A* **2017**, *703*, 116–123. [CrossRef]
7. Casatia, R.; Codurib, M.; Lecisa, N.; Andrianopolic, C.; Vedania, M. Microstructure and mechanical behavior of hot-work tool steels processed by Selective Laser Melting. *Mater. Charact.* **2018**, *137*, 50–57. [CrossRef]
8. Shamsaei, N.; Yadollahi, A.; Bian, L.; Thompson, S.M. An overview of direct laser deposition for additive manufacturing; Part II: Mechanical behavior, process parameter optimization and control. *Addit. Manuf.* **2015**, *8*, 12–35. [CrossRef]
9. Chan, K.S.; Koike, M.; Mason, R.L.; Okabe, T. Fatigue life of titanium alloys fabricated by additive layer manufacturing techniques for dental implants. *Metall. Mater. Trans. A* **2013**, *44*, 1010–1022. [CrossRef]

10. Wycisk, E.; Emmelmann, C.; Siddique, S.; Walther, F. High cycle fatigue (HCF) performance of Ti-6Al-4V alloy processed by selective laser melting. *Adv. Mater. Res.* **2016**, *816*, 134–139. [CrossRef]
11. Yan, J.J.; Zheng, D.L.; Li, H.X.; Jia, X.; Sun, J.F.; Li, Y.L.; Qian, M.; Yan, M. Selective laser melting of H13: Microstructure and residual stress. *J. Mater. Sci.* **2017**, *52*, 12476–12485. [CrossRef]
12. Santos, L.M.S.; Ferreira, J.A.M.; Jesus, J.S.; Costa, J.M.; Capela, C. Fatigue behaviour of selective laser melting steel components. *Theor. Appl. Fract. Mech.* **2016**, *85*, 9–15. [CrossRef]
13. Suryawanshi, J.; Prashanth, K.G.; Ramamurty, U. Tensile, fracture, and fatigue crack growth properties of a 3D printed maraging steel through selective laser melting. *J. Alloys Compd.* **2017**, *725*, 355–364. [CrossRef]
14. Mazur, M.; Brincat, P.; Leary, M.; Brandt, M. Numerical and experimental evaluation of a conformally cooled H13 steel injection mould manufactured with selective laser melting. *Int. J. Adv. Manuf. Technol.* **2017**, *93*, 881–900. [CrossRef]
15. Ackermann, M.; Šafka, J.; Voleský, L.; Bobek, J. Impact Testing of H13 Tool Steel Processed with Use of Selective Laser Melting Technology. *Mater. Sci. Forum* **2018**, *919*, 43–51. [CrossRef]
16. Cyr, E.; Asgari, H.; Shamsdini, S.; Purdy, M.; Hosseinkhani, K.; Mohammadi, M. Fracture behaviour of additively manufactured MS1-H13 hybrid hard steels. *Mater. Lett.* **2018**, *212*, 174–177. [CrossRef]
17. Ferreira, J.A.M.; Santos, L.M.S.; Jesus, J.S.; Costa, J.M.; Capela, C. Assessment of the fatigue life on functional hybrid laser sintering steel components. *Proc. Struct. Integr.* **2016**, *1*, 126–133. [CrossRef]
18. Campanelli, S.L.; Angelastro, A.; Signorile, C.G.; Casalino, G. Investigation on direct laser powder deposition of 18 Ni (300) marage steel using mathematical model and experimental characterization. *Int. J. Adv. Manuf. Technol.* **2017**, *89*, 885–895. [CrossRef]
19. ASTM E8/E8M-16a. *Standard Test Methods for Tension Testing of Metallic Materials*; ASTM International: West Conshohocken, PA, USA, 2016.
20. BS 7448-3:1991. *Fracture Mechanics Toughness tests, Part 1. Method for Determination of K1c, Critical CTOD and Critical J Values of Metallic Materials*; British Standards Institution: London, UK, 1991.
21. ASTM Standard E3, 2011. *Standard Guide for Preparation of Metallographic Specimens*; ASTM International: West Conshohocken, PA, USA, 2011. [CrossRef]
22. ASTM Standard E384, 2011e1. *Standard Test Method for Knoop and Vickers Hardness of Materials*; ASTM International: West Conshohocken, PA, USA, 2011. [CrossRef]
23. Isfahany, A.N.; Saghafian, H.; Borhani, G. The effect of heat treatment on mechanical properties and corrosion behavior of AISI420 martensitic stainless steel. *J. Alloys Compd.* **2011**, *509*, 3931–3936. [CrossRef]
24. Branco, R.; Costa, J.D.M.; Berto, F.; Mohammad, S.; Ferreira, J.A.M.; Capela, C.; Santos, L.; Antunes, F. Low-Cycle Fatigue Behaviour of AISI 18Ni300 Maraging Steel Produced by Selective Laser Melting. *Metals* **2018**, *8*, 32. [CrossRef]
25. Marashi, J.; Yakushina, E.; Xirouchakis, P.; Zante, R.; Foster, J. An evaluation of H13 tool steel deformation in hot forging conditions. *J. Mater. Process. Technol.* **2017**, *246*, 276–284. [CrossRef]
26. Antolovich, S.D.; Risbeck, T.R.; Saxena, A.; Kawabe, Y. The effect of microstructure on the fracture toughness of 300 and 350 grade maraging steels. *Eng. Fract. Mech.* **1980**, *13*, 717–739. [CrossRef]

applied sciences

MDPI

Article

Integrating the Shape Constants of a Novel Material Stress-Strain Characterization Model for Parametric Numerical Analysis of the Deformational Capacity of High-Strength X80-Grade Steel Pipelines

Onyekachi Ndubuaku [1,*], Michael Martens [2], J. J. Roger Cheng [1] and Samer Adeeb [1]

[1] Department of Civil and Environmental Engineering, University of Alberta, Edmonton, AB T6G 1R1, Canada; roger.cheng@ualberta.ca (J.J.R.C.); adeeb@ualberta.ca (S.A.)
[2] TransCanada Pipelines Ltd., Calgary, AB T2P 5H1, Canada; michael_martens@transcanada.com
* Correspondence: ndubuaku@ualberta.ca

Received: 11 December 2018; Accepted: 11 January 2019; Published: 17 January 2019

check for updates

Featured Application: A novel material stress-strain characterization model is presented herein, which is capable of precise and robust approximation of the full range of the nonlinear hardening portion of the stress-strain curve, notwithstanding the presence of a yield plateau. Parameterization of the stress-strain curve is, therefore, improved using two "shape" constants, and material behavior is easily incorporated in the test matrix for semi-empirical modeling of the mechanical response of structural systems under various loading conditions.

Abstract: Pipelines typically exhibit significant inelastic deformation under various loading conditions, making it imperative for limit state design to include considerations for the deformational capacity of pipelines. The methods employed to achieve higher strength of API X80 line pipe steels during the plate manufacturing process tend to increase the hardness of the pipe material, albeit at the cost of ductility and strain hardenability. This study features a simple and robust material stress-strain characterization model, which is able to mathematically characterize the shape of a diverse range of stress-strain curves, even for materials with a distinct yield point and an extended yield plateau. Extensive parametric finite element analysis is performed to study the relationship between relevant parameters and the deformational capacity of API X80 pipelines subjected to uniform axial compression, uniform bending, and combined axial compression and bending. Nonlinear regression analysis is employed to develop six nonlinear semi-empirical equations for the critical limit strain, wherein the shape constants of the material model are adapted as dimensionless parameters. The goodness-of-fit of the developed equations was graphically and statistically evaluated, and excellent predictive accuracy was obtained for all six developed equations.

Keywords: buckling; critical limit strain; deformational capacity; nonlinear regression equation; parametric analysis; semi-empirical model; shape constant; stress-strain model; X80 pipeline

1. Introduction

Pipelines are commonly used as the primary means for transmission of hydrocarbon fluids in the energy industry, and for transportation of water resources in the public utility industry. Previous investigations into the main causes of damages and loss of mechanical and structural integrity in pipelines have highlighted the detrimental impacts of geological conditions or environmental actions associated with various time-dependent thermomechanical phenomena, such as ground subsidence, soil liquefaction and land sliding, discontinuous frost heave/ or thaw settlement, fault movement due to seismic action, etc. [1–3]. Large displacements tend to induce significant strains in the wall of pipe segments, which may ultimately result in the failure of a pipeline either due to tensile rupture, especially at welded regions, or due to bulging or wrinkling deformation of the pipe under the influence of compressive stresses [4–6]. As a result of the inevitable requirement for pipelines to traverse large geographical areas, a majority of pipelines in the onshore energy and public utility industry are installed below the soil surface. It is observed that, compared to above-ground pipelines, the mechanical behavior of buried pipelines is significantly influenced by the geophysical properties of the surrounding soil. Hence, the intricacies of the mutual interaction between the pipe and the surrounding soil are regarded as an important consideration in the design of buried continuous pipelines [7–10].

The common consequence of ground movement and thermal buckling deformation on pipelines is the experience of high longitudinal stresses and strains. Conventional pipeline design procedures are generally based on an *"allowable stress"* concept, which aims to limit the resultant longitudinal and circumferential stresses in pipelines, either due to load-controlled or displacement-controlled conditions, to a prescribed fraction of the specified minimum yield strength (SMYS) of the pipe material. Unfortunately, the allowable stress design (ASD) approach tends to be severely limited by its inability to distinguish between stable and unstable failure modes, as well as between loss of serviceability and loss of containment [11]. A strain-based design (SBD) approach has therefore been regarded as more appropriate, and has been recently more adopted by various pipeline design standards (e.g., API RP 1111 [12], DNV-OS-F101 [13], ABS 2006 [14], CSA Z-662 [15]) for design of pipelines expected to experience high longitudinal strains caused by displacement-controlled environmental and operational conditions in service [16,17]. SBD circumvents the inherent conservativeness of the ASD approach by permitting a limited amount of plastic strain, while maintaining relevant ultimate and serviceability limit state considerations. However, rather than consider the SBD and ASD approaches as independent techniques, best design practice recommends the application of SBD as a complementary tool to the more traditional ASD procedure [18]. SBD employs a limit state design approach, whereby the in-service longitudinal tensile or compressive strain that a pipeline is expected to experience due to displacement-controlled loading conditions (i.e., strain demand) is evaluated and compared to the tolerable limit of strain which the pipeline has intrinsic ability to withstand (i.e., strain capacity). Estimation of strain demand in pipelines typically involves a rigorous and often complex process, which requires all-inclusive consideration of numerous interrelating factors related to various environmental and geological conditions, and the pipe's mechanical properties. The strain demand is also highly dependent on a number of peculiar, and sometimes transient, environmental conditions, making the development of deterministic prediction models difficult, and often necessitating the use of probabilistic and reliability-based estimation techniques for the SBD [19,20]. Unlike strain demand, evaluation of the strain capacity is essentially based on the inherent mechanical resistance of the pipe segments and has relatively less dependency on external factors, hence analytical evaluation and deterministic prediction of strain capacity is relatively more straightforward, as parameterization of influencing factors for incorporation in the SBD is relatively less complicated.

Extensive studies have been conducted by a number of researchers to investigate the limit load instability and deformational capacity of pipe segments, and cylindrical shells in general, subjected to various loading conditions. The earliest recorded investigations on the buckling behavior of pipe segments were analytical studies, known as the classical shell buckling theories, performed by Lorenz [21], Timoshenko [22], and Southwell [23] at the inception of the 20th century, which focused on deriving constitutive theoretical formulations for predicting the elastic critical buckling stress of perfect isotropic cylindrical shells subjected to pure axial compression. Subsequent experimental and analytical studies [24–28] pointed out some shortcomings of the classical shell buckling theories, and prescribed the extension of the classical equations to include nonlinear large-deflection considerations, as well as adequate characterization of material, geometric, and boundary characteristics. An analytical study by Brazier [29] on the circumferential flattening mode of distortion (referred to as *"ovalization"* or "*Brazier instability*") observed in an infinitely long, circular tube subjected to pure bending led to the derivation of an expression for the limit moment. Brazier's theory is based on the relationship between the strain energy per unit tube length and the change in axial curvature, such that the limit moment is directly related to the ovalization of the circular tube. Subsequent experimental studies [30–33], however, indicate that bifurcation instability (buckling) in the form of longitudinal wrinkling or bulging may precede the limit moment in pipes subjected to bending due to increase in axial stress on the compression side of the pipe. Corona and Kyriakides [34] explained that bifurcation instability is more likely to precede *Brazier* instability in "thinner" tubes, and vice versa in "thicker" tubes. They further speculated that the transition between one preceding instability phenomenon and the other occurs in the range of D/t (diameter/thickness) ratios between 35 and 45. Mathon and Limam [35] performed statistical evaluation of results obtained from (1) experimental tests of circular tubes under pure bending [36], (2) empirically-derived analytical formulations for cylinders under combined axial compression and bending [37], and (3) semi-empirical derivation of the critical buckling stress for pressurized circular tubes subjected to uniform bending [38]. They observed that the critical buckling stress of cylinders under pure bending is generally between 20 to 60% higher, depending on the diameter-to-thickness ratio (D/t ratio), than that of cylinders under pure axial compression.

Using a special-purpose non-linear finite element technique, Houliara and Karamanos [39] studied the structural stability of long uniformly pressurized thin elastic tubular shells subjected to in-plane bending. They observed that external overpressure reduces the buckling moment and corresponding curvature but increases the buckling wavelength, whereas internal overpressure tends to increase the buckling moment but reduce the cross-sectional ovalization and the buckling wavelength. Following a systematic series of experiments and by Limam et al. [40] involving stainless steel tubes bent to failure at fixed values of internal pressure, it was shown that cross-sectional ovalization and circumferential extension occur simultaneously, due to bending and internal pressure, respectively. Ref. [40] also performed computational simulation of the evolution of wrinkling and its eventual localization using a finite element (FE) shell model, and observed that internal pressure tends to stabilize the structure and increase the wavelength of the wrinkles, while also shifting the initiation of buckling towards the plastic range of the material. Ref. [40] indicates that accurate simulation of the structural stability of tubular shells requires accurate characterization of the nonlinear inelastic properties of the material stress-strain behavior, including yield anisotropies, as well as adequate representation of initial geometric imperfections. Highlighting the significant effect of the material properties on the moment-curvature relationship, Ref. [34] established that for pipes with D/t ratios less than 200, the mechanical and structural instabilities of the structure are strongly influenced by the interaction of the induced ovalization and the plastic and nonlinear characteristics of the material.

This study focuses on the deformational capacity of high-strength API-X80 [41] grade pipelines subjected to various loading conditions. The continuous increase in world energy demand has compelled the hydrocarbon industry to venture into more remote and more environmentally hostile environments in search of fossil fuels. This predicament necessitates the construction and operation of pipelines with the best possible transport efficiency and at the lowest possible cost, therefore initiating a tendency towards using line pipes of larger diameter or maintaining higher operational pressures. High-strength steel grades provide the advantage of forming line pipes with high material yield strength, thus reducing the wall thickness requirement for internal (or external; in the case of deep to ultra-deep offshore environments) pressure containment [42]. Suzuki and Toyoda [43] explained that the peak moment strain, and invariably the deformational capacity of line pipes, is strongly influenced by the strain-hardening characteristics of the pipe material. In a numerical study of API-X100 line pipes subjected to axial compression or bending moment, Suzuki et al. [44] pointed out that high-strength pipe steels tend to have inferior strain hardening properties, and consequently lower deformational capacity, compared to conventional pipeline steels of lower yield strength. However, results of a few experimental studies have shown that by carefully manipulating the thermomechanical control processes (TMCP) of slab reheating, rolling, and cooling during the manufacture of parent steel plates, the microstructural and mechanical properties of steel can be adjusted to produce line pipe steels with improved deformability, toughness, and resistance to strain-aging [45–47].

Kong et al. [48] explained that the basic morphology of most high-strength pipeline steels is a multiphase structure composed of ferrite, bainite, and martensite. While bainite and martensite ingredients naturally improve the strength and hardness of line pipe steel, ductile and plastic deformation capacities tend to be compromised. Also, variability in manufacturing processes of parent steel plates makes it inevitable for line pipes of the same grade classification to exhibit slight differences in material properties [49,50]. While manufacturing variability may be regarded as the primary cause of difference in the strain-hardening characteristics of materials with the same grade classification, other factors, such as strain aging, have also been observed to cause significant alteration of the material properties, leading to higher yield strength but lower deformational capacity [51]. Moreover, it has been established that the buckling response of pipes is highly sensitive to the material behavior, especially in the nonlinear range of the stress-strain relationship. Hence, it is imperative that the mathematical model that serves for characterization of the material stress-strain relationship should simultaneously provide reasonable simplicity and representative accuracy. Recent studies by Ndubuaku et al. [52–55] have led to the development of a novel stress-strain expression, referred to as the "Ndubuaku model", which has proven to be exceptionally convenient and effective for parameterizing the true stress-true strain relationship of any metallic material with a non-negative gradient throughout the stress-strain curve, including materials with a distinct yield point and an extended yield plateau. The "Ndubuaku model" approximates the material stress-strain curve over the full range of strains using only two constitutive model constants (or "shape" constants), referred to as the "knee" constant and the "heel" constant.

The procedures implemented in this study for estimating the peak moment strain and the compressive limit strain take a cue from the semi-empirical modeling approach adopted by various prominent pipeline design standards [12–15]. ABAQUS CAE [56] was employed to create FE (finite element) pipe models for this study and three different loading conditions were studied: uniform axial compression, uniform bending, and combined axial compression and bending. Four variations of a constant net-section compressive axial stress were applied to the pipe models prior to bending to simulate combined axial compression and bending. Four and five variations of the D/t ratio and applied internal pressure, respectively, were also applied to the pipe models. The "Ndubuaku model" was used to create two sets of stress-strain curves according to the two main classifications of stress-strain curves for metallic materials, i.e., round-house type (RHT) curves and yield-plateau type (YPT) curves. Four and five variations of RHT and YPT curves, respectively, were created so that the numerical analyses performed in this study culminated in a total of 1080 separate FEA (finite element analysis) runs; 180 runs for uniform axial compression, 180 runs for uniform bending, and 720 runs for combined axial compression and bending. Advanced nonlinear regression techniques were subsequently employed to develop mathematical expressions for predicting the critical limit strain (CLS) for each loading condition using a powerful computational package, Wolfram Mathematica [57]. The strain-hardening peculiarities of the stress-strain curves were parameterized by the model "shape" constants of the "Ndubuaku model" and incorporated in the derived semi-empirical models.

2. Characterization of Material Stress-Strain Behavior

The "Ndubuaku model" was adopted in this study to adequately capture the slight variations that are generally observed to exist in the stress-strain response of metallic materials. Details of the fundamental concept and successive steps for developing the adopted material model have been previously presented in recent publications by Ndubuaku et al. [52–55]. The mathematical form of the material model derives the true stress (σ) as a function of the true strain (ε), using two "shape" constants, the "knee" constant (k_{nm}) and the "heel" constant (h_{nm}):

$$\sigma = \begin{cases} E\varepsilon & \sigma \leq \sigma_{pl} \\ \sigma_{pl} + \left(\sigma_{up} - \sigma_{pl}\right)\left(\frac{\varepsilon - \varepsilon_{pl}}{\varepsilon_{up} - \varepsilon_{pl}}\right)^{k_{nm}\left(\frac{\varepsilon - \varepsilon_{pl}}{\varepsilon_{up} - \varepsilon_{pl}}\right)^{h_{nm}}} & \sigma > \sigma_{pl} \end{cases} \tag{1}$$

where E is Young's modulus of elasticity, σ_{pl} and ε_{pl} are the proportionality limit stress and the proportionality limit strain, respectively, and the ultimate proof stress and corresponding ultimate proof strain are represented by σ_{up} and ε_{up}, respectively. The first part of Equation (1) represents the linear (elastic) portion of the stress-strain curve, where the relationship between the stress and the strain is simply defined by the elastic modulus, E, while the second part characterizes the entire nonlinear portion of the stress strain curve, beginning at the proportionality limit stress (PLS) and terminating at the nominal ultimate proof stress (UPS).

The "Ndubuaku model" is easily applicable for generating any desired number of idealized strain curves by simply specifying two stress control points (σ_{c1} and σ_{c2}) and corresponding strain control points (ε_{c1} and ε_{c2}) according to the following expressions [54]:

$$k_{nm} = \frac{\varepsilon_{R1}{}^{(h_{nm})} In(\sigma_{R1})}{In(\varepsilon_{R1})} \; or \; \frac{\varepsilon_{R2}{}^{(h_{nm})} In(\sigma_{R2})}{In(\varepsilon_{R2})} \tag{2}$$

$$h_{nm} = \frac{In\left[\frac{In(\varepsilon_{R1}).In(\sigma_{R2})}{In(\sigma_{R1}).In(\varepsilon_{R2})}\right]}{\left[In(\varepsilon_{R1}) - In(\varepsilon_{R2})\right]} \tag{3}$$

where, $\sigma_{R1} = \frac{\sigma_{c1} - \sigma_{pl}}{\sigma_{up} - \sigma_{pl}}$, $\sigma_{R2} = \frac{\sigma_{c2} - \sigma_{pl}}{\sigma_{up} - \sigma_{pl}}$, $\varepsilon_{R1} = \frac{\varepsilon_{c1} - \varepsilon_{pl}}{\varepsilon_{up} - \varepsilon_{pl}}$, and $\varepsilon_{R2} = \frac{\varepsilon_{c2} - \varepsilon_{pl}}{\varepsilon_{up} - \varepsilon_{pl}}$.

To ensure conformity with the API 5L [41] specifications for X80 grade line pipe steels, the material model was employed such that all the stress-strain curves for this study (shown in Figure 1) are characterized by the same nominal yield stress (YS) and nominal ultimate proof stress (UPS), but different strain hardening properties.

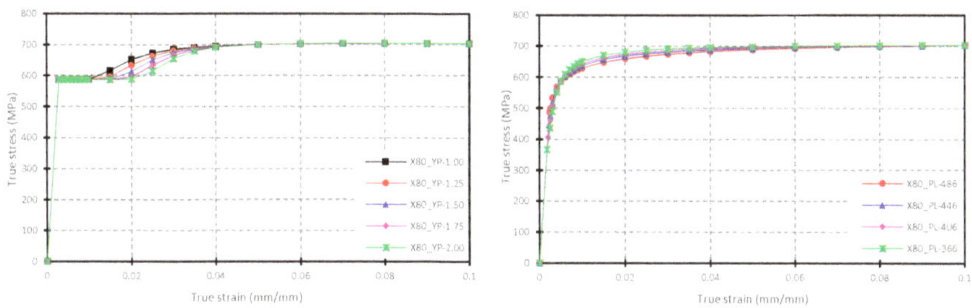

(**a**) (**b**)

Figure 1. Idealized variations for stress-strain curves of (**a**) X80-YPT, and (**b**) X80-RHT materials.

As indicated by the plots in Figure 1, the nominal yield stress (YS = 586 MPa) was maintained at the stress corresponding to a total strain of 0.5% (i.e., $\sigma_{0.5}$), while the nominal ultimate proof stress (703 MPa) was maintained at the stress corresponding to a total strain of 10% (i.e., σ_{10}). The value of the elastic modulus was selected as 205,800 MPa. The strain hardening for the YPT materials was varied by simply changing the yield plateau length (YPL) of the stress-strain curves, while the various strain hardening properties of the RHT curves were obtained by adjusting the proportionality limit stress (PLS) of the stress-strain curves. The numeric designations of the elements of the graph legends for the stress-strain curves in Figures 1a and 1b represent the YPL and the PLS, respectively. The YPLs are defined in terms of the total strain in percentage (%) units, while the PLSs are defined in megapascal (MPa) units.

The constitutive model constants which define the shapes, and invariably the strain-hardening characteristics, of the generated stress-strain curves are presented in Table 1.

Table 1. Model shape constants for the derivation of stress-strain curves.

YPT			RHT		
Curve Label	$K_f = 1/K_{NM}$	$H_f = 1/H_{NM}$	*Curve Label*	$K_f = 1/K_{NM}$	$H_f = 1/H_{NM}$
YP-1.00	83.7395	0.5187	*PL-486*	11.5448	3.9814
YP-1.25	115.2048	0.4221	*PL-446*	21.1099	2.7531
YP-1.50	162.9859	0.3510	*PL-406*	35.3391	2.1374
YP-1.75	240.2443	0.2953	*PL-366*	69.2445	1.5847
YP-2.00	374.3178	0.2501			

For better representation of the shape constants two new designations, the "knee" factor (K_f) and the "heel" factor (H_f) are defined in Table 1 as inverse functions of the "knee" constant, K_{nm}, and the "heel" constant, H_{nm}, respectively.

3. Methodology of Numerical Analysis

Pipelines made of API X80-grade line pipe steel were numerically simulated in this study using a tubular FE shell model developed with ABAQUS/Standard, in order to study the stability of steel pipelines when subjected to uniform axial compression, uniform bending, or combined axial compression + bending. The parameters investigated in the FE study were selected according to three factors considered to be most influential to the buckling response and respective limit strains of tubular shell structures, i.e., the dimensional factor (related to the D/t ratio), the load factor (related to internal or external overpressure), and the strain hardening factor (related to shape of stress-strain curve). Four variations of the D/t ratio were specified thus: DT1 = 41.152, DT2= 64.078, DT3 = 82.156, and DT4 = 104.622. The D/t ratio was varied by maintaining the pipe's outer diameter (OD) at a constant nominal pipe size (NPS) of 36 inches (914.4 mm), and respectively changing the wall thickness of the pipe models thus: t_1 = 22.22 mm (for DT1), t_2 = 14.27 mm (for DT2), t_3 = 11.13 mm (for DT3), and t_4 = 8.74 mm (for DT4). In addition to non-pressurized conditions, four levels of internal pressure corresponding to 20%, 40%, 60%, and 80% of the yield pressure in the circumferential direction were applied to the pipe models. To properly account for the influence of the material's strain hardening properties on the CLS, three material-related parameters each were derived for both YPT curves and RHT curves. The two parameters obtained for both RHT and YPT pipes comprised the two model "shape" constants (K_{nm} and H_{nm}), while the yield plateau length (YPL) was applied as the third parameter for YPT pipes and the ratio of the PLS to the UPS (herein referred to as the "PLUS" ratio) was considered as the third parameter for RHT pipes.

The pipe geometry was modeled as a 3D deformable shell structure, and for computational efficiency, symmetric boundary conditions were applied in the longitudinal and transverse directions of the pipe, so that only a quarter of the full pipe specimen (half of the specified length and half of the circumference) was modeled. The numerical analysis procedure conducted in this study is a geometric and material nonlinear type; hence, the boundary conditions applied to create the X- and Z-symmetries for the assumed quarter models are able to provide an accurate representation of the actual response of the full pipe models. With the application of the Z-symmetry boundary condition (symmetry across the X-Y plane), the quarter pipe models were therefore extruded to three times the pipe diameter (3.0*D). The assigned length of the pipe model was selected according to the recommendations of Liu et al. [58] to ensure that there are no interactions between the end boundary conditions and the strain and stress distribution at the pipe mid-length, where initiation of local buckling deformation is expected to occur. At the loading end of the pipe model (i.e., the end where the rotation or axial displacement was applied), a reference point was created at the center of the pipe cross-section and restrained such that only translation along the Z-axis was allowed for uniform axial compression, and rotation about the X-axis plus translation along the Z-axis was allowed for uniform bending and combined axial compression + bending. A kinematic coupling constraint was then assigned for connecting the nodes at the loading end to the reference point to allow the end of the pressurized pipe to expand in the radial direction throughout the applied loading. The kinematic coupling constraint was defined in the cylindrical coordinate system to permit only one degree of freedom, i.e., circumferential extension, thereby inducing a uniform stress field and consequently facilitating the initiation of buckling at the mid-length cross-section of the pipe segment.

The four-node reduced-integration shell elements (S4R) with hourglass control in the ABAQUS element library were used for all the analyses, enabling up to five integration points through the wall thickness of the pipe models. The S4R elements are suitable for the analysis of thin to moderately thick shell structures with material and geometric nonlinearities. A mesh convergence study was conducted to establish the optimum number of elements required for the FE simulation, and a mesh size approximately equal to 3% of the pipe diameter (0.03*D) was applied respectively to all the FE pipe models. A large-strain von Mises plasticity model with isotropic hardening was applied to the line pipe steel material. Calibration of the material response is assumed to be related to the true stress vs. true strain relationship from a uniaxial tensile test performed on a coupon specimen of the pipe materials, hence, respective stress-strain curves were assigned to the pipe models using the standard multilinear material definition facility in ABAQUS.

For the three different loading conditions considered in this study, two loading steps (an initial load-controlled step followed by a displacement-controlled step) were defined in addition to the default "Initial" step in ABAQUS CAE for each simulation. To adequately track the nonlinear equilibrium path of deformation at the reference critical portions of the pipe segment, recording of the resultant moment and resultant compressive force, as well as the corresponding limit strains, was performed at the displacement-controlled stage of each simulation. In the initial load-controlled step of the simulation for uniform axial compression, specified levels of internal pressure were applied to the pipe models, while a monotonically-increasing uniform axial displacement was applied to the loading-end reference point in the following displacement-controlled step. Internal pressure was applied in the load-controlled step for uniform bending, while a monotonically-increasing rotation was subsequently applied to the loading-end reference point in the displacement-controlled step. For the load-controlled phase of the combined axial compression and bending, a finite compressive axial force was applied to the loading-end reference point, while respective percentages of the circumferential yield pressure were simultaneously applied to the internal surface of the pipe models. A monotonically-increasing rotation was then applied to the loading-end reference point in the second step. For combined axial compression + bending, the compressive axial force applied to the pipe prior to the monotonically-induced curvature was measured as respective percentages (20%, 40%, 60%, and 80%) of the peak force under uniform axial compression.

The measurement scheme for the average compressive strain is adopted for this study in line with the recommendations of Liu et al. [58]. For uniformity of estimation, the same measurement scheme was employed to determine the average values of compressive strain in the pipe models. The measurement scheme was defined with reference to the pipe model under uniform bending by evaluating the average value of the compressive strain of all the meshed elements within the reference gauge area on the compressive side of the pipe model. The measurement area was selected immediately adjacent to both the Z-symmetry plane and the X-symmetry plane, and corresponded to a longitudinal distance equal to the pipe diameter (1.0*D), and a circumferential distance approximately equal to one-eighth of the pipe diameter (0.125*D), respectively (shown in Figure 2). The critical limit strain was derived from the results of the individual FE runs in this study based on a "peak load criterion" [59], which regards the "critical" value as the resultant compressive longitudinal strain measured at the onset of local buckling, and is derived as the average strain that corresponds to the onset of "softening" in the load-deformation response (i.e., the X-Y plot of the average compressive strain on the X-axis versus the loading-end reaction moment or the loading-end reaction force on the Y-axis). A cylindrical coordinate system (with the origin at the center of the pipe cross-section) was assigned to the pipe model for obtaining the average longitudinal strains, while the loading-end reaction moment and the loading-end reaction force were obtained in the default Cartesian coordinate system.

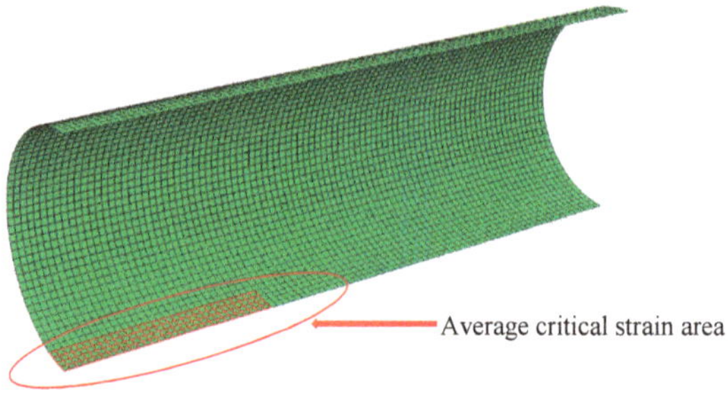

Figure 2. Meshed area for average strain measurement.

4. Numerical Results

Based on the stipulated parameters, and using the above numerical simulation techniques, the post-processing module in ABAQUS CAE was used to extract field output data after each run, such that the recorded incremental values for the reaction force ("RF3") or reaction moment ("RM1"), and the corresponding average compressive strains over the gauge length area were exported to a Microsoft Excel spreadsheet to estimate the critical limit strain of the API X-80 grade steel pipeline models. The individual variable functions that form the semi-empirical models generated in this study emanate from the observed trends of the CLS, with respect to the various parameters investigated.

4.1. Validation of Numerical Model

The pilot FE pipe model was validated by comparing the result obtained for pure bending of an unpressurized pipe model of D/t ratio = 64 to the result of an experimental study by Mohareb et al. [60], and satisfactory correspondence between the FE result and the experimental result was obtained for the end reaction bending moment vs. average induced curvature response, as well as the diamond-shaped wrinkling deformation at failure (Figure 3).

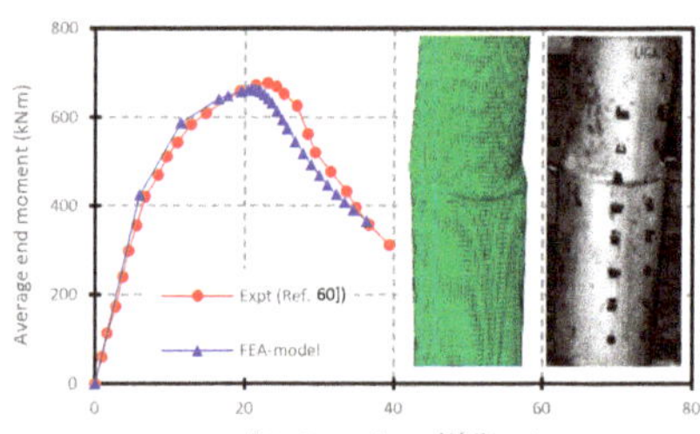

Figure 3. Comparison of FEA result with experimental result (UGA508) by Ref. [60].

4.2. Evolution of Stress in Pipes

The evolution of the resultant stress along the length of pipe segments due to applied loads and induced deformation is portrayed in the following figures using contour plots of the von Mises stress. The contour plots are captured at various points along the path of the load-displacement (or moment-curvature) curve, which coincide with three of the strategic stages of the load-deformation response, i.e., pre-buckling, limit load, and post-buckling. Figures 4a and 5a represent the axial load-displacement plots of YPT pipes and RHT pipes, respectively, under uniform axial compression. Figures 6a and 7a represent the moment-curvature plots of YPT pipes and RHT pipes, respectively, under uniform bending. The longitudinal distribution of the von Mises stress at the extreme of the compression side of the pipe's cross-section, which is aligned with the bending plane, is plotted in Figures 4b and 5b for YPT pipes and RHT pipes, respectively, under uniform axial compression.

The longitudinal stress distribution is plotted in Figures 6b and 7b for YPT pipes and RHT pipes, respectively, under uniform bending. Contour plots of the von Mises stress at specified stages of the load-deformation response are respectively presented in Figures 4c and 5c for YPT pipes and RHT pipes under uniform axial compression, while the contour plots for YPT pipes and RHT pipes under uniform bending are presented in Figures 6c and 7c, respectively. The results presented in Figures 4–7 are for DT4 pipes with $f_p = 0.4$; where the YPL = 1.0% for the YPT pipes and the PLUS ratio = 0.691 for the RHT pipes. The von Mises contour plots are longitudinally aligned with the longitudinal stress distribution plots such that the left end is the load end and the right end is mid-length cross-section of the pipe.

The plots presented in Figures 4–7 indicate that the von Mises stress is constant over the entire length at the compression zone of the pipe in the pre-buckling stage (i.e., from the beginning of load application until buckling occurs). Buckling is associated with the formation of a longitudinal wrinkle, and may either precede the limit load, as in Figures 5a, 6a and 7a, or coincide with the limit load, as in Figure 4a. The limit load is typically reached when the von Mises stress in the extreme compression fibers of the pipe reach the yield stress of the pipe material, after which localization of strains and resultant stresses commences at the mid-length region of the pipe, and the remaining portions of the pipe experience a stress relief. Attainment of the limit load is essentially followed by load collapse and softening of the load-deformation response. The above plots indicate that, beyond the limit load, the von Mises stresses in the compression zone evolve such that the stresses around the mid-length cross-section advance towards the UPS, while the stresses in the remaining portions of the pipe continue to decrease.

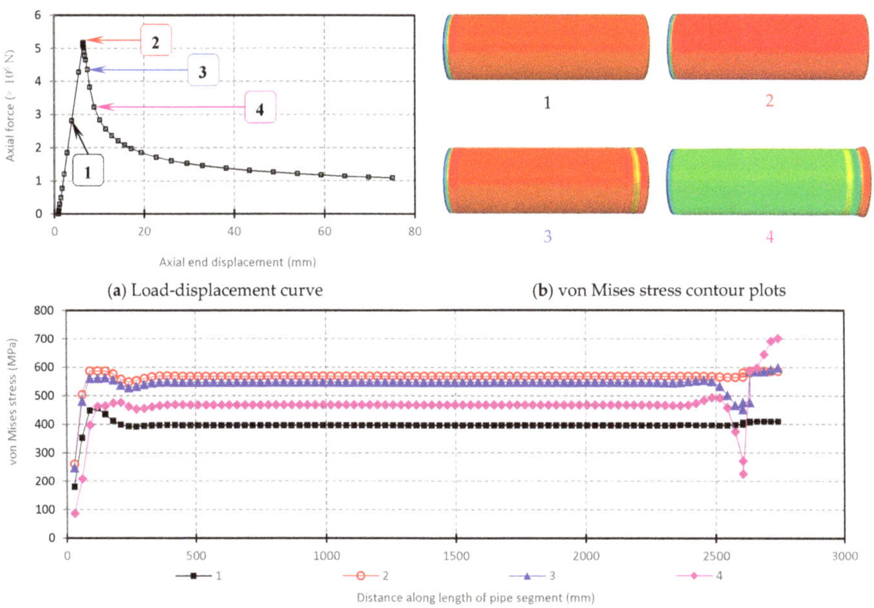

(**a**) Load-displacement curve (**b**) von Mises stress contour plots

(**c**) Longitudinal stress distribution plots

Figure 4. Axial load-displacement response and stress evolution in YPT DT4 pipes under uniform axial compression (YPL = 1.0%, $f_p = 0.4$).

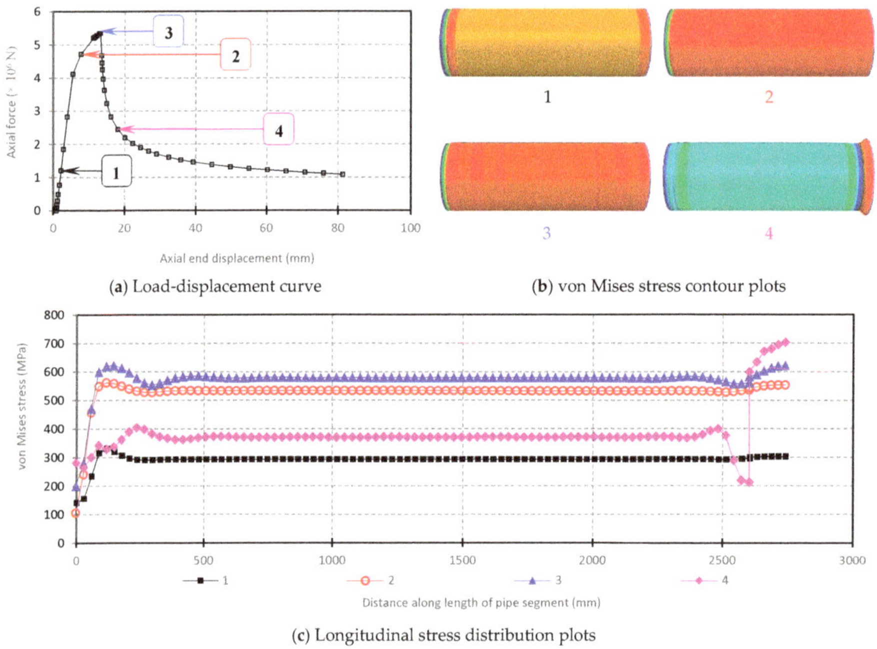

(a) Load-displacement curve (b) von Mises stress contour plots

(c) Longitudinal stress distribution plots

Figure 5. Axial load-displacement response and stress evolution in RHT DT4 pipes under uniform axial compression (PL/US = 0.691, f_p = 0.4).

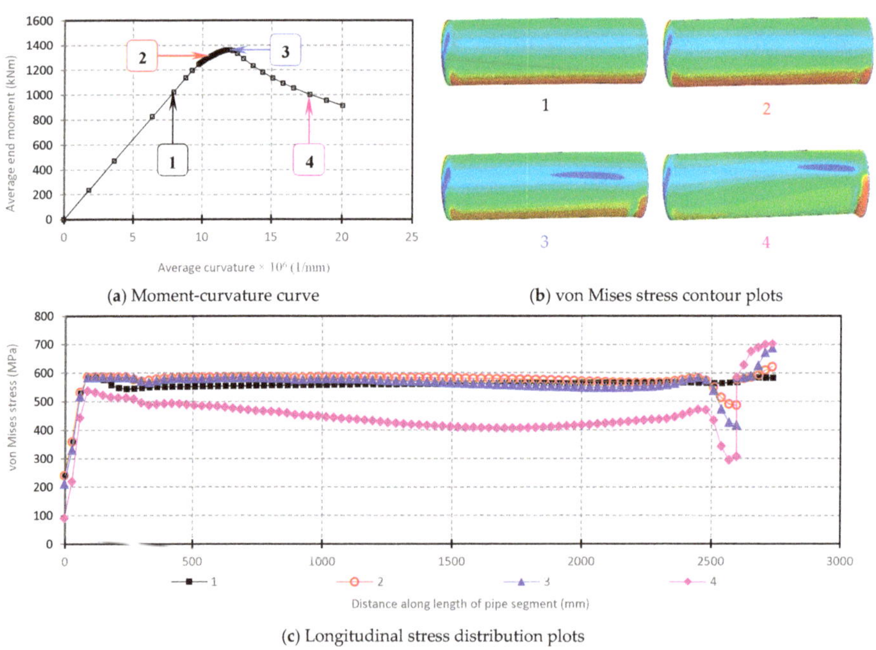

(a) Moment-curvature curve (b) von Mises stress contour plots

(c) Longitudinal stress distribution plots

Figure 6. Moment-curvature response and stress evolution in YPT DT4 pipes under uniform bending (YPL = 1.0%, f_p = 0.4).

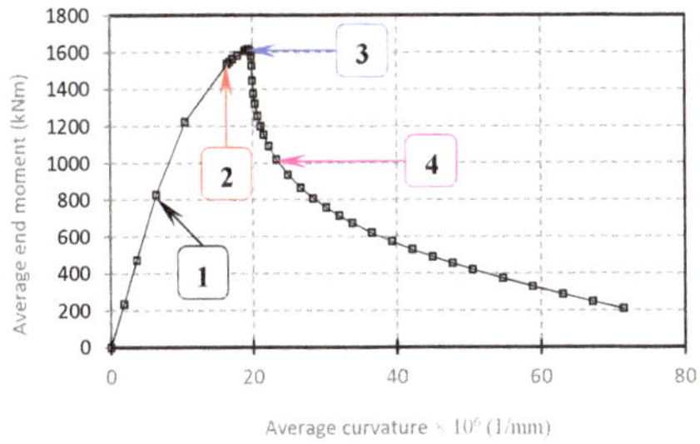

(**a**) Moment-curvature curve

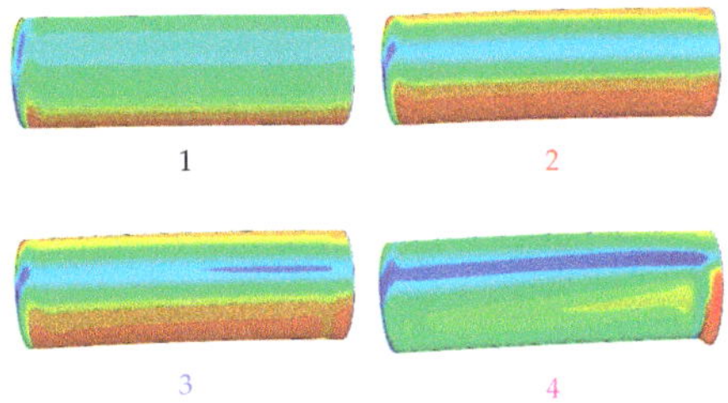

(**b**) von Mises stress contour plots

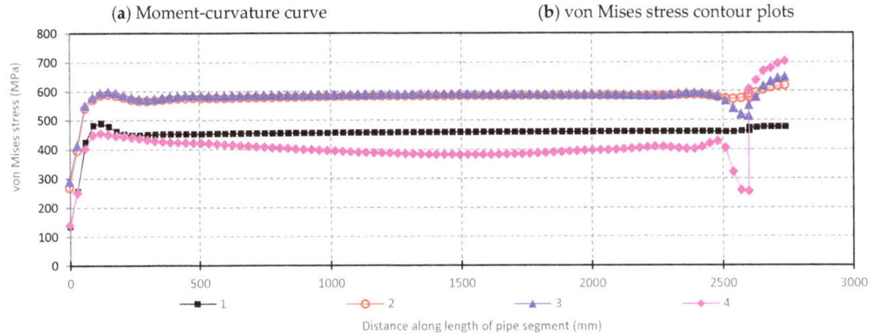

(**c**) Longitudinal stress distribution plots

Figure 7. Moment-curvature response and stress evolution in RHT DT4 pipes under uniform bending (PL/US = 0.691, f_p = 0.4).

4.3. Results of Parametric Analysis

The influence of the D/t ratio, internal pressure, and material strain-hardening (represented by the YPL for YPT pipes, and PLUS ratio for RHT pipes) on the CLS of X80 pipes is outlined herein for the three different loading conditions considered. The PLUS ratios, defined as the ratios of respective values of the proportionality limit stress (as indicated by the stress-strain curve labels in Figure 1) to the specified value of the ultimate proof stress (703 MPa), are obtained as: PL486 = 0.691, PL446 = 0.634, PL406 = 0.577, PL366 = 0.520. The ratio of applied internal pressure to the circumferential yield pressure is herein referred to as the "pressure factor", f_p, and various levels of internal pressurization are assigned numeric designations (0.0, 0.2, 0.4, 0.6, and 0.8) according to respective percentages (0%, 20%, 40%, 60%, and 80%) of the yield pressure. For combined axial compression + bending, the ratio of the applied stress to the limit stress of the pipe (obtained under uniform axial compression conditions) is herein referred to as the "compression factor", f_c, and various levels of axial compression are assigned numeric designations (0.2, 0.4, 0.6, and 0.8) according to respective percentages (20%, 40%, 60%, and 80%) of the limit stress.

4.3.1. Influence of D/t Ratio

The plots in Figures 8 and 9 illustrate the relationship between the CLS and the D/t ratio of YPT and RHT pipes, respectively. The plots in Figure 8 are obtained for YPT pipes with YPL = 1.50%, while the plots in Figure 9 are obtained for RHT pipes with PLUS ratio = 0.634. The plots for combined axial compression + bending in Figures 8c and 9c are both obtained for pipes subjected to compression factor, f_c = 0.6. The CLS trends are presented for all five levels of internal pressure considered.

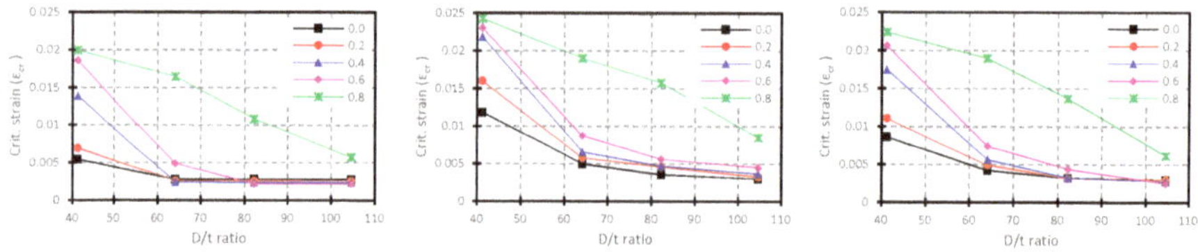

(a) (b) (c)

Figure 8. Plots of CLS vs. D/t ratio of YPT pipes with YPL = 1.50% for (**a**) uniform axial compression, (**b**) uniform bending, and (**c**) combined loading (f_c = 0.6).

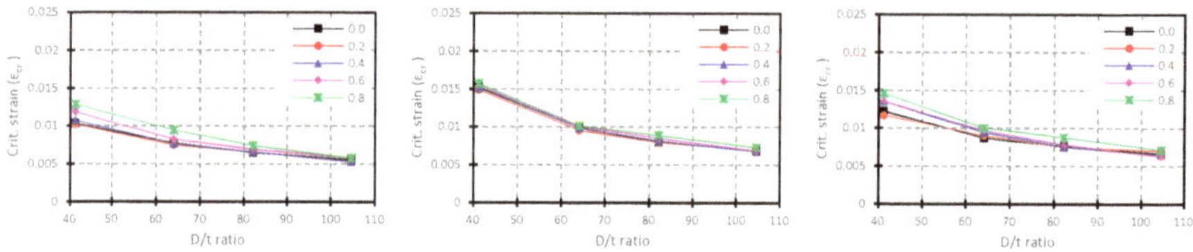

(a) (b) (c)

Figure 9. Plots of CLS vs. D/t ratio of RHT pipes with PL/US = 0.634 for (**a**) uniform axial compression, (**b**) uniform bending, and (**c**) combined loading (f_c = 0.6).

The most obvious aspect of the plots above is the nonlinear negative correlation between the CLS and the D/t ratio. The plots indicate the tendency for the CLS to reduce as the D/t ratio increases, and even more evident is this phenomenon in YPT pipes, especially at high levels of internal pressure. It is also reasonable to deduce that the influence of internal pressure on the CLS vs. D/t ratio trends is minimal in RHT pipes compared to YPT pipes, as the change in the slope of the CLS vs. D/t ratio trends for RHT pipes is observed to be generally much lower than for YPT pipes. The relationship between the CLS of YPT pipes and the D/t ratio becomes nearly linear at internal pressure = 80%. For intermediate D/t ratios (DT2 and DT3) and high D/t ratios (DT4), RHT pipes are observed to have a higher deformational capacity than YPT pipes at internal pressure $\leq$ 60% YS. However, at high internal pressure (f_p = 0.8) and/or low D/t ratio (DT1), the deformational performance of YPT pipes tends to supersede that of RHT pipes. The CLS of pipes under uniform bending is generally observed to surpass the CLS under uniform axial compression, while the CLS under combined axial compression and bending is somewhat bounded by the uniform axial and uniform bending CLS values.

4.3.2. Influence of Internal Pressure

The relationship between the CLS and internal pressure for YPT and RHT pipes is illustrated by the plots of CLS vs. f_p in Figures 10 and 11, respectively. The plots in Figure 10 are obtained for YPT pipes, with YPL = 1.25%, while the plots in Figure 11 are obtained for RHT pipes, with PLUS ratio = 0.691. Similar to the plots in Figures 8 and 9, the plots for combined axial compression + bending in Figures 10c and 11c are both obtained for pipes subjected to compression factor, f_c = 0.6. The CLS trends are presented for all four D/t ratios considered.

Unlike the relationship between the CLS and the D/t ratio, the correlation between the CLS and internal pressure is characteristically positive and is more evident in YPT pipes. In conformance with the deductions from the CLS vs. D/t ratio plots in Figure 9, variation of the internal pressure is observed to have a negligible influence on the CLS of RHT pipes for all D/t ratios. On the other hand, the influence of internal pressure on the CLS of YPT pipes is only minimal at intermediate D/t ratios (DT2 and DT3) and high D/t ratios (DT4) for internal pressures $\leq$ 60% YS. The most significant influence of internal pressure is observed in YPT pipes of low D/t ratios (DT1).

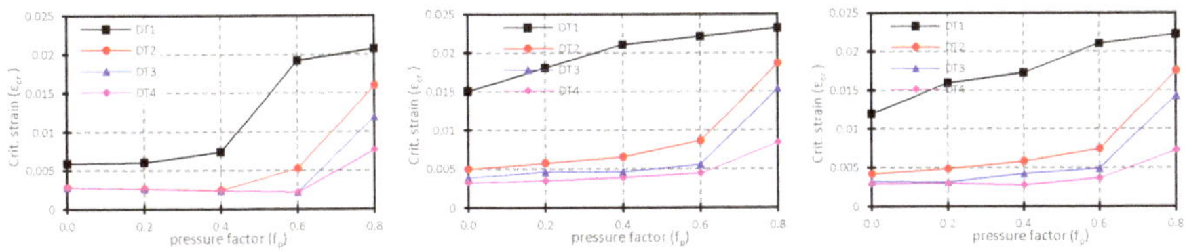

(a) (b) (c)

Figure 10. Plots of CLS vs. f_p of YPT pipes with YPL = 1.25% for (**a**) uniform axial compression, (**b**) uniform bending, and (**c**) combined loading (f_c = 0.6).

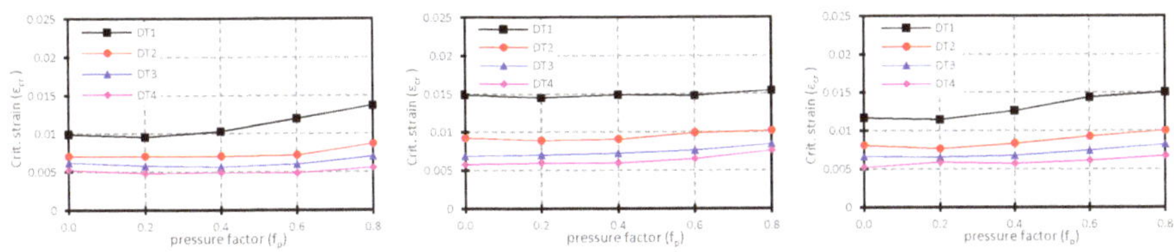

(a) (b) (c)

Figure 11. Plots of CLS vs. f_p of RHT pipes with PL/US = 0.691 for (**a**) uniform axial compression, (**b**) uniform bending, and (**c**) combined loading (f_c = 0.6).

4.3.3. Influence of Strain-Hardening Properties

The strain-hardening properties relate to the YPL of the stress-strain curve for YPT pipes, and the PLUS ratio for RHT pipes. The influence of variations in D/t ratio on the relationship between the CLS and the strain-hardening properties is considered to be significant, and is therefore included in the CLS-trend illustrations in Figures 12–15. The plots in Figures 12 and 13 are both obtained for YPT pipes, with D/t ratio = 41.15 and 82.16, respectively, while the plots in Figures 14 and 15 are both obtained for RHT pipes, with D/t ratio = 41.15 and 82.16, respectively. The plots for combined axial compression + bending in Figures 12c, 13c, 14c and 15c are all obtained for pipes subjected to compression factor, f_c = 0.2.

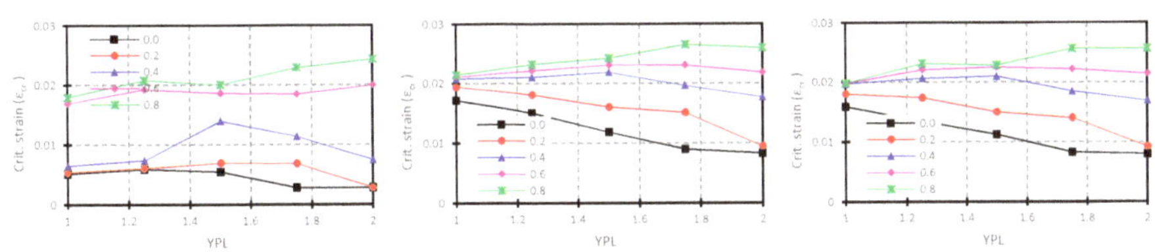

(a) (b) (c)

Figure 12. Plots of CLS vs. YPL of YPT DT1 pipes for (**a**) uniform axial compression, (**b**) uniform bending, and (**c**) combined loading (f_c = 0.2).

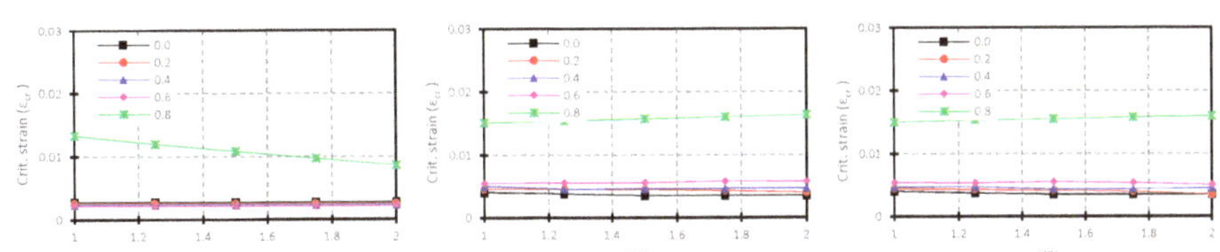

(a) (b) (c)

Figure 13. Plots of CLS vs. YPL of YPT DT3 pipes for (**a**) uniform axial compression, (**b**) uniform bending, and (**c**) combined loading (f_c = 0.2).

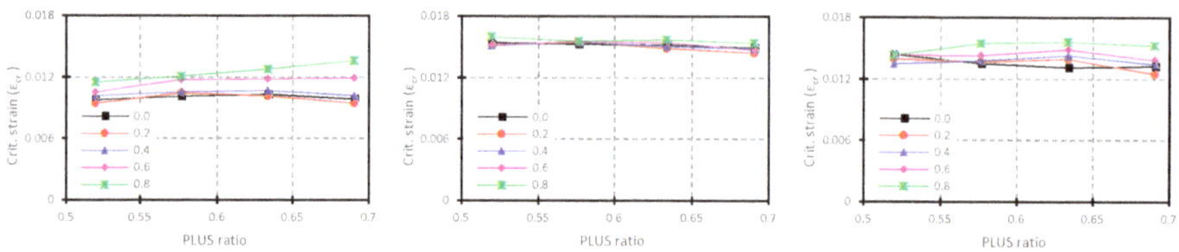

(a) (b) (c)

Figure 14. Plots of CLS vs. PLUS ratio of RHT DT1 pipes for (**a**) uniform axial compression, (**b**) uniform bending, and (**c**) combined loading (f_c = 0.2).

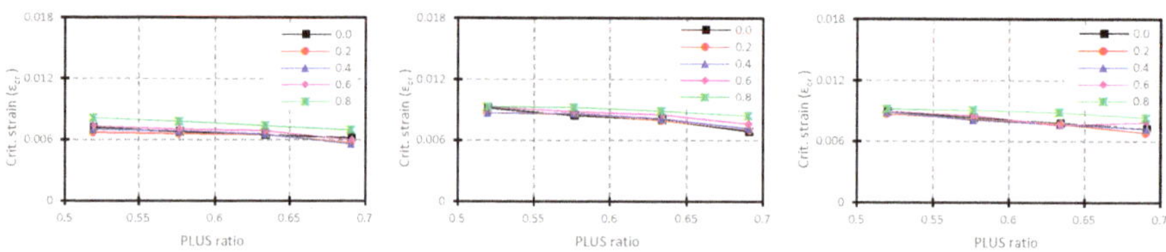

(a) (b) (c)

Figure 15. Plots of CLS vs. PLUS ratio of RHT DT3 pipes for (**a**) uniform axial compression, (**b**) uniform bending, and (**c**) combined loading (f_c = 0.2).

The plots above indicate that the influence of the strain hardening properties of the pipe material is more prevalent in pipes with low D/t ratios (DT1) for YPT pipes; as the D/t ratio of the pipe increases, the influence of the YPL on the CLS diminishes. For YPT pipes of low D/t ratio (DT1), a positive relationship is observed between the YPL and the CLS at a high internal pressure (f_p = 0.8). The YPL indicates a progressively negative correlation with the CLS as the internal pressure drops below 60% YS. On the other hand, a negative correlation between the PLUS ratio and the CLS is more apparent at higher D/t ratios and the influence tends to diminish as the D/t ratio decreases. As is the case for YPT pipes with low D/t ratio (DT1), a positive (but relatively less significant) relationship between the PLUS ratio and the CLS of RHT pipes with low D/t ratio (DT1) is also observed at high internal pressure (f_p = 0.8).

4.3.4. Influence of Compressive Net-Section Axial Force for Combined Loading

The effect of a compressive net-section axial force on the CLS of pipes subjected to monotonically-increasing curvature is depicted in Figures 16 and 17. The relationship between the CLS and the compression factor, f_c, is represented by the plots in Figure 16 for YPT pipes, with D/t ratio = 64.08 and YPL = 1.25%. Figure 17 comprises CLS vs. f_c plots for RHT pipes, with D/t ratio = 41.15 and PLUS ratio = 0.577. Each plot corresponds to a respective level of internal pressure, as indicated by the elements of the graph legends.

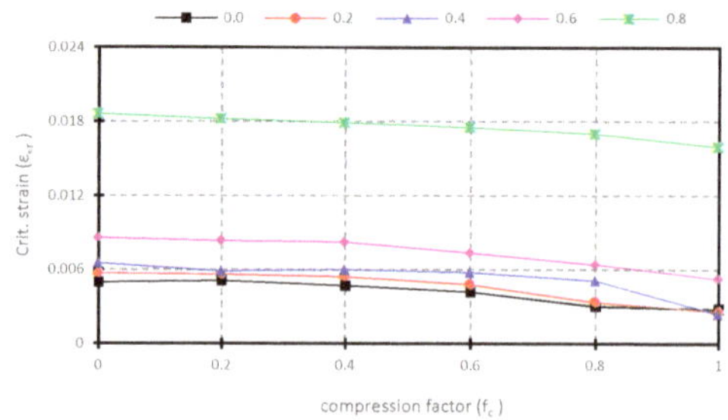

Figure 16. Plots of CLS vs. f_c for YPT DT2 pipes with YPL = 1.25%.

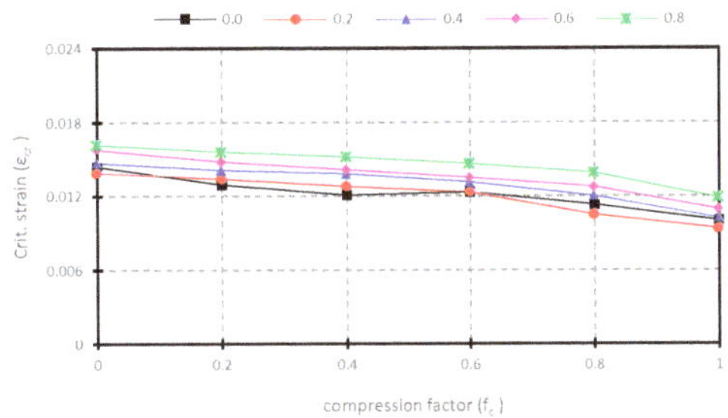

Figure 17. Plots of CLS vs. f_c for RHT DT1 pipes with PL/US = 0.577.

An inverse relationship is observed between the compression factor and the CLS for both YPT and RHT pipes. Observation of various parametric combinations of the factors investigated in this study shows a strong and general indication that the upper and lower bounds of the CLS for pipes subjected to combined axial compression + bending is coincident with the CLS under uniform bending and the CLS under uniform axial compression, respectively. The observed upper and lower bound phenomenon is portrayed accordingly on the horizontal axes of the plots in Figures 16 and 17. such that a $f_c = 0$ represents uniform bending and $f_c = 1$ represents uniform axial compression. It can therefore be inferred that for all combinations of investigated parameters, the CLS of both YPT and RHT pipes reduces progressively and somewhat nonlinearly from a state of uniform bending to a state of uniform axial compression as the compressive net-section axial force is increased from 0% to 100% of the limit axial stress.

5. Derivation of Semi-Empirical Models

5.1. CLS Derivation

The CLS trends obtained with respect to the various considered parameters were examined to derive appropriate individual variable functions for each parameter, while taking any significant inter-relationships between the constitutive factors into consideration. Taking a cue for successful development of semi-empirical models for prediction of mechanical performance in pipelines from recent numerical studies [58,59,61], a multiplicative approach was employed for development of six nonlinear mathematical expressions, each formed as a product of the individual variable functions for the respective constitutive parameters. The basic form of the nonlinear expressions for the CLS of YPT and RHT pipes under the three loading conditions investigated is:

$$\varepsilon_{cr} = f_{cr}[\pi_1, \pi_2, \pi_3, \pi_4, \pi_5, \pi_6] = f_1 . f_2 . f_3 . f_4 . f_5 . f_6 \quad (4)$$

where f_1, f_2, f_3, f_4, f_5, and f_6 represent the D/t ratio function (f_{dt}), the pressure factor function (f_{fp}), the strain hardening function (f_{sh}), the heel factor function (f_{hf}), the knee-to-heel ratio function (f_{kh}), and the compression factor function (f_{fc}), respectively. Here, π_1 represents the D/t ratio $\left(\frac{D}{t}\right)$, π_2 represents the pressure factor $\left(\frac{p}{p_y}\right)$, π_3 represents the yield plateau length (YPL) for YPT pipes or the PLUS ratio $\left(\frac{\sigma_{pl}}{\sigma_{us}}\right)$ for RHT pipes, π_4 represents the heel factor (h_f), π_5 represents the *"knee-to-heel"* ratio $\left(\frac{k_f}{h_f}\right)$, and π_6 represents the compression factor (f_c).

For simplicity of presentation, the three loading conditions investigated are hereafter assigned the following alphanumeric designations: LC1 for uniform axial compression, LC2 for uniform bending, and LC3 for combined axial compression + bending.

An iterative process was thoroughly implemented for deriving the final form of the individual variable functions by targeting the highest possible value of the coefficient of multiple determination

(R^2). An R^2 value greater than 0.95 was considered to be a satisfactory goodness-of-fit between the FEA-derived CLS values and the predictions of the derived nonlinear expressions.

The individual variable functions for YPT pipes are outlined thus:

$$\begin{array}{lll}
f_1 = f_{dt} & = a_1.(\pi_1)^{b_1} & \\
f_2 = f_{fp} & = \begin{bmatrix} a_2 + (b_2 + c_2\pi_1).Exp(\pi_2)^{(d_2+e_2\pi_4)} \\ a_2\pi_4 + (b_2 + c_2\pi_1).Exp(\pi_2)^{(d_2+e_2\pi_4)} \end{bmatrix} & \begin{array}{l}(LC1\ \&\ LC3) \\ (LC2\ only)\end{array} \\
f_3 = f_{sh} & = a_3\pi_4 + (b_3\pi_2 + c_3\pi_1).(\pi_3)^{d_3} & (LC3\ only) \\
f_4 = f_{hf} & = \begin{bmatrix} a_4 + (b_4\pi_1 + c_4\pi_2).(\pi_4)^{d_4} \\ a_4 + (b_4\pi_1).(\pi_4)^{c_4} \end{bmatrix} & \begin{array}{l}(LC1\ \&\ LC2) \\ (LC3\ only)\end{array} \\
f_5 = f_{kh} & = a_5 + (b_5\pi_3).(\pi_5)^{c_5} & \\
f_6 = f_{fc} & = a_6 + (b_6\pi_2 + c_6\pi_1).(\pi_6)^{d_6} & (LC3\ only)
\end{array} \tag{5}$$

The individual variable functions for RHT pipes are given by:

$$\begin{array}{lll}
f_1 = f_{dt} & = a_1.(\pi_1)^{b_1} & \\
f_2 = f_{fp} & = (a_2 + b_2.\pi_2).[c_2 + (d_2 + e_2\pi_1).Exp(\pi_2)^{(f_2)}] & \\
f_3 = f_{sh} & = a_3 + (b_3\pi_2 + c_3).(\pi_3)^{d_3} & \\
f_4 = f_{hf} & = a_4 + b_4.(\pi_4)^{c_4} & \\
f_5 = f_{kh} & = a_5 + b_5.(\pi_5)^{c_5} & \\
f_6 = f_{fc} & = a_6 + (b_6\pi_1).(\pi_6)^{c_6} & (LC3\ only)
\end{array} \tag{6}$$

The nonlinear regression coefficients and coefficients of multiple determination (R^2), obtained by advanced nonlinear regression analysis using the "NonlinearModelFit" command in the powerful computational package (Wolfram Mathematica [57]), are presented in Table 2.

Table 2. Nonlinear regression coefficients.

REG. COEFF.	YPT			RHT		
	LC1	*LC2*	*LC3*	*LC1*	*LC2*	*LC3*
a_1	0.07582	0.003171	0.7302	1.586	1.231	0.1829
b_1	−2.192	−2.386	−4.256	−0.5804	−1.112	−0.6736
a_2	0.06014	18.57	−386.2	2.919	0.4872	0.947
b_2	0.005009	−0.000358	−7.111	0.7857	−0.1282	−0.2245
c_2	−0.00003623	8.397×10^{-6}	0.02827	−218.1	−48.49	170.2
d_2	8.186	11.66	7.735	−34.63	−16.93	6.393
e_2	0.6897	3.537	−2.354	0.1272	−0.2596	−0.02223
f_2	-	-	-	3.565	1.335	2.82
a_3	-	-	72.59	−1.513	0.500779	7.783
b_3	-	-	182.2	−1.234	6.915×10^{-5}	14.71
c_3	-	-	1.116	3.225	−0.501086	1.263
d_3	-	-	3.632	0.2451	0.0002846	4.778
a_4	86.44	15.09	−30.34	−2.608	−1.272	3.37183
b_4	0.5966	0.004361	−1.788	1.392	2.396	−3.37182
c_4	−136.5	1.019	1.665	−0.1513	2.458	-1.127×10^{-6}
d_4	0.07415	−3.204	-	-	-	-
a_5	40.73	271.9	2.464×10^{-5}	−1.40902	−0.2618	−5.336
b_5	−0.0003071	−1.189	1.216	1.40905	0.4994	4.04
c_5	1.274	0.6593	−1.615	4.185×10^{-6}	0.9315	−0.1044
a_6	-	-	30.23	-	-	−18.19
b_6	-	-	−200.4	-	-	0.03894
c_6	-	-	3.893	-	-	2.456
d_6	-	-	−0.199	-	-	-
R^2	**0.962034**	**0.990746**	**0.983041**	**0.997271**	**0.998619**	**0.997224**

The applicable ranges for the constitutive factors of the developed semi-empirical models are determined by the range of the parameters used in the FE analyses, given in Table 3 as follows.

Table 3. Applicable range for dimensionless parameters.

Par.	YPT		RHT	
	≥	≤	≥	≤
π_1	41	105	41	105
π_2	0.0	0.8	0.0	0.8
π_3	1	2	0.520	0.691
π_4	0.250	0.519	1.58	3.98
π_5	161	1497	2.9	43.7
π_6	>0.0	<1.0	>0.0	<1.0

5.2. Limit Stress Derivation

The nonlinear regression procedures employed for deriving the above semi-empirical equations for the CLS were extended to obtain two nonlinear expressions for the limit stress, for YPT pipes and RHT pipes. The values of the limit stress were obtained based on the values of the peak load on the axial load vs. axial deformation plots for pipes subjected to uniform axial compression. The following regression equations are required for determining the appropriate fractions of the limit stress that feed into the semi-empirical equations for evaluating the CLS of pipes subjected to combined axial compression + bending:

For YPT pipes,

$$\begin{aligned}\sigma_{lim,YP} = {} & 6.16262(586) * [(8.49601 - 7.19159.\pi_2) * (\pi_1{}^{-0.307568})] \\ & *[(-8.36548) + 8.36562 * (Exp(\pi_2)^{5.59778*10^{-6}})] \\ & *[(1.52236.\pi_1) + 350.286 * (\pi_3{}^{-0.038848})]\end{aligned} \tag{7}$$

For RHT pipes,

$$\begin{aligned}\sigma_{lim,RH} = {} & 5.45103(586) * [(7.32071 - 6.48533.\pi_2) * (\pi_1{}^{-0.107001})] \\ & *[(8.45167) + 6.90582 * (Exp(\pi_2)^{0.843233})] \\ & *[(5.23197) + 7.78213 * (\pi_4{}^{-0.451323})] \\ & *[(-5.67028) + 5.67057 * (\pi_5{}^{-2.72452*10^{-6}})]\end{aligned} \tag{8}$$

6. Goodness-of-Fit

In addition to highlighting the coefficients of multiple determination, which indicate the level of correspondence between the predictions of the developed semi-empirical equations and the FEA results, the plots in Figures 18 and 19 are presented to provide graphical illustrations of the accuracy of the developed models. The model-predicted CLS values were plotted against the FEA-derived CLS values in Figures 18 and 19, respectively, for YPT pipes and RHT pipes subjected to (a) uniform axial compression, (b) uniform bending, and (c) combined axial compression + bending.

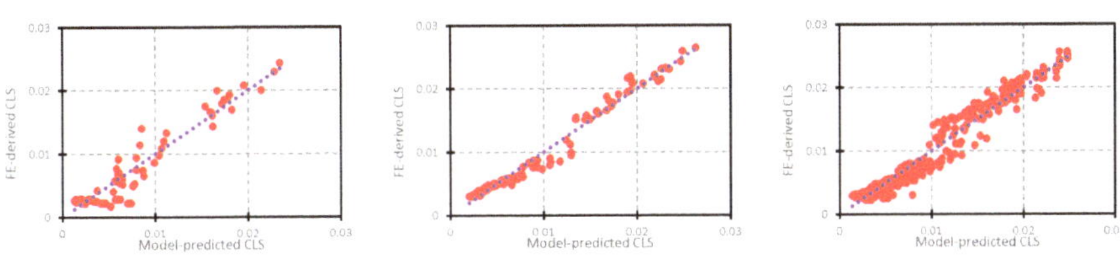

(a) (b) (c)

Figure 18. Model prediction vs. FEA results for YPT pipes for (**a**) LC1, (**b**) LC2, and (**c**) LC3.

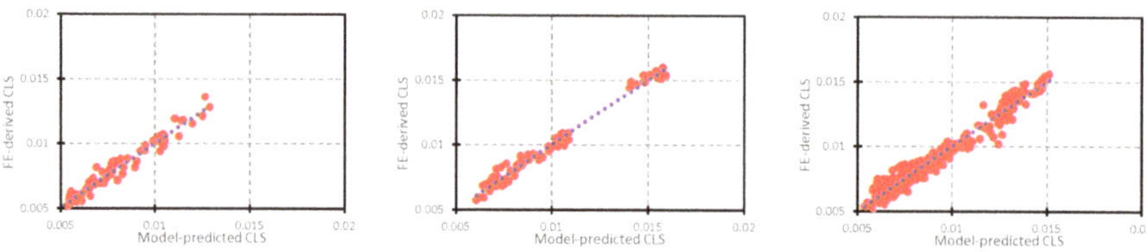

(a) (b) (c)

Figure 19. Model prediction vs. FEA results for RHT pipes for (**a**) LC1, (**b**) LC2, and (**c**) LC3.

7. Discussion of Results

The results obtained for YPT and RHT pipes indicate that the CLS varies inversely with the D/t ratio for all combinations of parameters. For the CLS of RHT pipes, however, the effect of varying the D/t ratio is observed to be mild for all cases considered. The observed deformation behavior, with respect to the D/t ratio, may be attributed to the somewhat dichotomous phenomenon associated with the buckling response of cylindrical shell structures as the initiation of buckling transitions from the elastic region of the material stress-strain curve (elastic buckling) to the nonlinear region of the material stress-strain curve (inelastic or plastic buckling). RHT pipes exhibit continuous strain-hardening, and invariably a gradual reduction in the tangential modulus of the stress-strain curve beyond the proportionality limit point, and therefore do not experience significant changes in deformational capacity due to the delay of buckling initiation. The effect of D/t variation on the CLS of RHT pipes can therefore be deduced to be predominantly influenced by the increase in axial or bending rigidity of the pipe due to decrease in slenderness of the pipe wall, and vice versa. YPT pipes, on the other hand, feature a sudden zeroing of the slope of the stress-strain curve at the proportionality limit, and subsequent strain hardening that commences toward the end of a significantly-extended yield plateau. The existence of a distinct yield point in the stress-strain curve tends to cause an automatic trigger of buckling instability in YPT pipes once the initiation of buckling extends beyond the elastic region of the pipe material. However, there exists the possibility of recovery from initial softening of the load-deformation or moment-curvature response if intrinsic properties (e.g., D/t ratio) or loading conditions (e.g., internal pressure) provide additional stiffness that sufficiently counteracts the detrimental impact of the yield plateau on the deformational response of the pipe.

The deformational response of a pipe to uniform axial compression is characteristically different from the response to bending, even with the additional application of a compressive force. Under uniform axial compression, the entire cross-section of the pipe is susceptible to buckling, and the pipe relies on geometric, material, loading, and boundary conditions for sustenance against induced deformation. Under bending, however, the pipe gains additional stiffness due to the counteractive interaction between tensile and compressive stresses in the pipe's cross-section. This can, therefore, be considered to be responsible for the more apparent effect of D/t variation on the CLS of YPT pipes subjected to uniform bending and combined bending, and axial compression at pressure levels between 0% and 60% YS. The deformational response of a YPT pipe subjected to all three loading conditions at a high level of internal pressure (f_p = 0.8) is analogous to the deformational response of RHT pipes, as an almost-linear relationship is observed between the CLS and the D/t ratio. This is attributable to the tendency that a pressure level of f_p = 0.8 significantly increases the stability of the pipe's response, such that catastrophic collapse leading to failure of the pipe inevitably takes advantage of the strain-hardening portion of the stress-strain curve at the end of the yield plateau. This also implies that the influence of D/t ratio variation on the CLS will be predominantly impelled by the stiffness of the pipe response associated with pipe wall slenderness factors.

The results presented for the influence of internal pressure on the CLS further substantiate the assertion that increase of the internal pressure in a pipe results in a corresponding increase in the stability of the pipe's response to loading. Internal pressure hinders the localization of wrinkling and bulging deformation, and also increases the tendency for buckling to be initiated further along the stress-strain curve. The results indicate a negligible influence of internal pressure variation on the CLS of RHT pipes at all levels of pressure. On the other hand, the CLS of YPT pipes tends to be unaffected by changes in internal pressure between 0% and 60% YS, whereas at higher pressure (f_p = 0.8), there is a spike in the CLS depending on the D/t ratio.

For YPT pipes, varying the length of the yield plateau is observed to have a higher effect on the CLS as the D/t ratio of the pipe decreases. For YPT pipes with D/t ratio = 41.15, the CLS is likely to be negatively affected by an increase in YPL at levels of internal pressure between 0% and 60% YS, whereas the YPL positively correlates with the CLS at a high level of internal pressure (f_p = 0.8). Unlike YPT pipes, varying the PLUS ratio tends to have a less significant influence on the CLS of RHT pipes as the D/t ratio decreases. At DT1, however, the influence of the PLUS ratio on the CLS of RHT pipes is observed to be negligible, except at a level of pressure equal to 80% YS. It can be inferred from the results for YPT pipes that improvement of the CLS for pipes with lower D/t ratio or high internal pressure is not necessarily automatic, as the length of the yield plateau exhibits a secondary, but significant, influence on the deformational capacity. Figure 12b,c further illustrate that subjection of a pipe to bending considerably enhances the stiffness response of the pipe, even at zero-to-low levels of internal pressure. Enhanced stiffness is, therefore, accompanied by an increased tendency for buckling to be initiated in the nonlinear region of the material stress-strain curve. The higher stiffness of a pipe under bending, compared to a pipe under uniform axial compression, leads to a relatively greater positive effect of a longer YPL on the CLS of a pipe under bending, especially at internal pressures as high as 80% YS. However, additional benefits of a longer yield plateau tend to diminish once the YPL exceeds 1.75%. For levels of internal pressure between 0% and 60% YS, increasing the length of the yield plateau results in a reduced ability of the pipe to recover from initial softening of the mechanical response before reaching the actual peak stress or moment. A decrease in the PLUS ratio for RHT pipes implies a lower proportionality limit point, and invariably, 'superior' strain hardening and vice versa. At higher D/t ratios, initiation of buckling tends to occur before the yield point of the material is reached, implying that materials with a lower proportionality limit earlier experience a softer inelastic response, which translates to higher deformability. As the D/t ratios become lower, initiation of buckling shifts towards the yield point of the material, where the stress-strain curves of materials with different PLUS ratios are likely to converge, hence the effect of PLUS ratio variation becomes negligible. However, beyond the yield point, the stress-strain curve of higher-PLUS ratio materials exhibits a softer post-YS response; hence, if the initiation of buckling is delayed beyond the yield point of the material (as is likely the case when internal pressure is as high as 80% YS), a higher PLUS ratio (i.e., inferior strain hardening) may have a positive effect on the deformational capacity of an RHT pipe, and vice versa.

8. Conclusions

The relationships between the CLS of pipelines and the relevant influencing parameters was studied herein, and a comparison was made between three different loading conditions; uniform axial compression, uniform bending, and combined axial compression and bending. The distinction in the nonlinear strain-hardening of YPT and RHT pipes has been observed in previous studies [62–66] to significantly impact on the deformational capacity of pipes, but marginally affect the limit stress and limit moment. An extensive parametric study was, therefore, conducted to derive the functional relationships between nondimensionalized parameters representing the relevant factors and the deformational performance of pipes made of API X80 grade line pipe steel, while separately considering the material behavior under YPT and RHT classifications.

The shape constants of a new material model were used to fully describe the shape of the nonlinear portion of the stress-strain curve of the material, and were included as dimensionless parameters in developing semi-empirical models for predicting the CLS. Individual variable functions were developed to mathematically represent the trends of the CLS with respect to the parameters that constitute the semi-empirical models. Advanced nonlinear multiple regression was performed using Wolfram Mathematica to obtain the nonlinear regression coefficients for each of the derived semi-empirical equations, and effort was made to achieve the highest possible R^2 value for each nonlinear expression by accounting for the complex interrelationships between the different parameters. The nonlinear regression analysis yielded high R^2 values (>0.95), indicating an excellent goodness-of-fit between the model prediction and the FEA-derived results of the CLS.

The semi-empirical models developed in this paper are considered as handy tools for evaluating the deformational capacity of both unpressurized and pressurized API X80 line pipes. The results clearly demonstrate that the CLS values for combined axial compression + bending typically exist in a spectrum between the CLS for uniform bending and the CLS for uniform axial compression. It is, however, the prerogative of the designer to determine the level of axial compression to consider, especially in the case of displacement-controlled loading conditions, where the axial stress is not readily derivable. For cases such as this, it may be necessary to conservatively adopt the models for uniform axial compression as a lower bound measure of the CLS.

Author Contributions: This research paper is a product of assiduous collaboration between the authors: conceptualization, O.N. and S.A.; methodology, O.N. and S.A.; validation, O.N., J.J.R.C., and S.A.; formal analysis, O.N.; investigation, O.N.; resources, M.M., J.J.R.C., and S.A.; writing—original draft preparation, O.N.; writing—review and editing, O.N. and S.A.; supervision, S.A.; project administration, M.M.; funding acquisition, M.M. and S.A.

Funding: This research and the APC for publication were funded by TransCanada Pipelines Ltd.

Acknowledgments: The authors wish to express their appreciation to TransCanada Pipelines Ltd. for providing the financial support and technical input necessary for conducting this research study.

Conflicts of Interest: The authors declare no conflict of interest.

References

1. Taylor, L.C. Fault displacement and ground deformation associated with surface faulting. In Proceedings of the Technical Council on Lifeline Earthquake Engineering Specialty Conference, Los Angeles, CA, USA, 30–31 August 1977; pp. 338–353.
2. Ariman, T.; Muleski, G.E. A review of the response of buried pipelines under seismic excitations. *Earthq. Eng. Struct. Dyn.* **1981**, *9*, 133–152. [CrossRef]
3. Selvadurai, A.P.S. Mechanics of buried pipelines induced by random ground movement. In Proceedings of the CSCE Annual Conference (Engineering Mechanics Symposium), Vancouver, BC, Canada, 29–31 May 1991; pp. 142–151.
4. Yun, H.; Kyriakides, S. On the beam and shell modes of buckling of buried pipelines. *Soil Dyn. Earthq. Eng.* **1990**, *9*, 179–193. [CrossRef]
5. Dama, E.; Karamanos, S.A.; Gresnigt, A.M. Failure of Locally Buckled Pipelines. *J. Press. Vessel Technol.* **2007**, *129*, 272. [CrossRef]
6. Trifonov, O.V.; Cherniy, V.P. Elastoplastic stress–strain analysis of buried steel pipelines subjected to fault displacements with account for service loads. *Soil Dyn. Earthq. Eng.* **2012**, *33*, 54–62. [CrossRef]
7. American Lifelines Alliance. *Guidelines for the Design of Buried Steel Pipe (with Addenda through February 2005)*; American Society of Civil Engineers: Reston, VA, USA, 2005.
8. Vazouras, P.; Karamanos, S.A.; Dakoulas, P. Finite element analysis of buried steel pipelines under strike-slip fault displacements. *Soil Dyn. Earthq. Eng.* **2010**, *30*, 1361–1376. [CrossRef]
9. Vazouras, P.; Karamanos, S.A.; Dakoulas, P. Mechanical behavior of buried steel pipes crossing active strike-slip faults. *Soil Dyn. Earthq. Eng.* **2012**, *41*, 164–180. [CrossRef]
10. Vazouras, P.; Dakoulas, P.; Karamanos, S.A. Pipe–soil interaction and pipeline performance under strike–slip fault movements. *Soil Dyn. Earthq. Eng.* **2015**, *72*, 48–65. [CrossRef]
11. Kan, W.C.; Weir, M.; Zhang, M.M.; Lillig, D.B.; Barbas, S.T.; Macia, M.L.; Biery, N.E. Strain-Based Pipelines: Design Consideration Overview. In Proceedings of the Eighteenth International Offshore and Polar Engineering Conference, Vancouver, BC, USA, 6–11 July 2008; International Society of Offshore and Polar Engineers: Vancouver, BC, USA, 2008.

12. *API 1111 Design, Construction, Operation, and Maintenance of Offshore Hydrocarbon Pipelines*; American Petroleum Institute (API): Washington, DC, USA, 1999.
13. Det Norske Veritas. *Submarine Pipeline Systems—DNV-OS-F101*; DNV: Oslo, Norway, 2010.
14. American Bureau of Shipping. *Guide for Building and Classing Subsea Pipeline Systems*; ABS: Houston, TX, UAS, 2006.
15. *Canadian Standards Association Oil and Gas Pipeline Systems*; CSA-Z662-2007; CSA Group: Mississauga, ON, Canada, 2007.
16. Macia, M.L.; Kibey, S.A.; Arslan, H.; Bardi, F.; Ford, S.J.; Kan, W.C.; Cook, M.F.; Newbury, B. Approaches to Qualify Strain-Based Design Pipelines. In Proceedings of the 2010 8th International Pipeline Conference, Calgary, AB, Canada, 27 September–1 October 2010; Volume 4, pp. 365–374.
17. Liu, B.; Liu, X.J.; Zhang, H. Strain-based design criteria of pipelines. *J. Loss Prev. Process Ind.* **2009**, *22*, 884–888. [CrossRef]
18. Wang, Y.-Y.; Liu, M.; Horsley, D.; Salama, M.; Sen, M. Overall Framework of Strain-Based Design and Assessment of Pipelines. In Proceedings of the 10th International Pipeline Conference, Calgary, AB, Canada, 29 September–3 October 2014; p. V004T11A023.
19. Kenny, S.; Barrett, J.; Phillips, R.; Popescu, R. Integrating Geohazard Demand And Structural Capacity Modelling Within a Probabilistic Design Framework For Offshore Arctic Pipelines. In Proceedings of the The Seventeenth International Offshore and Polar Engineering Conference, Lisbon, Portugal, 1–6 July 2007; International Society of Offshore and Polar Engineers: Lisbon, Portugal, 2007.
20. Nessim, M.; Zhou, W.; Zhou, J.; Rothwell, B. Reliability Based Design and Assessment for Location-Specific Failure Threats With Application to Natural Gas Pipelines. *J. Press. Vessel Technol.* **2009**, *131*, 041701. [CrossRef]
21. Lorenz, R. Achsensymmetrische Verzerrungen in dünnwandigen Hohlzylindern. *Zeitschrift des Vereines Deutscher Ingenieure* **1908**, *52*, 1706–1713.
22. Timoshenko, S.P. Einige stabilitäts probleme der elastizitäts theorie. *Z. Math. Phys.* **1910**, *58*, 337–385.
23. Southwell, R.V. On the General Theory of Elastic Stability. *Philos. Trans. R. Soc. A Math. Phys. Eng. Sci.* **1914**, *213*, 187–244. [CrossRef]
24. Robertson, A. *The Strength of Tubular Struts*; Report No.: 23-9003-1-29; British Aeronautical Research Committee: Bedford, UK, 1929.
25. Wilson, W.M.; Newmark, N.M. The strength of thin cylindrical shells as columns. In Proceedings of the Civil Engineering Classics, The Engineering Experiment Station, University of Illinois, Urbana, IL, USA, 28 February 1933; pp. 1–42.
26. Lundquist, E.E. *Strength Tests of Thin-Walled Duralumin Cylinders in Compression*; NACA Technical Note, No 473; National Advisory Committee for Aeronautics: Washington, DC, USA, 1934.
27. Von Kármán, T.; Tsien, H.S. The Buckling of Thin Cylindrical Shells under Axial Compression. In *Collected Works of Hsue-Shen Tsien (1938–1956)*; Elsevier: Oxford, UK, 2012; Volume 8, pp. 165–181, ISBN 9780123982773.
28. Koiter, W.T. The Stability of Elastic Equilibrium. Ph.D. Thesis, Cornell University, New York, NY, USA, 1945.
29. Brazier, L.G. On the Flexure of Thin Cylindrical Shells and Other "Thin" Sections. *Proc. R. Soc. A Math. Phys. Eng. Sci.* **1927**, *116*, 104–114. [CrossRef]
30. Jirsa, J.O.; Lee, F.H.; Wilhoit, J.C.; Merwin, J.E. Ovaling Of Pipelines Under Pure Bending. In Proceedings of the Offshore Technology Conference, Houston, TX, USA, 1–3 May 1972.
31. Sherman, D.R. Test of Circular Steel Tubes in Bending. *J. Struct. Div.* **1976**, *102*, 2181–2195.
32. Tugcu, P.; Schroeder, J. Plastic deformation and stability of pipes exposed to external couples. *Int. J. Solids Struct.* **1979**, *15*, 643–658. [CrossRef]
33. Reddy, B.D. An experimental study of the plastic buckling of circular cylinders in pure bending. *Int. J. Solids Struct.* **1979**, *15*, 669–683. [CrossRef]
34. Corona, E.; Kyriakides, S. On the collapse of inelastic tubes under combined bending and pressure. *Int. J. Solids Struct.* **1988**, *24*, 505–535. [CrossRef]
35. Mathon, C.; Limam, A. Experimental collapse of thin cylindrical shells submitted to internal pressure and pure bending. *Thin-Walled Struct.* **2006**, *44*, 39–50. [CrossRef]
36. Lundquist, E.E. *Strength Tests of Thin-Walled Duralumin Cylinders in Pure Bending*; NACA Technical Note, No 479; National Advisory Committee for Aeronautics: Langley Field, VA, USA, 1933.

37. Donnell, L.H. A new theory for the buckling of thin cylinders under axial compression and bending. *Trans. ASME* **1934**, *56*, 795–806.
38. Suer, H.S.; Harris, L.A.; Skene, W.T.; Benjamin, R.J. The Bending Stability of Thin-Walled Unstiffened Circular Cylinders Including the Effects of Internal Pressure. *J. Aerosp. Sci.* **1958**, *25*, 281–287. [CrossRef]
39. Houliara, S.; Karamanos, S.A. Buckling and post-buckling of long pressurized elastic thin-walled tubes under in-plane bending. *Int. J. Non. Linear. Mech.* **2006**, *41*, 491–511. [CrossRef]
40. Limam, A.; Lee, L.-H.; Corona, E.; Kyriakides, S. Inelastic wrinkling and collapse of tubes under combined bending and internal pressure. *Int. J. Mech. Sci.* **2010**, *52*, 637–647. [CrossRef]
41. American Petroleum Institute. *API 5L: Specification for Line Pipe*, 44th ed.; American Petroleum Institute: Washington, DC, USA, 2007.
42. Bai, Q.; Bai, Y. Use of High-Strength Pipeline Steels. In *Subsea Pipeline Design, Analysis, and Installation*; Gulf Professional Publishing: Oxford, UK, 2014; pp. 675–694.
43. Suzuki, N.; Toyoda, M. Critical Compressive Strain of Linepipes Related to Workhardening Parameters. In Proceedings of the 21st International Conference on Offshore Mechanics and Arctic Engineering, Oslo, Norway, 23–28 June 2002; Volume 3, pp. 217–224.
44. Suzuki, N.; Muraoka, R.; Glover, A.; Zhou, J.; Toyoda, M. Local Buckling Behavior of X100 Linepipes. In Proceedings of the ASME 2003 22nd International Conference on Offshore Mechanics and Arctic Engineering, Cancun, Mexico, 8–13 June 2003; pp. 67–76.
45. Kang, K.-B.; Yoo, J.-Y.; Ahn, S.-S.; Cho, W.Y.; Yoon, T.-Y. Buckling Behavior of API-X80 Linepipe. In Proceedings of the Seventeenth International Offshore and Polar Engineering Conference, Lisbon, Portugal, 1–6 July 2007; International Society of Offshore and Polar Engineers: Lisbon, Portugal, 2007; pp. 3254–3260.
46. Ishikawa, N.; Okatsu, M.; Endo, S.; Shikanai, N.; Muraoka, R.; Kondo, J. Mechanical And Metallurgical Properties of Grade X80 High Strain Linepipe Produced By Heat Treatment On-line Process. In Proceedings of the Eighteenth International Offshore and Polar Engineering Conference, Vancouver, BC, Canada, 6–11 July 2008; International Society of Offshore and Polar Engineers: Vancouver, BC, Canada, 2008; pp. 13–20.
47. Yoo, J.-Y.; Ahn, S.-S.; Seo, D.-H.; Song, W.-H.; Kang, K.-B. New Development of High Grade X80 to X120 Pipeline Steels. *Mater. Manuf. Process.* **2011**, *26*, 154–160. [CrossRef]
48. Kong, L.Z.; Zhou, X.Y.; Chen, L.Q.; Shuai, J.; Huang, K.; Yu, G.J. True Stress–Strain Curves Test and Material Property Analysis of API X65 and API X90 Gas Pipeline Steels. *J. Pipeline Syst. Eng. Pract.* **2018**, *9*, 04017030. [CrossRef]
49. Porter, D.; Laukkanen, A.; Nevasmaa, P.; Rahka, K.; Wallin, K. Performance of TMCP steel with respect to mechanical properties after cold forming and post-forming heat treatment. *Int. J. Press. Vessel. Pip.* **2004**, *81*, 867–877. [CrossRef]
50. Mouriño, N.S.; Petrov, R.; Bae, J.-H.; Kim, K.; Kestens, L.A.I. Texture Dependent Mechanical Anisotropy of X80 Pipeline Steel. *Adv. Eng. Mater.* **2010**, *12*, 973–980. [CrossRef]
51. Hines, J.R.; Timms, C.M.; DeGeer, D.D. Thermal Ageing Effects on Thickwalled Line Pipe. In Proceedings of the ASME 2007 26th International Conference on Offshore Mechanics and Arctic Engineering, San Diego, CA, USA, 10–15 June 2007; pp. 573–579.
52. Ndubuaku, O.; Martens, M.; Cheng, R.; Ahmed, A.; Adeeb, S. A Novel Approach for True Stress-True Strain Material Characterization of Metallic Materials Using the Product-Log (Omega) Function. In Proceedings of the 6th International Conference on Engineering Mechanics and Materials, Vancouver, BC, Canada, 31 May–3 June 2017.
53. Ndubuaku, O.; Martens, M.; Cheng, J.J.R.; Adeeb, S. Expression of a Generic Full-Range True Stress-True Strain Model for Pipeline Steels Using the Product-Log (Omega) Function. In Proceedings of the ASME 2017 Pressure Vessels and Piping Conference, Waikoloa, HI, USA, 16–20 July 2017; p. V06BT06A050.
54. Ndubuaku, O.; Liu, X.; Martens, M.; Roger Cheng, J.J.; Adeeb, S. The effect of material stress-strain characteristics on the ultimate stress and critical buckling strain of flat plates subjected to uniform axial compression. *Constr. Build. Mater.* **2018**, *182*, 346–359. [CrossRef]
55. Ndubuaku, O. A New Material Characterization Approach for Evaluating the Deformational Capacity of Onshore Pipelines. Ph.D. Thesis, University of Alberta, Edmonton, AB, Canada, 2019.
56. Hibbitt, D.; Karlsson, B.; Sorensen, P. *ABAQUS Standard User's and Reference Manuals*, version 6.14; Dassault Systèmes: Providence, RI, USA, 2014.

57. Wolfram Research Inc. *Mathematica*, version 11.3; Wolfram Research Inc.: Champaign, IL, USA, 2018.
58. Liu, M.; Wang, Y.-Y.; Zhang, F.; Kotian, K. *Realistic Strain Capacity Models for Pipeline Construction and Maintenance*; Prepared for US Department of Transportation; Contract No. DTPH56-10-T-000016; Pipeline and Hazardous Materials Safety Administration, Office of Pipeline Safety: Dublin, OH, USA, 2013.
59. Dorey, A.B.; Murray, D.W.; Cheng, J.J.R. Critical Buckling Strain Equations for Energy Pipelines—A Parametric Study. *J. Offshore Mech. Arct. Eng.* **2006**, *128*, 248. [CrossRef]
60. Mohareb, M.E.; Elwi, A.E.; Kulak, G.L.; Murray, D.W. *Deformation Behavior of Line Pipe*; Structural Engineering Report No. 202; University of Alberta: Edmonton, AB, Canada, 1994.
61. Liu, X.; Zhang, H.; Han, Y.; Xia, M.; Zheng, W. A semi-empirical model for peak strain prediction of buried X80 steel pipelines under compression and bending at strike-slip fault crossings. *J. Nat. Gas Sci. Eng.* **2016**, *32*, 465–475. [CrossRef]
62. Dorey, A.B.; Murray, D.W.; Cheng, J.J.R. Material Property Effects on Critical Buckling Strains in Energy Pipelines. In Proceedings of the 2002 4th International Pipeline Conference, Calgary, AB, Canada, 30 Septembe–3 October 2002; pp. 475–484.
63. Gresnigt, A.M.; Karamanos, S.A. Local Buckling Strength And Deformation Capacity of Pipes. In Proceedings of the Nineteenth International Offshore and Polar Engineering Conference, Osaka, Japan, 21–26 July 2009; International Society of Offshore and Polar Engineers: Osaka, Japan, 2009; pp. 212–224.
64. Neupane, S.; Adeeb, S.; Cheng, R.; Ferguson, J.; Martens, M. Modeling the Deformation Response of High Strength Steel Pipelines—Part I: Material Characterization to Model the Plastic Anisotropy. *J. Appl. Mech.* **2012**, *79*, 051002. [CrossRef]
65. Neupane, S.; Adeeb, S.; Cheng, R.; Ferguson, J.; Martens, M. Modeling the Deformation Response of High Strength Steel Pipelines—Part II: Effects of Material Characterization on the Deformation Response of Pipes. *J. Appl. Mech.* **2012**, *79*, 051003. [CrossRef]
66. Adeeb, S.; Zhou, J.; Horsley, D. Investigating the Effect of UOE Forming Process on the Buckling of Line Pipes Using Finite Element Modeling. In Proceedings of the 2006 International Pipeline Conference, Calgary, AB, Canada, 25–29 September 2006; Volume 2006, pp. 169–174.

applied sciences

Review

Welding Capabilities of Nanostructured Carbide-Free Bainite: Review of Welding Methods, Materials, Problems, and Perspectives

Aleksandra Królicka [1,*], Andrzej Ambroziak [1] and Andrzej Żak [1]

Faculty of Mechanical Engineering, Wroclaw University of Science and Technology, Wybrzeże Wyspiańskiego 27, 50-370 Wrocław, Poland
* Correspondence: aleksandra.krolicka@pwr.edu.pl

Received: 12 July 2019; Accepted: 9 September 2019; Published: 11 September 2019

Featured Application: Expanding the applications of nanostructured carbide-free bainitic steels by enabling them to be joined. This particularly applies to structures operating in difficult conditions (e.g., abrasive wear, contact fatigue, dynamic loads).

Abstract: This article presents state-of-the-art welding methods and the weldability aspect of steels, particularly high-carbon nanobainitic (NB) steels, without carbide precipitates (CFB—carbide-free bainite). On the basis of research conducted to date, all welding methods with parameters and weld metals for NB CFB are presented. It was found that the process parameters significantly affected the mechanical properties of the welds, which were determined primarily by the properties of the low-temperature heat-affected zone. The microstructures of welded joints in the heat-affected and fusion zones are also described. The general requirements for welding processes, as well as problems and perspectives for further research, are presented.

Keywords: nanobainite; NB; CFB; high-carbon steels; high-strength steels; Si-rich steels; welding; HAZ; fusion zone

1. Introduction

Bhadeshia and Edmonds [1,2] were the first to introduce methods for designing steel that is characterized by a structure consisting of bainitic ferrite and austenite with high mechanical properties. A carbide-free structure is possible to obtain with a concentration of 2% Si [3,4]. In addition, the nanometric widths of the bainitic ferrite laths and film-like austenite (even with 20-40 nm, [5]) allow for an ultimate tensile strength (UTS) of 2.5 GPa to be reached with satisfactory ductility [6]. Austenite can be said to play an important role. It has a film-like morphology and is characterized by a high carbon concentration and high dislocation density and stability in comparison to blocky austenite [5,7]. Such materials are called nanostructured bainitic (NB) steels without carbide precipitates (carbide-free bainite (CFB)) and their development continues to this day (among others, recently published [8–10]). The high mechanical properties of nanostructured carbide-free bainite create the possibility of a significant reduction in mass, and also increase the durability of structural parts in many branches of industry. In addition to the high price, which is mainly associated with complex, long-term heat treatment and demanding chemical composition, the main limitation of the common use of NB and CFB in industry is their difficult welding [11]. A high carbon content in the range of 0.6/1.0 wt.%, silicon of 1.5/3.0 wt.%, and manganese of 0.6/2.0 wt.% allows for the refinement of the microstructure to nanometric values using a low temperature for the isothermal heat treatment [12], which is the main reason for the poor weldability of these steels. The fundamental problems that reduce the mechanical properties of the welded joints are the formation of brittle martensite in the fusion zone, cementite precipitation in the heat-affected zone, and cold cracking [11].

2. Design of Weld Metals

Obtaining the structure of nanocrystalline bainite in an entire welded joint area determines the most favorable mechanical properties. The homogeneous bainitic structure is difficult to achieve during conventional welding methods. In addition, the NB CFB structure can be obtained for materials characterized by a specific chemical composition, which causes problems with the selection of weld metals. At present, there are no commercially available weld metals that allow for the structure of NB CFB in the weld zone. In Reference [13], the chemical composition of an electrode was proposed, which allows such a structure in the weld zone to be obtained with a small amount of martensite and retained austenite (Table 1). After the welding process, as a result of the high hardenability associated with the designed chemical composition of the weld material, the typical Widmanstätten structure was not achieved. High ultimate tensile strength (approximately 950 MPa) and yield strength (approximately 830 MPa) was not achieved in the joint. It was found that silicon does not significantly affect strength parameters but does have an influence on impact toughness. The increased concentration of silicon caused a decrease in the impact energy, which was explained by the slow development of martensite tempering in relation to the material with a lower content of silicon. Krishna Murthy et al. [14] proposed weld metals with a higher carbon content (0.32%), which, according to the authors, can replace the austenitic welds when welding armor steels. The used weld metals allowed the desired structure of CFB to be obtained with blocky austenite in the inter-dendritic areas. The presence of unstable blocky austenite reduces the impact toughness of the welds due to the induction of martensitic transformation under loads [15].

Table 1. Review of weld metals that allow a carbide-free bainite (CFB) structure to be obtained.

Chemical Composition of Weld Metals wt.%	Process Parameters	Mechanical Properties	Microstructure and Comments	Reference(s)
(1) 0.10% C; 2.2% Mn; 0.86% Si; 2.1% Ni; 0.2% Mo (2) 0.12% C; 2.3% Mn; 1.38% Si; 2,1% Ni; 0,2% Mo (3) 0.1% C; 2.2% Mn; 1.63% Si; 2.1% Ni; 0.2% Mo	TIG Interpass temperature: 250 °C; Electrode ⌀4 mm; 30 runs; Current: 174 ± 1 A; Voltage: 25 ± 1 V; Input energy: 1.08 kJ/mm.	YS = 830 MPa UTS = 950 MPa The impact toughness decreases with increasing silicon content. No significant influence of silicon on weld strength was found.	Bainitic ferrite, austenite, and martensite in the weld. The structure is relatively homogeneous despite many weld beads.	[13]
0.32% C; 1.6% Si; 1.6% Mn; 1.1% Ni; 1.1% Co; 1.0% Cr; 0.3% Mo	Shielded metal arc welding; Pre-heat: 350 °C; Current: 175 A; Voltage: 23 V; Input energy: 1.2 kJ/mm; Speed: 180 mm/min; Electrode ⌀4 mm; Electrode polarity: DCEN; Regeneration: 350 °C/6 h.	YS = 1010 MPa UTS = 1200 MPa Elongation: 14% Charpy Energy: 15 J	In the fusion zone FZ- the structure consists of bainitic ferrite and austenite (CFB). Blocky austenite in inter-dendritic areas.	[14] [16]
YS—Yield Strength	UTS—Ultimate Tensile Strength	DCEN—Direct Current Electrode Negative		

It was also found that the weld material with the structure of CBF was resistant to hydrogen cold cracking [14] and hot cracking [16]. The obtained ultimate tensile strength was 1200 MPa [14], but this is still insufficient in the context of high-carbon nanostructured bainitic steels (even at 2500 MPa [6]).

Therefore, weld materials with a higher strength should be designed. On the other hand, increasing the strength requires the carbon content to be increased, which causes worse weldability. Thereat. Therefore, medium and high carbon bainitic steels were welded without weld metals in order to obtain high-strength parameters for the joints.

3. Review of Welding Methods

Over the last decade, attempts have been made to weld nanostructured CFB bainite. In Table 2, a review of welding processes of these steels that have so far been conducted is presented. The welding methods, chemical composition of the base materials, process parameters, mechanical properties, and a brief description of the microstructures are specified; in addition, comments are also provided. Furthermore, some welding methods involve steels with a high silicon content and a relatively low carbon concentration (0.34 wt.%). The table also includes some of the welding methods of steels after Quenching and Partitioning (Q&P) with regards to the chemical composition that favors the preparation of CBF structures.

Welding of high carbon bainitic steels (0.55/0.82%) was carried out, inter alia, using the TIG method [17–21]. It was found that the process parameters significantly influenced the microstructure and mechanical properties. A higher amount of delivered heat caused a higher volume of retained austenite in the weld (in which martensite was also found) [17]. The authors of Reference [17] applied preheating and also analyzed the welding process for steel in the softened delivery state. Due to the occurrence of cracks in welded sheets with a final bainitic structure, they proposed that it should be welded in the delivery state and that the heat treatment process should be carried out afterwards. Fang et al. [19–21] carried out a welding process using the regeneration technique. Regeneration involves controlled cooling of the welded joint to the isothermal heat treatment temperature and annealing at this temperature for a sufficient time to allow completion of bainite transformation. After welding with regeneration, a bainitic structure with coarse retained austenite was obtained [19]. No cold cracks or martensitic structures were found in the weld zone. After a tensile strength test, fractures were located in the heat-affected zone, where the presence of cementite was identified [19]. In order to accelerate the beginning of the bainite transformation, use of deformation in the welded joint was suggested by applying the rotary impacting trailed welding (RITW) method [22]. In this method, the welding and impacting occurs synchronously, which makes it possible to introduce compressive and shear stresses in the material at the same time during the welding process. Bainitic steel after heat treatment was also welded using the TIG method. The rotary impacting head was placed 30 mm behind the welding gun and 5 mm away from the welding centerline, which, in this area, corresponds to a temperature of 600°C. However, in this method, heat treatment (regeneration) is necessary after the welding process. It was found that the RITW method allows for a reduction in the time necessary for bainite transformation in the deformed austenite zone, but the used regeneration time, however, was not long enough to complete the transformation. In addition, the deformation of austenite before phase transformation affected the morphology of bainite—the bainitic ferrite laths were arranged in the direction of deformation. In Reference [23], two-pass welding was used to refine the austenite grains. The authors proposed such a process due to the fact that fine austenite grains will accelerate the nanobainitic transformation [24,25], such as in the case of the material used in this study [23]. Acceleration of bainitic transformation will shorten the regeneration time, which will reduce the cost of heat treatment and increase the process' technological efficiency. As with the previous method, the impacting trailed welding (ITW) method uses a first impacting head placed 30 mm behind the welding gun, and the second welding gun is 60 mm behind the first welding gun in the welding direction (Figure 1). Regeneration of the weld joint involved two stages, and the cycles are shown in Figure 2. Different recrystallization temperatures were used, ranging from 700 °C, 750 °C, and 800 °C with times of 10 s, 30 s, 50 s, and 100 s. It was found that the conducted static recrystallization process in the impact zone reduced the grain size from 106 ± 42 µm to 36 ± 13 µm. The recrystallized grains were also smaller than the base material, where the grain size was around 50 µm. In addition, it was found that a higher volume of bainitic ferrite was obtained in the recrystallized grains, which also confirmed the reduction of the bainite transformation time and the same time of regeneration. Welded joints after the ITW process showed a higher ultimate tensile strength (2100 MPa) when compared to the process without ITW (1400 MPa) [23].

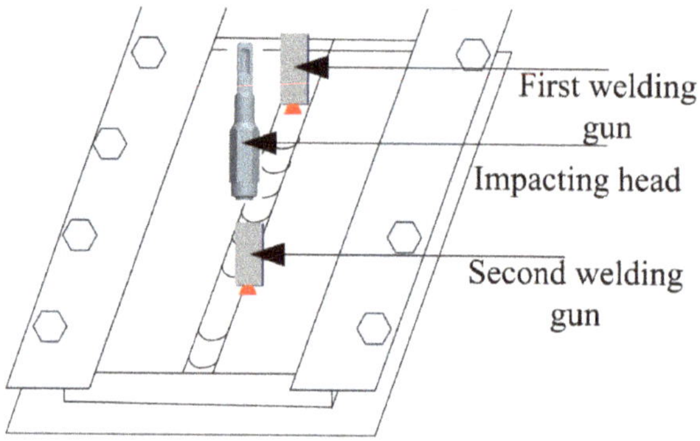

Figure 1. Scheme of the welding process using the two-pass impacting Trailed welding (ITW) method [23].

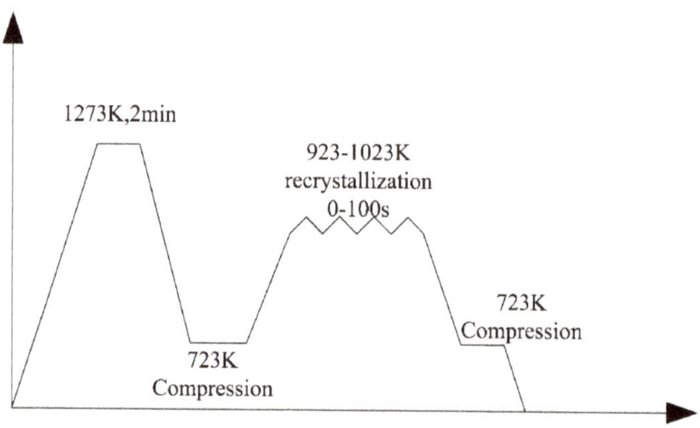

Figure 2. Two-pass regeneration with static recrystallization [23].

In Reference [26], for welding high carbon bainitic steels, laser welding with post-weld rapid heat treatment (PWRHT) is proposed. When the welded joint after the welding process approaches the beginning of the martensitic transformation (Ms), it is quickly heated to a temperature lower than A_1 (in less than 10 seconds). As a consequence, disadvantageous, brittle martensite in the welds and cold cracks are avoided. The structure after laser welding and PWRTH consisted of ferrite, austenite, and cementite, and its hardness was comparable to the base material. The obtained structure was different to the correct bainitic structure and, despite a comparable hardness level, it showed lower mechanical properties. Fang et al. [21] proposed a laser beam welding process with regeneration, similar to the TIG method. It was found that low input energy (60 kJ/m) in laser welding, when compared to the TIG method where high input energy was used (908 kJ/m), allows for the achievement of high ultimate tensile strength (33 MPa lower than the strength of the base material). The high tensile strength of welded joints is due to the high stability of the microstructure in the heat-affected zone at low input energy processes.

Another welding method used for nanostructured CFB steels is friction stir welding (FSW) [27,28]. The joining process takes place in a solid state, which reduces the formation of distortions and defects associated with thermal contraction. In the FSW method, a rotating tool containing the shoulder and pin is plunged into the joint between the two front-mounted plates. As a result of friction and plastic deformation of the pin on the surface of the plates, heat is generated, which causes the softening of the material in the area of the tool. The softened material then flows to the rear edge of the shoulder, where it is stirred. The material after cooling forms a joint between the plates [29]. In References [27,28], it was found that the parameters of the FSW process affect the mechanical properties and microstructure of the obtained joints. The research concerned silicon-rich steels with a relatively low carbon content (0.34 wt.%). The increase in rotational speed increases the hardness of the zone, defined in the work as the stir-zone (SZ), located in the area of the moving shoulder. However, the hardness obtained at the lowest rotational speed was also high when compared to the thermo-mechanical affected zone (TMAZ) and base material. This increase in hardness is explained by the appearance of martensite under the influence of the thermo-mechanical effect (TRIP) and suggests the use of heat treatment after the joining process. However, the presence of martensite was not confirmed and the change in the volume of retained austenite in this zone was not determined. Scanning electron microscopy does not allow microstructures to be unequivocally identified. Carbide precipitates, which are likely in the TMAZ zone, were not analyzed.

In Reference [18], an attempt was made to friction weld NB CFB steel with C45 steel. The parameters of the welding process are presented in Table 2. No cracking was found in the Fusion zone, while microcracks were identified in the heat-affected zone (HAZ). When joining C45 steel with bainitic steel in the delivery state (perlite and ferrite), no microcracks were found [18]. However, welding steel in the delivery state requires heat treatment after the process in order to obtain a bainitic structure, which is an additional technological process (it will also change the mechanical properties of the C45 steel).

Wang and Speer [30] described the welding capabilities of commercial QP980 steel after quenching and partitioning (Q&P) heat treatment. The Q&P process was proposed by Speer et al. [31] and allows a high-strength martensite–ferrite–austenite structure for C–Si–Mn/C–Si–Mn–Al steels to be obtained. Heat treatment consists of complete or partial austenitization and cooling to the temperature between the beginning and the end of the martensitic transformation (M_s and M_f), followed by isothermal anealing in the higher temperature range of martensitic or bainitic transformation [32,33]. Due to the high concentrations of silicon and aluminum in these steels, precipitation of cementite is inhibited, and the structures are therefore carbide free. Carbon enriches (above all) austenite and ferrite [33]. Heat treatment parameters determine the proportions of phases in the structure in which CFB bainite occur [33]. For this reason, this review also included welding methods for steel after Q&P treatment. The microstructure of the tested QP980 steel contained martensite, austenite, and ferrite, and no carbide-free bainite was identified [30]. However, the chemical composition of QP980 steel favors the structure of carbide-free bainite Therefore, the presented welding methods can be useful for planning the welding process of bainitic steels with a similar chemical composition. Steel after Q&P treatment was welded using resistance spot welding (RSW), laser welding, and the MAG method (Table 1). The authors assessed the weldability of QP980 steel as good and all welding methods were successfully carried out (i.e., no excessive welding defects and good mechanical properties) [29]. Resistance spot welding (with thermal simulation) was also performed for steel with a higher carbon concentration (42SiCr) after Q&P treatment. The satisfactory mechanical properties (shear strength) of the welded joints were not obtained and, in order to reconstruct the mechanical properties of the base material, the authors proposed the future use of a regenerative treatment in conditions similar to the Q&P parameters [34]. This result confirms the difficulty of welding steel with a higher carbon concentration.

Table 2. Review of welding methods for nanostructured carbide-free bainite and for selected steels after quenching and partitioning(Q&P) heat treatment.

Welding Methods	Chemical Composition of Base Material wt.%	Process Parameters	Mechanical Parameters	Microstructure and Comments	Reference(s)
TIG	0.82% C; 1.2% Si; 2.5% Mn; 0.8% Mo; 1.8% Cr; 1.5% Al; 1.0% Ni	1. Current: 160 A; Voltage: 16 V; Speed: 22 mm/min; Sheet thickness: 6.0 mm 2. Current: 160 A; Voltage: 15 V; Speed: 50 mm/min; Sheet thickness: 6.0 mm 3. Current: 150 A; Voltage: 15 V; Speed: 80 mm/min; Sheet thickness: = 4.0 mm	UTS up to 60% of the base material; 1. UTS = 850 MPa 2. UTS = 1200 MPa 3. UTS = 1578 MPa	A higher amount of heat caused a higher volume of retained austenite. A small volume of martensite in the weld and HAZ. No preheating and regeneration.	[17]
TIG	(1) 0.61% C; 1.5% Mn; 1.7% Si; 1.3% Cr. (2) 0.55% C; 1.9% Mn; 1.8 %Si; 1.3% Cr; 0.8% Mo.	Pre-heat = 230 °C; Current: 100 A (impulse 160 A); Impulses frequency: 30 Hz; Speed: 100 mm/min; Sheets thickness: 5.0 and 8.0 mm Cover gas: 100%Ar; Gas flow: 15 L/min.	Base material: Hardness: 640 HV. Softened state + heat treatment: 590/610 HV. Maximum value of nanobainitic welds: 720 HV in the HAZ	The sheets welded in the softened state did not show cold racks. Cold cracks were identified in welded joints of steel with a nanobainitic structure.	[18]
TIG + regeneration	0.76% C; 1.0% Si; 1.3% Cr; 1.0% Mn;	Current: 140 A; Voltage: 20 V; Speed: 185 mm/min; Samples: 2 × 40 × 100 mm Regeneration: 250 °C /5 days	Base material: UTS = 1950 MPa, A = 2% After welding: UTS = 1410 MPa, A = 0.8%	No cold cracks in the weld. After regeneration, the weld had a bainitic structure and a small amount of retained austenite. Cementite precipitates were identified in the HAZ.	[19] [20]
TIG + regeneration	0.82% C; 1.7% Si; 2.0% Mn; 0.2% Cr; 0.4% Mo; 1.1% Ni;	Input energy: 908 kJ/m Speed: 18.5 mm/min Sheets thickness: 2 mm Regeneration: 250 °C/5 days	Base material: UTS = 1877 MPa After welding: UTS = 1680 MPa	Precipitates of cementite in HAZ, which increased the width of bainitic ferrite laths, austenite decomposition in the LHAZ.	[21]

Table 2. *Cont.*

Welding Methods	Chemical Composition of Base Material wt.%	Process Parameters	Mechanical Parameters	Microstructure and Comments	Reference(s)
RITW (TIG + rotary impacting head) + regeneration	0.87% C; 1.2% Si; 1.5% Mn; 0.3% Mo; 0.5% Ni; 1.1% Al	Current: 140 A; Voltage: 18 V; Speed: 90 mm/min Samples: 10 × 80 × 100 mm Regeneration: 1. 250 °C/1.5 h 2. 250 °C/2.5 h		The RITW process accelerated the bainite transformation. Bainitic ferrite in the deformed austenite zone was arranged in accordance with the direction of deformation.	[22]
ITW (TIG + two-pass impacting head) + regeneration	0.87% C; 1.2% Si; 1.5% Mn; 0.3% Mo; 0.5% Ni; 1.1% Al	Current: 220 A; Voltage: 18 V; Speed: 2 mm/s Samples: 10 × 80 × 100 mm Argon flow: 2 L/min Recrystallization: 700 °C; 750 °C; 800 °C 10 s; 30 s; 50 s; 100 s. Regeneration: 250 °C/2 h	Without ITW: UTS = 1400 MPa Elongation: 2% With ITW: UTS = 2010 MPa Elongation: 3/4%	Grain size: Base material: ~50 μm Without ITW: 106 ± 42 μm With ITW, recrystallization area: 36 ± 13 μm. The ferrite volume was higher for recrystallized grains and the bainite transformation time was reduced.	[23]
Laser Welding + PWRHT	0.78% C; 1.0% Si; 1.5% Mn	PWRHT: Cooling of the welded joint (T > Ms); Heating with speed 10 s (T < Ac1); cooling to RT.		No cold crack in the weld. HAZ: ferrite, cementite, and austenite.	[26]
Laser Beam Welding + regeneration	0.82% C; 1.7% Si; 2.0% Mn; 0.2% Cr; 0.4% Mo; 1.1% Ni;	Input energy: 60 kJ/m Speed: 100 mm/min Sheet thickness: 2 mm Regeneration: 250 °C/5 days	Base material: UTS = 1877 MPa After welding: UTS = 1844 MPa	Slight changes in LHAZ. A small amount of cementite, austenite partly decomposed.	[21]
Friction Stir Welding	0.34% C; 1.8% Mn; 1.5% Si; 0.9% Cr;	Stir speed: 80; 100, 150, 200 rpm Feed rate: 35 mm/min	Hardness of base material: 425 HV Hardness of stir-zone: 650–725 HV	In the stir zone, there was an increase of hardness in relation to the thermo-mechanical affected zone. Thermoplastic deformations caused the transformation of austenite into martensite.	[27,28]

Table 2. *Cont.*

Welding Methods	Chemical Composition of Base Material wt.%	Process Parameters	Mechanical Parameters	Microstructure and Comments	Reference(s)
Quenching and Partitioning + Resistance Spot Welding	QP980 steel 0.15–0.30% C; 1.5–3.0% Mn; 1.0–2.0%Si; Electrode: ISO 5821:2009 type B; ⌀4 mm	Weld force: 5.8 kN, 1305 lbf Weld pulse: 3 Sheet thickness: 1.6 mm Cooling: 2 L/min; 0.5 gal/min	Hardness of base material: 300 HV. After welding: 473–512 HV. Shear strength increases with current.	Welding defects such as cracks, pores or shrinkage cavities were not identified. The authors assessed the overall shear strength of welded joints as good.	[30]
Quenching and Partitioning + Laser Welding	QP980 steel 0.15–0.30% C; 1.5–3.0% Mn; 1.0–2.0% Si;	Energy: 3 kW Speed: 5 m/mm Cover gas: He Gas flow: 15 L/min Sheet thickness: 1.6 mm	After welding: UTS = 1081 MPa (Fracture in area of base material) Elongation: 7.3% (base material: 10.3%)	No welding defects. Weld area: low-carbon martensite. HAZ: martensite and ferrite.	[30]
Quenching and Partitioning + MAG	QP980 steel 0.15–0.30% C; 1.5–3.0% Mn; 1.0–2.0% Si; Electrode: ER110S	Energy: 3.6 kJ/cm Speed: 35 cm/min Cover gas: 80%Ar + 20% CO_2 Gas flow: 14 L/min Sheet thickness: 1.6 mm	After welding: UTS = 991 MPa Hardness: <500 HV	No visible welding defects or softened areas.	[30]
Quenching and Partitioning + Resistance Spot Welding+ simulation	0.42% C; 2.0% Si; 1.3% Cr; 0.8% Mn;	Welding Energy: 3.13/3.39 J Nugget diameter: 7.26/7.33 mm Sheet thickness: 1.2 mm	Base material: UTS = 1841 MPa YS = 1030 MPa Hardened zone: 700 HV. Softened zone in HAZ: 370 HV. After welding: Shear strength: 7–10 kN	The maximum process temperature had a stronger effect on hardness than heating speed and heating time.	[34]
UTS—Ultimate Tensile Strength	YS—Yield Strength	HAZ—Heat-Affected Zone	PWRHT—Post-Weld Rapid Heat Treatment	RITW—Rotary Impacting Trailed Welding ITW—Impacting Trailed Welding	

4. Analysis of Microstructure Changes after the Welding Process

The mechanical properties of welded joints are related to the microstructure of welds and the heat-affected zone. During welding processes, both the microstructure and grain size of the prior austenite change as a function of the distance from the weld face. A thorough analysis of structural changes may enable optimization of the welding parameters, which in turn will minimize zones with adverse mechanical properties.

The following zones were distinguished in welded joints (Figure 3):

- Low-temperature heat-affected zone (LHAZ): between A_1 temperature and the base material;
- High-temperature heat-affected zone (HHAZ): between the fusion line and A_1 temperature;
- Fusion zone (FZ): located at the weld area.

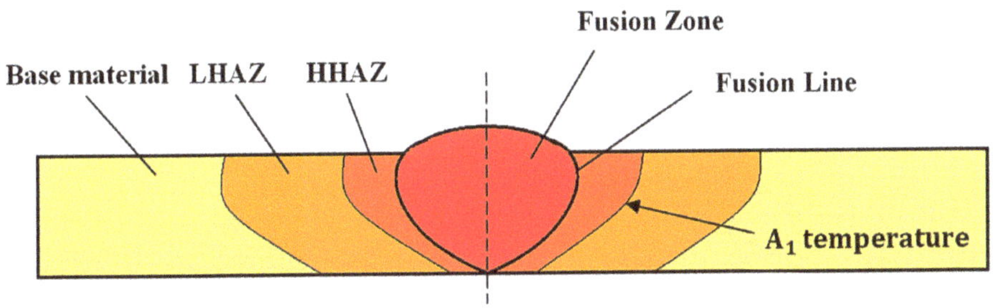

Figure 3. Scheme of welded joint.

4.1. *Low-Temperature Heat-Affected Zone*

One of the main problems of welding nanostructured bainitic steels is the precipitation of cementite in low-temperature heat-affected zones (LHAZs). Analogous to tempering processes, there are areas located in the LHAZ that correspond to tempering at different temperature ranges. In Reference [6], it was found that tempering at temperatures of up to 500 °C did not cause significant changes in the hardness and microstructure of nanobainitic steels. However, during longer period of heating, precipitation of cementite may occur at temperatures above 450 °C [35–38]. Areas of blocky austenite at higher temperatures (above 550 °C) were decomposed to fine dispersive perlite, and austenite with a film-like morphology was decomposed to ferrite and cementite due to the limited volume for pearlite growth [38]. In this temperature range, the austenite completely decomposed [38]. Further increases of the tempering temperature resulted in a ratio of ferrite and cementite that corresponded to the equilibrium state and an increase in the width of the ferrite plates [39], which reduced the mechanical properties. However, according to Podder and Bhadeshia [37], long-term tempering also causes a significant decrease in the volume of austenite, which becomes less stable by the precipitate of cementite. Then, austenite with a lower concentration of carbon was transformed to martensite after cooling to an ambient temperature. There was also no segregation of alloying elements between the ferrite/austenite phases during tempering [36]. In addition to research on the nanostructured CFB bainite stability during tempering, in situ tests were carried out during the continuous heating of two silicon-rich steels [40]. It was found that for a steel of 0.84C–2.26Mn–1.78Si–0.25Mo–1.55Co–1.47Cr, austenite began to decompose at 400 °C, while for the alloy 1.04C–1.97Mn–3.89Si–0.24Mo–1.43Al, decomposition began to occur at 580 °C. Therefore, the higher contents of silicon and aluminum in the steel caused a higher stability in the austenite. In addition, tests performed during continuous heating were closer to welding conditions than tempering over longer time periods. The differences in the obtained results indicate the need for further research, especially during continuous heating. In the context of welding processes, cooling to an ambient temperature or a regeneration temperature after subsequent heating stages is also important due to the fact that the mechanisms of austenite decomposition can be significantly different.

Fang et al. [21] conducted research using a thermomechanical simulation of the LHAZ zone at 400 °C for low- and high-energy welding with regeneration (Figure 4). The samples after reaching 400 °C were cooled to the regeneration temperature (250 °C). In the initial steel consisting of 0.82C–1.66Si–2.05Mn–0.22Cr–0.36Mo–1.06Ni, a typical nanobainitic structure consisting of thin ferrite laths and austenite with a film-like morphology (and a small fraction of blocky austenite) was also identified (Figure 5a). For the low-energy process, no significant differences were found when compared to the base material. It was found that only short austenite structures with a film-like morphology were decomposed to spheroidal precipitations of austenite (Figure 5b). In the case of high-energy welding processes, the microstructure changed substantially. The short film-like austenite completely decomposed, while the areas of long filmy austenite became discontinuous. Many spheroidal precipitations also formed in the bainitic ferrite matrix (Figure 5c). According to the authors, the blocky austenite in a high-energy welding process decomposes into a fine nanostructured bainite with a different crystallographic orientation than the nearby bainite (Figure 5d). On the basis of investigations using transmission electron microscopy, no precipitates of cementite were found in the nanostructured bainite in the base material (Figure 6a). For the low-energy process, a small amount of cementite located at the ferrite–austenite interface was identified (Figure 6b). For the high-energy process, a large amount of cementite was found in the bainitic ferrite matrix with clearly increased widths of the laths (Figure 6c) [21]. However, in the presented research, the identification of cementite and its crystallographic orientation to the matrix using electron diffraction was not determined. This would have provided knowledge on the precipitation mechanism in steels with high carbon concentrations.

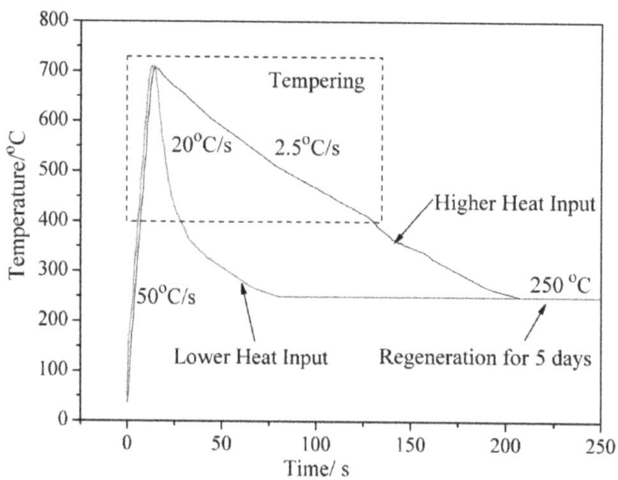

Figure 4. Thermal cycles simulation of the low-temperature heat-affected zone for low- and high-energy welding processes [21].

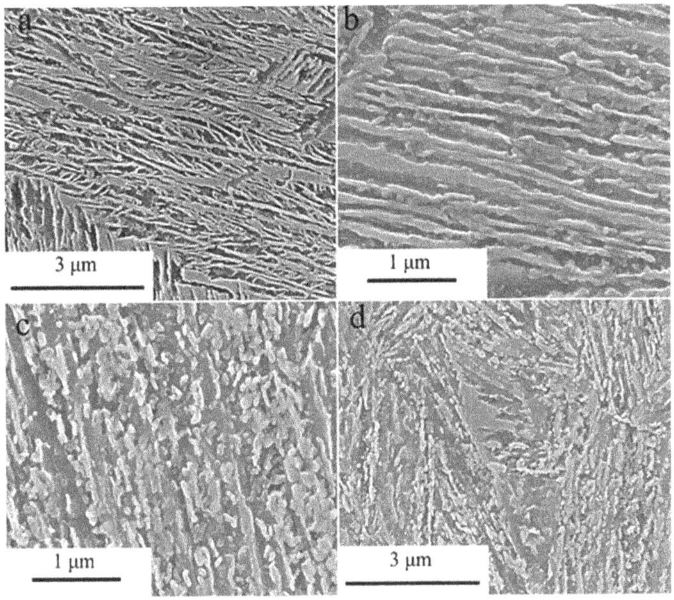

Figure 5. Microstructure of 0.82C–1.66Si–2.05Mn–0.22Cr–0.36Mo–1.06Ni steel. (**a**) Initial microstructure; (**b**) film-like morphology of the microstructure of the simulated low-energy welding process; (**c**) degraded microstructure for the simulated high-energy welding process; (**d**) area of blocky austenite for the high-energy welding process. Images produced via SEM [21].

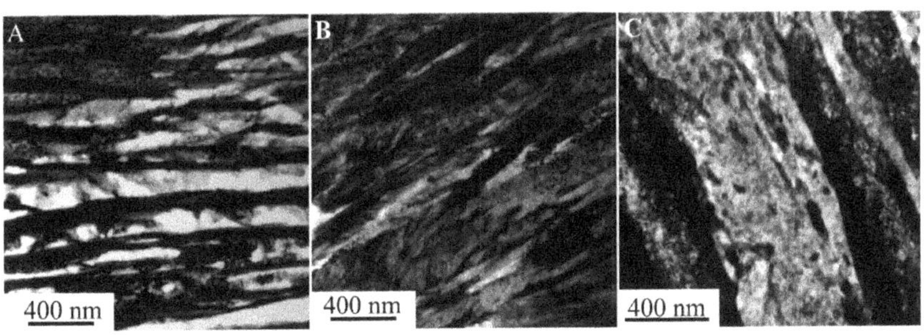

Figure 6. Bright field image of 0.82C–1.66Si–2.05Mn–0.22Cr–0.36Mo–1.06Ni steel. (**a**) Base material; (**b**) low-energy welding process; (**c**) high-energy welding process. Images produced via TEM [21].

In References [19,21], it was found that the fracture after tensile strength tests was located in the LHAZ, which indicates a significant impact of structural changes on this area. Welding and regeneration parameters significantly affect the phase morphology after the welding process. Fang et al. [20] stated that low-input energy processes are more advantageous when compared to high-input energy processes, which is confirmed, apart from significant changes in the microstructure, by the results of tensile strength tests of welds. For this reason, further research is necessary in order to reduce the most unfavorable zones and to develop analyses of the mechanisms of decomposition in nanocrystalline CFB structures in real welding process conditions.

4.2. High-Temperature Heat-Affected Zone and Fusion Zone

In the high-temperature heat-affected zone and fusion zone, as a result of high temperatures (above A_1), recrystallized austenite is formed, which during the rapid cooling after the welding process is transformed into brittle martensite. This is the main reason for the cold cracking of these steels [19]. Thus, the following analysis includes welds after regeneration aimed at reconstructing the structure of the base material.

In the fusion zone, the columnar solidified structure characteristics for welding processes were identified. The visible structure in the inter-dendritic areas differs from the structure of the base material due to the segregation of carbon, chromium, and manganese [20]. In References [14,20], it was found that unstable blocky austenite occurs in the inter-dendritic areas. It was also determined that in the FZ zone, in addition to the coarse inter-dendritic structure, there is also a typical fine nanobainite structure and blocky austenite [20]. However, the volume of austenite in the weld is lower than in the base material [20], which can be explained by the reduction of the carbon content during welding [41] and, thus, by the increased rate of bainite transformation.

The presence of less stable, coarse blocky austenite in both the FZ and HHAZ is the reason for a significant reduction of impact toughness in relation to the base material [15,42,43]. On the other hand, blocky austenite increases elongation due to the TRIP effect [6]. A more preferred morphology is stable film-like austenite because it increases strength and fracture toughness. For this reason, a large amount of blocky retained austenite in the FZ and HHAZ is undesirable. The temperature of regeneration can affect the amount of retained austenite. It is known that increasing the temperature of bainitic transformation increases the amount of retained austenite [44,45], in which case the welded joints may increase the volume of unstable blocky austenite [20].

5. Conclusions

On the basis of current scientific knowledge, the methods of welding nanostructured bainitic steels commonly used today were determined. This study focused on obtaining high strength parameters of welded joints, often comparable with the base material. Problems and prospects for further research were also presented.

5.1. Welding Process

The conducted research on the possibility of welding high-carbon nanostructured CFB steels allowed for the determination of the requirements for welding parameters in order to achieve the high-strength parameters of welded joints without weld metals:

- Before welding processes, pre-heating should be used at a temperature close to that of the bainitic transformation. This prevents cold cracking and allows for control of the cooling process after the welding process is finished.
- Welding process parameters should be characterized by low-input energy, which results in more advantageous structural changes in the HAZ, less blocky austenite in the weld (fusion zone and high-temperature affected-zone), and higher strength compared to high-input energy processes.
- In order to obtain high mechanical parameters, comparable to the base material, heat treatment should be performed after the welding process which will reconstruct the NB CFB structure in the welded joint. Regeneration should be designed similar to the previously conducted heat treatment on the base material. It should be noted that an appropriate regeneration design should be based on dilatometric research, because the regeneration technique requires the determination of real phases of transformation temperatures and times.
- Long-term regeneration can be reduced by introducing deformation into the FZ, HHAZ, and grain refinement. However, the implementation of these methods requires further research and the design of appropriate equipment.
- The highest mechanical properties of welded joints can be obtained after welding in the delivery (softened) state, and then after the conducted isothermal heat treatment that is aimed at obtaining a nanocrystalline structure. There are no problems with the precipitation of cementite in the heat-affected zone due to the complete recrystallization of the weld and the base material. However,

this requires an additional technological process. In addition, due to the fact of their dimensions, not all welded constructions can be heat treated.

5.2. Problems and Perspectives

Welding of high-strength and high-carbon nanobainitic steels is difficult in terms of many factors, including welding parameters, heat treatment, choice of weld-metals, and microstructure. The main problems, as well as prospects resulting from the conducted research, were identified:

- The weld materials so far proposed have shown satisfactory resistance for cold cracking when using pre-heating, an ultimate tensile strength of up to 1200 MPa, and a structure consisting only of austenite and ferrite phases in the weld. However, welding of high-carbon steels requires stronger weld metals. Designing stronger weld metals, including those with a higher carbon concentration, would allow for easier joining and a wider range of welding methods for nanobainitic steels, which, until now, were welded without weld materials.
- The weakest zone of welded joints is the low-temperature affected-zone (LHAZ), where softening occurs due to the decomposition of austenite. Welding parameters directly affect the resumption of bainitic structures to the equilibrium state. However, the mechanisms of the austenite decomposition process and the precipitation of cementite require further research. This research should also involve real welding conditions. The development of a theory regarding these processes could allow for the reduction of unfavorable zones.
- A disadvantage which is difficult to avoid is the grain growth due to the influence of high temperatures, which results in lower mechanical properties. Analysis of the grain growth of bainitic steels (containing silicon and the lack of elements forming carbides inhibiting the growth of austenite grains) has not yet been described in detail.
- The required regeneration time to complete the bainite transformation is long (up to a few days), which is a problem in industrial applications. Shortening the time requires further research regarding the design of the chemical composition of base materials and weld metals, and also improvement of the deformation process (austempering) and the kinetics of phase transformations that depend on the regeneration parameters and grain size.
- To date, no residual stress research [46] on NB CFB welded joints has been presented. Due to the complex welding process and the focus on maximizing mechanical properties, the effect of residual stressors can be significant.
- Investigations on the mechanical properties of NB CFB welded joints did not include the aspect of fatigue. Due to the possibility of using welded structures during fatigue conditions, such research should be carried out in the future.

Author Contributions: A.K.; writing—original draft preparation, A.Ż.; writing—review and editing, A.A.; supervision.

Funding: POWR.03.02.00-00-I003/16.

Acknowledgments: Figures 1 and 2 copyright © Institute of Materials, Minerals and Mining reprinted with permission of Taylor & Francis Ltd. (http://www.tandfonline.com) on behalf of the Institute of Materials, Minerals and Mining. Song, K.J.; Fang, K.; Yang, J.G.; Ma, R.; Liu, X.S.; Wang, J.J.; Fang, H.Y., "Acceleration of regeneration treatment for nanostructured bainitic steel welding by static recrystallization", Materials Science and Technology. Figures 3–5 copyright © reprinted with permission of the publisher Taylor & Francis Ltd. Fang, K.; Yang, J.G.; Song, K.J.; Liu, X.S.; Wang, J.J.; Fang, H.Y., "Study on tempered zone in nanostructured bainitic steel welded joints with regeneration", Science and Technology of Welding and Joining, 26 June 2014.

Conflicts of Interest: The authors declare no conflict of interest.

References

1. Bhadeshia, H.K.D.H.; Edmonds, D.V. Bainite in silicon steels: New composition–property approach Part 1. *Met. Sci.* **1983**, *17*, 411–419. [CrossRef]

2. Bhadeshia, H.K.D.H.; Edmonds, D.V. Bainite in silicon steels: New composition-property approach Part 2. *Met. Sci.* **1983**, *17*, 420–425. [CrossRef]
3. Caballero, F.G.; Bhadeshia, H.K.D.H.; Mawella, K.J.A.; Jones, D.G.; Brown, P. Design of novel high strength bainitic steels: Part 1. *Mater. Sci. Technol.* **2001**, *17*, 512–516. [CrossRef]
4. Caballero, F.G.; Bhadeshia, H.K.D.H. Very strong bainite. *Curr. Opin. Solid State Mater. Sci.* **2004**, *8*, 251–257. [CrossRef]
5. Bhadeshia, H.K.D.H. Nanostructured bainite. *Proc. R. Soc. A Math. Phys. Eng. Sci.* **2010**, *466*, 3–18. [CrossRef]
6. Dong, B.; Hou, T.; Zhou, W.; Zhang, G.; Wu, K. The Role of Retained Austenite and Its Carbon Concentration on Elongation of Low Temperature Bainitic Steels at Different Austenitising Temperature. *Metals* **2018**, *8*, 931. [CrossRef]
7. Caballero, F.G.; Bhadeshia, H.K.D.H.; Mawella, K.J.A.; Jones, D.G.; Brown, P. Very strong low temperature bainite. *Mater. Sci. Technol.* **2002**, *18*, 279–284. [CrossRef]
8. Caballero, F.G.; Rementeria, R.; Morales-Rivas, L.; Benito-Alfonso, M.; Yang, J.R.; de Castro, D.; Poplawsky, J.D.; Sourmail, T.; Garcia-Mateo, C. Understanding Mechanical Properties of Structural Analysis. *Metals* **2019**, *9*, 426. [CrossRef]
9. Chu, C.; Qin, Y.; Li, X.; Yang, Z.; Zhang, F.; Guo, C.; Long, X.; You, L. Effect of Two-Step Austempering Process on Transformation Kinetics of Nanostructured Bainitic Steel. *Materials* **2019**, *12*, 166. [CrossRef] [PubMed]
10. Zhao, J.; Ji, H.; Wang, T. High-cycle, push-pull fatigue fracture behavior of high-C, Si-Al-rich nanostructured bainite steel. *Materials* **2018**, *11*, 54. [CrossRef] [PubMed]
11. Fang, K.; Yang, J.G.; Zhao, D.L.; Song, K.J.; Yan, Z.J.; Fang, H.Y. Review of Nanobainite Steel Welding. *Adv. Mater. Res.* **2012**, *482–484*, 2405–2408. [CrossRef]
12. Caballero, F.G.; Garcia-Mateo, C.; Miller, M.K. Design of novel bainitic steels: Moving from ultrafine to nanoscale structures. *JOM* **2014**, *66*, 747–755. [CrossRef]
13. Bhadeshia, H.K.D.H.; Load, M.; Svensson, L. Silicon-rich bainitic steel welds. *Trans. JWRI* **2003**, *43–52*, 2003.
14. Krishna, M.N.; Janaki, R.G.D.; Murty, B.S.; Reddy, G.M.; Rao, T.J.P. Carbide-Free Bainitic Weld Metal: A New Concept in Welding of Armor Steels. *Metall. Mater. Trans. B Process. Metall. Mater. Process. Sci.* **2014**, *45*, 2327–2337. [CrossRef]
15. Tsai, Y.T.; Chang, H.T.; Huang, B.M.; Huang, C.Y.; Yang, J.R. Microstructural characterization of Charpy-impact-tested. *Mater. Charact.* **2015**, *107*, 63–69. [CrossRef]
16. Krishna, M.N.; Janaki, R.G.D. Hot cracking behavior of carbide-free bainitic weld metals. *Mater. Des.* **2016**, *92*, 88–94. [CrossRef]
17. Kirbiš, P.; Vuherer, T.; Irgolič, T.; Anžel, I. Metallographic Analysis of Kinetically Activated Bainite (KAB) Welds. *Metallogr. Microstruct. Anal.* **2015**, *4*, 570–578. [CrossRef]
18. Węglowski, M.S.; Grobosz, W.; Marcisz, J.; Garbarz, B. Characteristics of Fusion Welded and Friction Welded Joints Made in High-Carbon Nanobainitic Steels. *Biul. Inst. Spaw.* **2018**, *62*, 7–15. [CrossRef]
19. Fang, K.; Yang, J.G.; Liu, X.S.; Song, K.J.; Fang, H.Y.; Bhadeshia, H.K.D.H. Regeneration technique for welding nanostructured bainite. *Mater. Des.* **2013**, *50*, 38–43. [CrossRef]
20. Fang, K.; Yang, J.G.; Song, K.J.; Liu, X.S.; Dong, Z.B.; Fang, H.Y. Microstructure and mechanical properties of nanostructured bainite weld with regeneration. *Met. Mater. Int.* **2014**, *20*, 923–928. [CrossRef]
21. Fang, K.; Yang, J.G.; Song, K.J.; Liu, X.S.; Wang, J.J.; Fang, H.Y. Study on tempered zone in nanostructured bainitic steel welded joints with regeneration. *Sci. Technol. Weld. Join.* **2014**, *19*, 572–577. [CrossRef]
22. Fang, K.; Yang, J.G.; Song, K.J.; Liu, X.S.; Fang, H.Y. Acceleration of regeneration treatment for nanostructured bainitic steel by rotary impacting trailed welding. *J. Mater. Process. Technol.* **2014**, *214*, 2935–2940. [CrossRef]
23. Song, K.J.; Fang, K.; Yang, J.G.; Ma, R.; Liu, X.S.; Wang, J.J.; Fang, H.Y. Acceleration of regeneration treatment for nanostructured bainitic steel welding by static recrystallization. *Mater. Sci. Technol.* **2015**, *31*, 835–842. [CrossRef]
24. Garcia-Mateo, C.; Caballero, F.G.; Bhadeshia, H.K.D.H. Acceleration of Low-temperature Bainite. *ISIJ Int.* **2003**, *42*, 1821–1825. [CrossRef]
25. Rees, G.I.; Bhadeshia, H.K.D.H. Bainite transformation kinetics Part 2 Non-uniform distribution of carbon. *Mater. Sci. Technol.* **1992**, *8*, 985–993. [CrossRef]
26. Hong, S.G.; Cho, H.M.; Lee, J.S. Effects of Post Weld Rapid Heat Treatment on Microstructural Evolution in High-Carbon Bainitic Steel Welds. In Proceedings of the 63rd Annual Assembly & International Conference of the International Institute of Welding, Istambul, Turkey, 11–17 July 2010; pp. 47–54.

27. Ramakrishna, R.V.S.M.; Gautam, J.P.; Reddy, G.M.; Sankara, R.K.B. Recent Advances in Welding of High-Strength Bainitic Steel for Automotive Applications. In *Advanced High. Strength Steel. Processing and Applications*; Roy, T.K., Bhattacharya, N., Ghosh, C., Ajmani, S.K., Eds.; Springer: Singapore, 2018; pp. 79–85.
28. Ramakrishna, R.V.S.M.; Sankara, R.K.B.; Reddy, G.M.; Gautam, J.P. Friction stir welding of advanced high strength (bainitic) steels for automotive applications. *Mater. Today Proc.* **2018**, *5*, 17139–17146. [CrossRef]
29. Thomas, W.; Nicholas, E. Friction stir welding for the transportation industries. *Mater. Des.* **2002**, *18*, 269–273. [CrossRef]
30. Wang, L.; Speer, J.G. Quenching and Partitioning Steel Heat Treatment. *Metallogr. Microstruct. Anal.* **2013**, *2*, 268–281. [CrossRef]
31. Speer, J.; Matlock, D.K.; De Cooman, B.C.; Schroth, J.G. Carbon partitioning into austenite after martensite transformation. *Acta Mater.* **2003**, *51*, 611–2622. [CrossRef]
32. Tan, Z.L.; Wang, K.K.; Gao, G.H.; Gui, X.L.; Bai, B.Z.; Weng, Y.Q. Mechanical properties of steels treated by Q-P-T process incorporating carbide-free-bainite/martensite multiphase microstructure. *J. Iron Steel Res. Int.* **2014**, *21*, 191–196. [CrossRef]
33. Speer, J.G.; Edmonds, D.V.; Rizzo, F.C.; Matlock, D.K. Partitioning of carbon from supersaturated plates of ferrite, with application to steel processing and fundamentals of the bainite transformation. *Curr. Opin. Solid State Mater. Sci.* **2004**, *8*, 219–237. [CrossRef]
34. Dittrich, F.; Kaars, J.; Masek, B.; Jenicek, S.; Wagner, M.F.X.; Mayr, P. HAZ characterization of welded 42SiCr steel treated by quenching and partitioning. *J. Mater. Process. Technol.* **2019**, *268*, 37–46. [CrossRef]
35. Caballero, F.G.; Garcia-Mateo, C.; de Andrés, C.G. Dilatometric Study of Reaustenitisation of High Silicon Bainitic Steels: Decomposition of Retained Austenite. *Mater. Trans.* **2005**, *46*, 581–586. [CrossRef]
36. Caballero, F.G.; Miller, M.K.; Garcia-Mateo, C. The approach to equilibrium during tempering of a bulk nanocrystalline steel: An atom probe investigation. *J. Mater. Sci.* **2008**, *43*, 3769–3774. [CrossRef]
37. Saha Podder, A.; Bhadeshia, H.K.D.H. Thermal stability of austenite retained in bainitic steels. *Mater. Sci. Eng. A* **2010**, *527*, 2121–2128. [CrossRef]
38. Garcia-Mateo, C.; Peet, M.; Caballero, F.G.; Bhadeshia, H.K.D.H. Tempering of hard mixture of bainitic ferrite and austenite. *Mater. Sci. Technol.* **2004**, *20*, 814–818. [CrossRef]
39. Caballero, F.G.; Miller, M.K.; Garcia-Mateo, C.; Capdevila, C.; Babu, S.S. Redistribution of alloying elements during tempering of a nanocrystalline steel. *Acta Mater.* **2008**, *56*, 188–199. [CrossRef]
40. Hulme-Smith, C.N.; Lonardelli, I.; Peet, M.J.; Dippel, A.C.; Bhadeshia, H.K.D.H. Enhanced thermal stability in nanostructured bainitic steel. *Scr. Mater.* **2013**, *69*, 191–194. [CrossRef]
41. Zhang, C.; Song, X.; Lu, P.; Hu, X. Effect of microstructure on mechanical properties in weld-repaired high strength low alloy steel. *Mater. Des.* **2012**, *36*, 233–242. [CrossRef]
42. Avishan, B.; Yazdani, S.; Hossein Nedjad, S. Toughness variations in nanostructured bainitic steels. *Mater. Sci. Eng. A* **2012**, *548*, 106–111. [CrossRef]
43. Garcia-Mateo, C.; Caballero, F.G.; Sourmail, T.; Kuntz, M.; Cornide, J.; Smanio, V.; Elvira, R. Tensile behaviour of a nanocrystalline bainitic steel containing 3 wt% silicon. *Mater. Sci. Eng. A* **2012**, *549*, 185–192. [CrossRef]
44. Amel-Farzad, H.; Faridi, H.R.; Rajabpour, F.; Abolhasani, A.; Kazemi, S.; Khaledzadeh, Y. Developing very hard nanostructured bainitic steel. *Mater. Sci. Eng. A* **2013**, *559*, 68–73. [CrossRef]
45. Yang, J.; Wang, T.S.; Zhang, B.; Zhang, F.C. Microstructure and mechanical properties of high-carbon Si-Al-rich steel by low-temperature austempering. *Mater. Des.* **2012**, *35*, 170–174. [CrossRef]
46. Withers, P.J.; Bhadeshia, H.K.D.H. Residual stress part 2—Nature and origins. *Mater. Sci. Technol.* **2001**, *17*, 366–375. [CrossRef]

MDPI
St. Alban-Anlage 66
4052 Basel
Switzerland
Tel. +41 61 683 77 34
Fax +41 61 302 89 18
www.mdpi.com

Applied Sciences Editorial Office
E-mail: applsci@mdpi.com
www.mdpi.com/journal/applsci

www.ingramcontent.com/pod-product-compliance
Lightning Source LLC
Chambersburg PA
CBHW061439210926
51850CB00008B/1550
9783036520131